SECOND EDITION

ENERGY MANAGEMENT
AND
CONSERVATION

HANDBOOK

SECOND EDITION

ENERGY MANAGEMENT AND CONSERVATION

HANDBOOK

EDITED BY

FRANK KREITH

D. YOGI GOSWAMI

CRC Press
Taylor & Francis Group
Boca Raton London New York

CRC Press is an imprint of the
Taylor & Francis Group, an **informa** business

CRC Press
Taylor & Francis Group
6000 Broken Sound Parkway NW, Suite 300
Boca Raton, FL 33487-2742

© 2017 by Taylor & Francis Group, LLC
CRC Press is an imprint of Taylor & Francis Group, an Informa business

Printed on acid-free paper
Version Date: 20160726

International Standard Book Number-13: 978-1-4665-8516-4 (Hardback)

Visit the Taylor & Francis Web site at
http://www.taylorandfrancis.com

and the CRC Press Web site at
http://www.crcpress.com

Printed and bound in the United States of America by Publishers Graphics, LLC on sustainably sourced paper.

Contents

Preface

Since the first edition of the *Energy Management and Conservation Handbook* was published in 2007, awareness of the need to shift from an economy based on fossil fuels to one that relies mostly on low- or zero-carbon sustainable sources has received global acceptance. This recognition for the need to change our energy system was most recently addressed at the United Nations Climate Change Conference (COP21) in Paris in late 2015. The conference was attended by representatives of 185 countries and the European Union. The primary goal of the conference was to create an agreement limiting global warming to less than 1.5°C compared to preindustrial levels. Toward this end, countries must find ways to achieve zero net anthropogenic emissions over the next 35 years. While these goals were uniformly supported by the attendees of the conference, adequate funding to achieve them is still lacking.

Updating the first edition of this handbook after 10 years is important because research into and development of the tools necessary to achieve cost-effective reduction in fossil fuel consumption has been ongoing. Furthermore, the need to take effective action is apparent from the continued increase in global temperature, severe health-threatening air pollution in major cities such as Beijing and Delhi, and rising ocean levels, which threaten the very existence of low-lying countries.

This book presents updates on the most important tools to manage energy conservation and reduction in the use of CO_2-generating technologies. Chapter 1 presents an overview of sustainability management. Chapter 2 deals with transportation issues from the perspective of a well-to-wheel analysis. Chapter 3 covers economic methods, including risk assessment. Chapters 4 through 8 cover energy conservation in buildings. Chapter 9 deals with heat pump technology. Chapters 10 and 11 deal with industrial energy management and electric motors. Chapter 12 covers energy storage, and Chapter 13 addresses demand-side management.

With the exception of Chapter 1, all other chapters have been extracted from the more expansive coverage in the *Energy Efficiency and Renewable Energy Handbook*. Energy conservation is still the least expensive means of reducing the detrimental environmental effects of anthropogenic burning of fossil fuels. The updated versions in this handbook will be useful for engineering design of more efficient buildings, industrial technology, and energy planning. The editors express their appreciation to the contributors for updating the material published in 2007 and providing the most effective available tools for sustainable energy management. In a work of this type that covers such a variety of topics, errors and omissions are unavoidable. The editors would, therefore, appreciate feedback from the readers to rectify any of these issues they may discover.

Frank Kreith
D. Yogi Goswami

Editors

Professor Frank Kreith is an internationally known energy consultant and professor emeritus of engineering at the University of Colorado, Boulder, Colorado. In 1945, after graduation from the University of California, Berkeley, he accepted a position at the Jet Propulsion Laboratory of the California Institute of Technology, where he developed a heat transfer laboratory and conducted research on building heat transfer. He received his MS in engineering in 1949 from the University of California, Los Angeles, and in 1950 was awarded a Guggenheim Fellowship to Princeton University, followed in 1951 by an appointment to the faculty of the University of California, Berkeley. From 1953 to 1959, he was associate professor of mechanical engineering at Lehigh University, where he did research on heat transfer in rotating systems and wrote the first edition of *Principles of Heat Transfer*, now in its seventh edition. In 1958, he received the Robinson Award from Lehigh University for excellence in teaching. In 1959, he joined the University of Colorado where he held appointments as professor of mechanical and chemical engineering. In 1962, he published a text on the design of solar power plants based on his consulting for the NASA space program. During his 20 years of tenure at the University of Colorado, he did research on heat transfer in biological systems and renewable energy. In 1964, he was awarded a doctorate in engineering from the University of Paris.

Dr. Kreith served as chief scientist and ASME legislative fellow at the National Conference of State Legislatures (NCSL) from 1988 to 2001, providing professional advice and assistance to all 50 state legislatures on energy and the environment. Prior to joining NCSL, he was senior research fellow at the Solar Energy Research Institute (now the National Renewable Energy Laboratory) where he participated in the Presidential Domestic Energy Review and served as energy advisor to the governor of Colorado. From 1974–1977, he was president of Environmental Consulting Services. Dr. Kreith has been an energy consultant to NATO, the U.S. Agency of International Development, and the United Nations. He has published more than a hundred peer-reviewed technical articles and more than 15 books, many translated into foreign languages and used extensively in engineering programs around the world. His books include *Principles of Heat Transfer* (now in its seventh edition), *Principles of Solar Engineering* (with J.F. Kreider), *Nuclear Impact* (with C.B. Wrenn), the *Handbook of Solid Waste Management*, the *CRC Handbook of Mechanical Engineering*, and *Principles of Sustainable Energy*, which is now in its second edition. He is a fellow of AAAS and was promoted to honorary member of ASME in 2004. Dr. Kreith's work has received worldwide recognition, including the Washington Award, Charles Greeley Abbot Award from ASES, the Max Jacob Award from ASME-AIChE, and the Ralph Coats Roe Medal from ASME for "significant contributions…through provision of information to legislators about energy and the environment." In 2004, ASME recognized Dr. Kreith's lifelong contributions to heat transfer and renewable energy by establishing the Frank Kreith Energy Award. He has recently completed an autobiography, *Sunrise Delayed: A Personal History of Solar Energy*, which is available on Amazon.com.

Dr. D. Yogi Goswami is a university distinguished professor, the John and Naida Ramil Professor, and director of the Clean Energy Research Center at the University of South Florida, Tampa, Florida.

He conducts fundamental and applied research on solar thermal energy, thermodynamics, heat transfer, HVAC, photovoltaics, hydrogen, and fuel cells.

Dr. Goswami has served as an advisor and given testimonies on energy policy and the transition to renewable energy to the U.S. Congress and the Government of India, as well as provided consultant expertise to the U.S. Department of Energy, USAID, World Bank, and NIST, among others.

Professor Goswami is the editor-in-chief of the *Solar Energy* and *Progress in Solar Energy* journals. Within the field of RE, he has published as author/editor 16 books, 6 conference proceedings, and 393 refereed technical papers. He has delivered 52 keynote and plenary lectures at major international conferences. He holds 18 patents.

A recognized leader in professional scientific and technical societies, Professor Goswami has served as a governor of ASME-International (2003–2006), president of the International Solar Energy Society (2004–2005), senior vice president of ASME (2000–2003), vice president of ISES and president of the International Association for Solar Energy Education (IASEE, 2000–2002).

Dr. Goswami is a fellow of AAAS, ASME International, ASHRAE, the American Solar Energy Society, the National Academy of Inventors and a member of the Pan American Academy of Engineers. He is a recipient of the following awards:

- Technical Communities Globalization Medal, (ASME) 2013.
- Theodore and Venette Askounes-Ashford Distinguished Scholar Award, Univ. South Florida, 2011.
- Frank Kreith Energy Award, ASME, 2007
- Farrington Daniels Award, ISES, 2007 (highest award of ISES)
- Hoyt Clark Hottel Award, ASES, 2007
- Charles Greely Abbott Award for Outstanding Scientific, Technical and Human Contributions to the Development and Implementation of Solar Energy (highest award of the American Solar Energy Society), 1998.
- John Yellott Award for Outstanding Contributions to the Field of Solar Energy, ASME Solar Energy Division, 1995 (highest solar energy award from ASME).

He has also received more than 50 other awards and certificates from major engineering and scientific societies for his work in renewable energy.

Contributors

Aníbal T. de Almeida
Department of Electrical and Computer
    Engineering
University of Coimbra
Coimbra, Portugal

Andrea L. Alstone
Energy Efficiency Standards Group
Lawrence Berkeley National Laboratory
Berkeley, California

Barbara Atkinson
Energy Efficiency Standards Group
Lawrence Berkeley National Laboratory
Berkeley, California

Peter Biermayer
Pacific Gas & Electric Co.
San Francisco, California

Barney L. Capehart
Department of Industrial and Systems
    Engineering
University of Florida
Gainesville, Florida

Jeffrey P. Chamberlain
Argonne National Laboratory
Lemont, Illinois

David E. Claridge
Department of Mechanical Engineering
Texas A&M University
College Station, Texas

Charles H. Culp
Energy Systems Laboratory
Texas A&M University
College Station, Texas

Karina Garbesi
California State University, East Bay
Hayward, California

Clark W. Gellings
Electric Power Research Institute
Palo Alto, California

Brian F. Gerke
Energy Efficiency Standards Group
Lawrence Berkeley National Laboratory
Berkeley, California

Steve F. Greenberg
Lawrence Berkeley National Laboratory
University of California, Berkeley
Berkeley, California

Roel Hammerschlag
Hammerschlag & Co., LLC
Olympia, Washington

Eric Kleinert
FORTIS Colleges and Institutes
Lake Worth, Florida

Moncef Krarti
Civil, Environmental and Architectural
    Engineering Department
University of Colorado
Boulder, Colorado

Jan F. Kreider
K&A, LLC
Boulder, Colorado

Frank Kreith (Emeritus)
Mechanical Engineering Department
University of Colorado
Boulder, Colorado

Alex Lekov
Lawrence Berkeley National Laboratory
University of California, Berkeley
Berkeley, California

James Lutz (Retired)
Lawrence Berkeley National Laboratory
University of California, Berkeley
Berkeley, California

James E. McMahon
Better Climate Research and Policy
    Analysis
Moraga, California

Stephen Meyers
Lawrence Berkeley National Laboratory
University of California, Berkeley
Berkeley, California

Kelly E. Parmenter
Applied Energy Group, Inc.
Walnut Creek, California

Terry Penney (Retired)
National Renewable Energy Laboratory
Golden, Colorado

Prakash Rao
Lawrence Berkeley National Laboratory
University of California, Berkeley
Berkeley, California

Bryan P. Rasmussen
Department of Mechanical Engineering
Texas A&M University
College Station, Texas

Wesley M. Rohrer, Jr. (Deceased)

Greg Rosenquist
Lawrence Berkeley National Laboratory
University of California, Berkeley
Berkeley, California

Rosalie Ruegg (Retired)
TIA Consulting, Inc.
Emerald Isle, North Carolina

Christopher P. Schaber (Retired)
Institute for Lifecycle Environmental
    Assessment
Seattle, Washington

Walter Short (Retired)
National Renewable Energy Laboratory
Golden, Colorado

Craig B. Smith
Dockside Consultants, Inc.
Newport Beach, California

Herbert W. Stanford, III (Retired)
Stanford White, Inc.
Raleigh, North Carolina

Alex J. Valenti
Energy Efficiency Standards Group
Lawrence Berkeley National Laboratory
Berkeley, California

Vagelis Vossos
Energy Analysis and Environmental
    Impacts Division
Lawrence Berkeley National Laboratory
Berkeley, California

1

Planning for Sustainability

Frank Kreith

CONTENTS

1.1 Sustainability Principles in Management

The historic growth in all forms of energy use has led to unsustainable growth in population and the use of natural resources and land, as well as adverse environmental impact. Figure 1.1 shows the kind of interrelated issues that an engineer must consider in managing energy system investments. It is important to keep in mind that energy is not only an engineering challenge but also a politically and socially charged management field.

The energy choices made in the near future are among the most important of any choices in human history. Sustainability considerations reflect priorities in our society as well as our attitude toward future generations. Management of a sustainable energy future is only possible if we develop an overall technical, social, and political strategy that combines renewable energy development, energy conservation, and adaptation of our lifestyle to greatly reduce energy consumption (Kreith & Krumdieck, 2013; Rojey, 2009). One of the most important international studies of sustainable energy was the World Commission on Energy and the Environment headed by Gro Harlem Brundtland, the former prime minister of Norway, in 1983. The goal of this commission was to formulate a realistic proposal that allows human progress, but without depriving future generations of the resources they will need. The outcome of this study was summarized in an important book entitled *Our Common Future* (Brundtland, 1987); it concluded that the current development of human progress in both developed and developing countries is unsustainable because it uses an increasing amount of environmental resources, especially fossil fuels. The Brundtland Commission's definition of sustainable development is as valid today as it was in 1983: "Sustainable developments should meet the needs of the present without compromising the ability of future generations to meet their own needs." The traditional assumption of economists that when we run short of any one resource or material, engineers will always find a substitute is no longer valid for future planning.

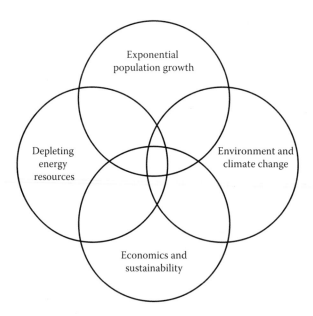

FIGURE 1.1
Complex and interrelated nature of engineering, social, and environmental issues. (From Alliance for Water Efficiency and American Council for an Energy-Efficient Economy, *Addressing the Energy–Water Nexus: A Blueprint for Action and Policy Agenda*, May 2011; ASME, *ETP: Energy–Water Nexus–Cross-cutting Impacts*.)

1.2 Management of Common Resources

Key elements for the management of common resources are the allocation of resources that might become in short supply and the regulations that limit environmental impacts to protect public health. Allocations of limited water resources in California and Australia and punishment for people who pollute rivers and lakes are examples of such regulations. The last 30 years have seen a number of federal environmental legislations in the United States, for example, the Clean Air Act (1972) and the Safe Drinking Water Act (1974) have led to improvement in air and water quality. At present, many countries are considering legislation to limit the exhaust of carbon dioxide to avoid dangerous global warming that could adversely impact the environment. Energy engineers will be responsible for providing energy and services with current technology, while at the same time transitioning to more suitable systems. An example of the potential for such an approach is the 1987 Montreal Protocol on Substances that Deplete the Ozone Layer, substances that created an ozone hole above the Arctic. The protocol, which was ratified by most developed nations, required corporations to phase out production and use of hydrochlorofluorocarbons (HCFC), refrigerants that created the ozone hole. Engineers under pressure developed substitutes for HClF that have stopped the ozone hole from growing.

The most important lesson to learn from each experience is that in the past environmental regulations are developed *after a problem arises* when a technical alternative is available. There are many examples for this, such as the emission of pesticide DT and sulfur dioxide from coal fire power plants. The companies that were innovators of a transition technology, for example, electrostatic precipitators or catalytic converters, ahead of legal requirements were in a superior business position rather than those who fought the changes. One of the guiding principles for future sustainability management is to seek out scientific evidence and begin work on changing the existing processes before environmental regulations are enacted. Today, there are developing social businesses and B Corps that are for-profit companies that have social and environmental well-being as part of their objectives. Some corporations have hired sustainability managers even when there is no requirement for corporations to consider sustainability or social welfare. But consideration of sustainability is increasing in business and industry, and engineers capable of innovative and creative solutions are in short supply (Winston, 2014).

1.3 Water, Population, and Food Issues

Population growth, lack of food supply, and limited water resources are three issues that grossly affect the resource management challenges of the future. Although as shown in Figure 1.2 the rate of growth of world population has gone down, the actual number of people on the globe continues to increase. The United Nations predicts that global population will reach 9 billion by the year 2050, and even today with a global population of 7 billion, a large percentage of people, particularly in Africa and the Middle East, face lack of clean water and inadequate food supply. The world food production has increased substantially in the past century as has the calorie intake per capita, but the absolute number of undernourished people has increased considerably. It is estimated that almost a trillion are undernourished, and a continued increase in global population will increase

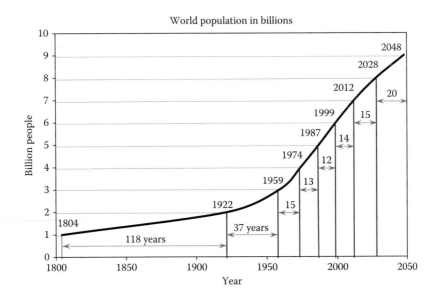

FIGURE 1.2
Annual additions and the annual growth rate of global population. (From Population Division of the Department of Economic and Social Affairs of the United Nations Secretariat, World population prospects: The 2008 revision, March 2009, http://esa.un.org/unpp.)

food demand. Increased fertilizer application as well as water usage has been responsible for more than 70% of the crop yield increase in the recent past according to the UNEP (GRID-Arendal, n.d.), but cereal yield has nearly stabilized now and fisheries landings have decreased in the past decade mainly as a result of overfishing and unsound fishing methods. About 30 million tons of fish needed to sustain the growth in aquaculture correspond roughly to the amount of fish discarded at sea with current fishing methods. An extensive study providing recent data and findings from a range of international collaborations and studies can be found in Martindale (2014).

Although there is an enormous amount of water on the planet, less than 1% is usable for drinking and agriculture; the rest is salty, brackish, or frozen. Freshwater refers to rivers or lakes fed by seasonal precipitations. Aquifers are underground freshwater reservoirs in permeable gravel or sand. Some of these, called unconfined, are replenished by surface precipitation. But others, called confined aquifers, such as the massive Ogallala Aquifer under the Great Plains of the United States, are actually finite and were deposited over a million years ago. Confined aquifers have a finite lifetime and some of them have already run dry. A recent study found that at least 30% of the southern Great Plains will exhaust their ground water reserve within the next 30 years (Scanlon et al., 2012).

The World Health Organization stipulates that the basic requirement for water is 20 L per day per person and that it be accessible within 1 km of the user. In industrial societies, personal water consumption is considerably larger, and if industrial and energy production are added, freshwater usage can exceed 5000 L per day per capita, and water scarcity in a developed country is equivalent to an annual availability of less than 1000 m^3 per person (World Bank, 2003).

The biggest user of freshwater from lakes and rivers are cooling towers of electrothermal power plants. A majority of power plants operate on the Rankine cycle and they need cooling water in the condensers to operate. Although much of the water used by power plants is returned to rivers or lakes, the returned water is warmed to between 4°C and 10°C and the upper regulation of temperature rise is necessary to protect aquatic ecosystems. But 3% of all US water consumption is evaporated from the cooling towers of power plants and lost as water vapor into the atmosphere. Figure 1.3 shows a breakdown of the water withdraws and consumption according to a recent study (ASME, 2011).

Although the technology for desalination is well known, the process requires a very large amount of energy. There are more than 7000 desalination projects in operation, with 60% located in the Middle East. The levelized cost of water from desalination plants is

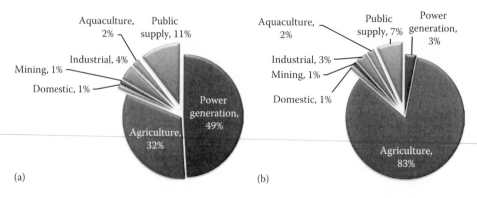

FIGURE 1.3
(a) Freshwater withdrawals. (b) Freshwater consumption.

about \$0.60 per m³ for large plants and considerably higher for smaller ones. Thermal processes like multistage desalination (MFD) use 10–15 kWh/m³ of water, while reverse osmosis (RO) uses between 1 and 2.5 kWh/m³ for brackish water and 4–13 kW/m³ of electricity for seawater (Loupasis, 2002). Water desalination can be a source of fresh drinking water in parts of the world that have excess energy, but it is not a solution to the forthcoming water crisis as the population grows.

References

ASME. 2011. *ETP: Energy-Water Nexus—Cross-Cutting Impacts and Addressing the Energy-Water Nexus: A Blueprint for Action and Policy Agenda.* Alliance for Water Efficiency and American Council for an Energy-Efficient Economy, New York.

Brundtland, G. H. 1987. *Our Common Future.* World Commission on Environmental and Development. New York: Oxford University Press.

GRID-Arendal. n.d. World food supply. http://www.grida.no, Web June 12, 2015.

Kreith, F. and Krumdieck, S. 2013. *Principles of Sustainable Energy Systems*, 2nd edition. CRC Press, Boca Raton, FL.

Loupasis, S. 2002. Technical analysis of existing RES desalination schemes—RE Driven Desalination Systems, REDDES. Report, Contract # 4.1030/Z/01-081/2001.

Martindale, W. December 2014. *Global Food Security and Supply.* Chichester, U.K.: Wiley.

Rojey, A. 2009. *Energy & Climate: How to Achieve a Successful Energy Transition.* London, U.K.: SCI.

Scanlon, B., C. Faunt, L. Longuevergne, R. Reedy, W. Alley, V. McGuire, and P. McMahon. June 2012. Groundwater depletion and sustainability of irrigation in the US high plains and central valley. *Proceedings of the National Academy of Sciences*, 109(24):9320–9325.

Winston, A. S. 2014. *The Big Pivot, Radically Practical Strategies for a Hotter, Scarcer and More Open World.* Boston, MA: Harvard Business Review Press.

World Bank. 2003. Development data and statistics. www.worldbank.org/data (accessed May 2007).

2

Transportation

Terry Penney and Frank Kreith

CONTENTS

2.1 Introduction

A viable transportation system is a crucial part of a sustainable energy future. Transportation is a complex interdisciplinary topic, which really deserves a book unto itself. However, some of the main issues related to transportation have to do with fuels and energy storage, which are the topics covered in this book. This section does not purport to be exhaustive, but examines some of the key issues related to a viable transportation future.

Gasoline and diesel not only are very convenient fuels for ground transportation, but also have a high energy density that permits storage in a relatively small volume—an important asset for automobiles. For example, these liquid fuels have a volumetric specific energy content of about 10,000 kWh/m^3, compared to hydrogen compressed to 100 bar at about 300 kWh/m^3. But known petroleum resources worldwide are being consumed rapidly, and future availability of these resources is bound to decline. At present, more than 97% of the fuel used for ground transportation in the United States is petroleum based. Importation of fossil fuels in the United States has recently been decreasing, primarily because of domestic exploration and production of oil and natural gas. The increase in the cost of gas and oil has become of growing concern to average citizens, and the emission from current transportation systems is a major component of CO_2 pollution that produces global warming.

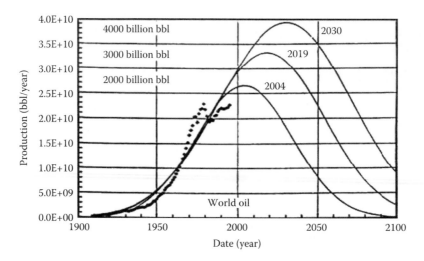

FIGURE 2.1
Oil production vs. time for various recoverable amounts of petroleum. (According to Bartlett, A.A., *Math. Geol.*, 32(1), 2000. With permission.)

There is worldwide agreement among oil experts that global oil production will reach a peak sometime between 2020 and 2030. The predictions for the date of peak world oil production according to various estimates [1] are demonstrated in Figure 2.1. The oil production is shown as a function of time for three total amounts of recoverable oil that span the entire range of assumptions by experts. Although new oil fields may be discovered in the future, it is not expected that they will substantially increase the total recoverable amounts. Hence, it is believed that the total amounts lie somewhere between 3000 and 4000 billion barrels of oil (bbl). The obvious conclusion to be drawn from these predictions is that the production peak is imminent and the price of oil will continue to escalate as supplies decline. Thus, planning for a sustainable transportation system that does not depend entirely on petroleum resources is an imperative segment of a sustainable energy future.

2.2 Alternative Fuels

Alternative fuels available to supplement oil as well as their feedstock are shown in Table 2.1 [2]. Inspection of the table shows that biodiesel, electricity, ethanol, and hydrogen (via electricity) are the fuels potentially independent of a petroleum resource such as oil or natural gas. The potential of producing liquid fuel from biomass has been treated elsewhere in this handbook. There is every reason to believe that biomass will provide an increasing percentage of the future transportation fuel, especially if ethanol can be produced from cellulosic materials such as switchgrass or bio-waste, or diesel from algae. The reason for this is that ethanol produced by traditional methods from corn kernels has only an energy return on energy investment on the order of 1.25, whereas the EROI for cellulosic ethanol is claimed to be about 6. Diesel from algae may be even better. As shown in Figure 2.2, even cellulosic ethanol would not be able to replace oil, because growing it would require too large a percentage of all the arable lands in the United States, and the

TABLE 2.1

Feedstocks for Alternative Fuels

Fuel	Feedstock
Propane (LPG)	Natural gas (NG), petroleum
CNG	NG
Hydrogen	NG or (water + electricity)
FT diesel	NG, coal, or algae
Methanol (M85)	NG or coal
Ethanol (E85)	Corn, sugarcane, or cellulosic biomass
Electricity	NG, coal, uranium, or renewables

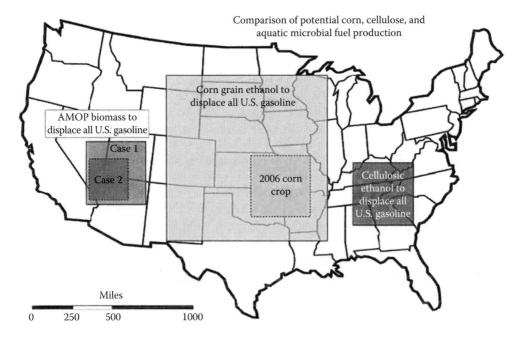

FIGURE 2.2
Relative land area requirement for various liquid fuel biosources. (From Dismukes, C. et al., *Curr. Opin. Biotechnol.*, 19, 235, June 2008. With permission.)

production of large amounts of ethanol would compete with the production of food, which is of increasing importance for a socially sustainable energy system.

It has recently been proposed [3] to use biofuels derived from aquatic microbial oxygenic photoautotrophs (AMOPs), commonly known as algae. In this study, it was shown that AMOPs are inherently more efficient solar collectors, use less or no land, can be converted to liquid fuels using simpler technologies than cellulose, and offer secondary uses that fossil fuels do not provide. AMOPs have a 6- to 12-fold energy advantage over terrestrial plants because of their inherently higher solar energy conversion efficiency, which is claimed to be between 3% and 9%. Figure 2.2 compares the area needed for three different biomass sources. The data are for corn grain, switchgrass, mixed prairie grasses, and AMOPs. Each box superimposed on the map of the United States represents the area needed to produce a sufficient amount of biomass to generate enough liquid fuel to displace all the gasoline

TABLE 2.2

Comparison of Some Sources of Biodiesel

Crop	Oil Yield (L/ha)	Land Area Needed (Mha)[a]	Percentage of Existing U.S. Cropping Area[a]
Corn	172	1540	846
Soybean	446	594	326
Canola	1,190	223	122
Oil palm	5,950	45	24
Microalgae[b]	136,900	2	1.1
Microalgae[c]	58,700	4.5	2.5

Source: Abstracted from Christi, Y., *Biotechnol. Adv.*, 25, 294, 2007.

[a] For meeting 50% of all transport fuel needs of the United States.

[b] 70% oil (by wt) in biomass.

[c] 30% oil (by wt) in biomass.

used in the United States in the year 2007. The two boxes for AMOPs are for 30% and 70% conversion efficiency. The overall solar energy conversion to biofuels works out to about 0.05% for solar to ethanol from corn grain and roughly 0.5% from switchgrass to ethanol. Comparatively, this value is about 0.5%–1% for AMOPs to ethanol or biodiesel.

An even more favorable assessment for the potential of algae to produce biodiesel is presented in Ref. [4]. According to this study, microalgae may be a source of biodiesel that has the potential to displace fossil fuel. According to Ref. [4], microalgae grow extremely rapidly and are exceedingly rich in oil. Some microalgae double their biomass every day, and the oil content of microalgae can exceed 80% by weight. Table 2.2 shows a comparison of some sources of biodiesel that could meet 50% of all of the transportation needs in the United States. According to estimates in Ref. [4], only a small percentage of U.S. cropping areas would be necessary to supply 50% of the entire transport fuel needs in the United States. However, no full-scale commercial algae biodiesel production facility has been built and operated for a sufficient time to make reliable predictions regarding the future of algae for a sustainable transportation system.

2.3 Well-to-Wheel Analysis

Rather than simply looking at the efficiency of an engine or a given fuel, a more comprehensive way to determine overall efficiency when evaluating the potential of any new fuel for ground transportation is to use what is called a *well-to-wheel analysis*. The approach to a well-to-wheel analysis is shown schematically in Figure 2.3 [5,14]. The well-to-wheel approach of a fuel cycle includes several sequential steps: feedstock production; feedstock transportation and storage; fuel production; transportation, storage, and distribution (T&S&D) of fuel; and finally vehicle operation. This approach is essential for a fair comparison of different options because each step entails losses. For example, whereas a fuel cell has a much higher efficiency than an internal combustion (IC) engine, for its operation, it depends on a supply of hydrogen, which must be produced by several steps from other sources. Moreover, there is no infrastructure for transporting and storing hydrogen, and this step in the overall well-to-wheel analysis has large losses and contributes to much larger energy requirements compared to a gasoline or electrically driven vehicle.

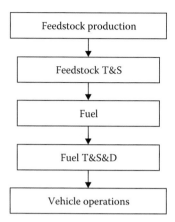

FIGURE 2.3
Steps in a well-to-wheel analysis for ground transportation vehicles.

2.4 Mass Transportation

An opinion held widely among state governments and environmentalists is that the mass transportation would greatly reduce the total energy consumption for the transportation sector. However, as shown in Table 2.3, based upon data collected by the U.S. Transportation Department in this country, the energy intensity of intercity rail and transit buses, that is, the energy spent per passenger-mile traveled, is virtually the same as the energy intensity of automobiles with current use. This is due to the urban sprawl in major cities that makes it difficult to reach outlying areas by a mass transport network. In other words, unless there are incentives for mass transport or disincentives to use the automobile, thereby achieving higher mass transport load factors—that is, more passengers per mile—on transit buses and light rail, the availability of mass transport systems will not materially change the overall energy use by the transportation sector. Moreover, installing new light rail is very capital-intensive and may not always be worth the energy and/or the money invested in its construction. Although passenger mass transit does not appear to offer large untapped opportunities, using ship or rail instead of air for shipment of freight and commercial goods is a source of enormous fuel reduction potential.

2.5 Hybrid Electric Vehicles

Another obvious approach to ameliorating the expected increase in price and lack of availability of petroleum fuel is to increase the mileage of the vehicles. This can be achieved by improving the efficiency of the IC engine, for instance, by using advanced diesel engines that have a higher compression ratio than spark ignition (SI) engines, or by using hybrid electric vehicles (HEVs). Increasing the efficiencies of IC or diesel engines is a highly specialized topic and is not discussed here. However, HEVs offer a near-term option for utilizing improved battery technology.

An HEV is powered by the combination of a battery pack and electric motor—like that of an electric vehicle—and a power generation unit (PGU), which is normally an IC or

TABLE 2.3

Passenger Travel and Energy Use, 2002

	No. of Vehicles (Thousands)	Vehicle-Miles (Millions)	Passenger-Miles (Millions)	Load Factor (Persons/Vehicle)	Energy Intensities (Btu/Vehicle-Mile)	(Btu/Passenger-Mile)	Energy Use (Trillion Btu)
Automobiles	135,920.7	1,658,640	2,604,065	1.57	5,623	3,581	9,325.9
Personal trucks	65,268.2	698,324	1,201,117	1.72	6,978	4,057	4,872.7
Motorcycles	5,004.2	9,553	10,508	1.22	2,502	2,274	23.9
Demand response	34.7	803	853	1.1	14,449	13,642	11.6
Vanpool	6.0	77	483	6.3	8,568	1,362	0.7
Buses	a	a	a	a	a	a	191.6
Transit	76.8	2,425	22,029	9.1	37,492	4,127	90.0
Intercity[b]	a	a	a	a	a	a	29.2
School[b]	617.1	a	a	a	a	a	71.5
Air	a	a	a	a	a	a	2,28.9
Certified route[c]	a	5,841	559,374	95.8	354,631	3,703	2071.4
General aviation	211.2	a	a	a	a	a	141.5
Recreation boats	12,409.7	a	a	a	a	a	187.2
Rail	18.2	1,345	29,913	22.2	74,944	3,370	100.8
Intercity[d]	0.4	379	5,314	14.0	67,810	4,830	25.7
Transit[e]	8.5	682	15,095	22.1	72,287	3,268	49.3
Commuter	5.3	284	9,504	33.5	90,845	2,714	25.8

Source: Davis, S. and Diegel, S., *Transportation Energy Data Book*, Oak Ridge National Laboratory/U.S. Department of Energy, Oak Ridge, TN, 2004.

[a] Data are not available.
[b] Energy use is estimated.
[c] Includes domestic scheduled service and half of international scheduled service. These energy intensities may be inflated because all energy use is attributed to passengers; cargo use not taken into account.
[d] Amtrak only.
[e] Light and heavy rail.

diesel engine. Unlike electric vehicles, however, HEV batteries can be recharged by an onboard PGU, which can be fueled by existing fuel infrastructure.

HEVs can be configured in a parallel or a series design. The parallel design enables the HEV to be powered by both the PGU and the motor, either simultaneously or separately. The series design uses the PGU to generate electricity, which recharges the HEV battery pack and produces power via an electric motor. The key element of either design is that the battery pack, as well as the PGU, can be much smaller than those of a typical electric vehicle or a vehicle powered by an IC engine because the IC engine can be operated at its maximum efficiency nearly all the time.

Currently available HEVs, such as the Toyota Prius, use a parallel configuration, as shown in Figure 2.4. A parallel HEV has two propulsion paths: one from the PGU and one from the motor, while computer chips control the output of each. A parallel-configuration HEV has a direct mechanical connection between the PGU and the wheels, as in a conventional vehicle (CV), but also has an electric motor that can drive the wheels. For example, a parallel vehicle could use the electric motor for highway cruising and the power from

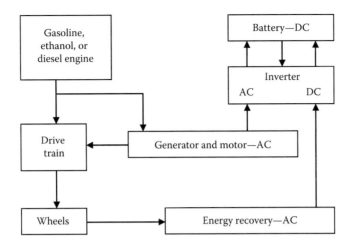

FIGURE 2.4
Schematic of a parallel-configuration hybrid electric vehicle.

(a)   (b)

FIGURE 2.5
Fuel-efficient cars (a) 2010 Toyota Prius: Toyota's third-generation Prius is a *parallel* hybrid; the 80 hp (60 kW) electric motor and gas engine work together to produce an average 50 mpg. (b) 2010 Ford Fusion: The Fusion can travel up to 47 mph (75 kph) on electric power alone. The motor is powered by a nickel metal hydride battery pack, and the car's engine is a 2.5 L Atkinson Cycle I-4 gasoline engine. The Fusion gets 41 mpg in the city and 36 mpg highway. (Courtesy of *Solar Today*, November/December 2009.)

IC engine for accelerating. The power produced by the PGU also drives the generator, which in turn can charge the battery as needed. The system to transfer electricity from the generator to the battery pack is exactly like that of an electric vehicle, with alternating current converted to DC by the inverter. HEV parallel designs also use a regenerative braking feature that converts energy stored in the inertia of the moving vehicle into electric power during deceleration (see Figure 2.5).

Some benefits of a parallel configuration versus a series configuration include the following:

- The vehicle has more power because both the engine and the motor provide power simultaneously.
- Parallel HEVs do not need a separate generator.
- Power is directly coupled to the road, thus operating the vehicle more efficiently.

Energy and cost savings from an HEV depend on many factors, such as the overall car design, cost of fuel, and the cost and efficiency of the batteries. Preliminary estimates indicate that over 5–8 years, the reduced size of the motor and savings in gasoline could pay for the additional cost of the batteries. Economically, an HEV would be beneficial when the price per gallon of gasoline or diesel exceeds $3.00/gal, but the life cycle of the batteries will also have to be considered. If batteries have a life cycle of 150,000 miles, as claimed by Toyota, they will not need to be replaced during the life of an average vehicle. On the other hand, if the battery life is considerably less and battery replacement is necessary during the expected 10-year life of the car, the operation cost over the life of an HEV would be considerably higher.

2.6 Plug-In Hybrid Electric Vehicles

Given the current state of technology, probably the most promising near-term solution to the ground transportation crisis is the use of plug-in HEVs (PHEVs). PHEVs have the potential of making the leap to the mainstream consumer market because they require neither a new technology nor a new distribution infrastructure. Like hybrids, which are already widely available, PHEVs have a battery and an IC engine for power, but the difference is that a PHEV has a larger battery capacity and a plug-in charger with which the battery can be recharged whenever the car is parked near a 110 or 220 V outlet. A so-called series PHEV40 can travel the first 40 miles on grid-supplied electric power when fully charged. When that charge is depleted, the gas or diesel motor kicks in, and the vehicle operates like a conventional hybrid. Because most commuter trips are less than 40 miles, it is estimated that a PHEV could reduce gasoline usage by 50% or more for many U.S. drivers. Moreover, using electric energy is cheaper and cleaner than using gasoline in automobile-type ground transportation. Figure 2.6 shows a Prius with

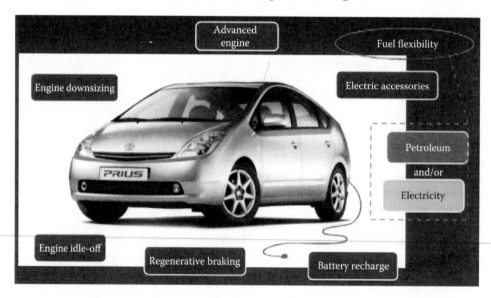

FIGURE 2.6
Plug-in hybrid electric vehicle. (Courtesy of NREL, Golden, CO.)

plug-in potential added. These early plug-in additions are available as aftermarket conversions, but because they are not in mass production, their cost is high, and Toyota does not honor their warranty when plug-in features are added aftermarket. More recently, all major original equipment manufacturers (OEMs) have introduced plug-in vehicles with various battery storage capacities.

An important feature of PHEVs is that their batteries can be charged at night when utilities have excess power available. Utilities have taken notice of the potential energy charging and storage capabilities of PHEVs because off-peak charging would help utilities to use low-cost baseload generation more fully. Furthermore, more advanced vehicle-to-grid concepts would allow utilities to buy back energy from the batteries of vehicle owners during peak demand periods and thus make a fleet of PHEVs into a large distributed storage-generation network. The arrangement would also enable renewable energy storage by charging PHEVs using solar- or wind-generated excess capacity. As mentioned in the previous section on batteries, lithium-ion and lithium-polymer batteries have the potential to store large charges in a lightweight package, which would make the HEVs even more attractive than the current technology, which uses nickel metal hydride (NiMH) batteries.

The efficiency of a PHEV depends on the number of miles the vehicle travels on liquid fuel and electricity, respectively, as well as on the efficiency of the prime movers according to the following equation:

$$\eta_{PHEV} = \frac{\text{Energy to wheels}}{\text{Energy from primary source}} = f_1\eta_1\eta_2 + f_2\eta_3\eta_4 \tag{2.1}$$

where
  η_1 is the efficiency from the primary energy source to electricity
  η_2 is the efficiency of transmitting energy to the wheels
  f_1 is the fraction of energy supplied by electricity
  f_2 is the fraction of energy supplied by fuel = $(1 - f_1)$
  η_3 is the well-to-wheel efficiency
  η_4 is the tank-to-wheel efficiency

PHEVs can be designed with different all-electric ranges. The distance, in miles, that a PHEV can travel on batteries alone is denoted by a number after PHEV. Thus, a PHEV20 can travel 20 miles on fully charged batteries without using the gasoline engine. According to a study by the Electric Power Research Institute (EPRI) [7], on average, one-third of the annual mileage of a PHEV20 is supplied by electricity and two-thirds by gasoline. The percentage depends, of course, on the vehicle design and the capacity of the batteries on the vehicle. A PHEV60 can travel 60 miles on batteries alone, and the percentage of electric miles will be greater as will the battery capacity and weight. The tank-to-wheel (more appropriately battery-to-wheel) efficiency for a battery all-electric vehicle according to EPRI [7] is 0.82.

Given the potentials for plug-in hybrid vehicles, the EPRI [7] conducted a large-scale analysis of the cost, the battery requirements, and the economic competitiveness of plug-in vehicles today and within the near-term future. Table 2.4 presents the net present value of life cycle costs over 10 years for a midsized IC engine vehicle such as the Ford Focus [IC], HEVs such as the Prius [HEV], and a future PHEV20 plug-in electric vehicle. The battery module cost in dollars per kWh is the cost at which the total life cycle costs of all three vehicles would be the same. According to projections for the production of NiMH battery

TABLE 2.4

Net Value of Life Cycle Costs over 117,000 Miles/10 Years for Conventional Gasoline (IC),
HEV and PHEV20 Midsize Vehicles with Gasoline Costs at $1.75/gal

Vehicle Type	IC	HEV	PHEV 20
Battery unit cost ($/kWh)		385[a]	316[a]
Incremental vehicle cost ($)		547	224
Battery pack cost ($)	60	3,047	3,893
Fuel costs ($)	5,401	3,725	2,787
Maintenance costs ($)	5,445	4,733	4,044
Battery salvage costs ($)		54	43
Total life cycle costs ($)	10,906	10,904	10,905

Source: Extracted from EPRI, Advanced batteries for electric drive vehicles: A technology and cost-effectiveness assessment for battery electric vehicles, power assist hybrid electric vehicles and plug-in hybrid electric vehicles, EPRI Tech. Report 1009299, EPRI, Palo Alto, CA, 2004.

[a] Battery module price at which life cycle parity with CV occurs.

modules, a production volume of about 10,000 units per year would achieve the necessary cost reduction to make both an HEV and a PHEV20 economically competitive. The EPRI analysis was conducted in 2004 and is, therefore, extremely conservative because it assumed a gasoline cost of $1.75/gal. A reevaluation of the analysis based upon a gasoline cost of $2.50/gal showed that the permitted battery price at which the net present value of conventional IC vehicles and battery vehicles are equal for battery module costs $1135 for an HEV and $1648 for a PHEV20, respectively.

Table 2.5 shows the electric and plug-in hybrid vehicle battery requirements (module basis for the cost estimates in Table 2.4), and it is apparent that, even with the currently available NiMH batteries, the cost of owning and operating HEVs and PHEVs is competitive with that of an average IC engine vehicle.

TABLE 2.5

NiMH Battery Cost Assumptions for Table 2.4

Assumption	ARB 2000 Report for BEVs		EPRI Assumptions
	2003	Volume	
Module cost ($/kWh)[a]	300	235	Varied[b]
Added cost for pack ($/kWh)	40	20	680 + 13
Multiplier for manufacturer and dealer markup	1.15	1.15	Varies[c]
Battery life assumptions (years)	6	10	10

[a] Equivalent module costs for an HEV 0 battery is $480 for 2003 and $384 for volume. Equivalent module costs for a PHEV 20 battery is $376 for 2003 and $301 for volume. HEV 0 and PHEV 20 batteries have a higher power-to-energy ratio and are more costly. These figures are based on data shown in the biomass chapter of this handbook.

[b] Battery module costs were varied in this analysis to determine the effect of battery module cost on life cycle cost.

[c] Manufacturer and dealer mark-up for HEV 0 battery modules estimated at $800, PHEV 20 battery modules $850, pack hardware mark-up assumed to be 1.5. Method documented in 2001 EPRI HEV report.

A cautionary note in the expectation of future ground transportation systems is the reduction in the rate of petroleum consumption that can be expected as HEVs and PHEVs is introduced into the fleet [8]. In these estimates, a rate of new vehicle sales of 7% of the fleet per year, a retirement rate of 5% per year resulting in a net increase in total vehicles of 2% per year was assumed. This increase is in accordance with previous increase rates between 1966 and 2003. Based upon the existing mileage for IC engine, HEVs, and PHEVs, it was estimated that *even if all new cars were HEVs or PHEVs*, after 10 years, the annual gasoline savings as a percentage of the gasoline usage by an all-gasoline fleet in the same year would only be about 30% for the HEVs and 38% for the PHEVs. These relatively small reductions in the gasoline use are due to the fact that despite introduction of more efficient vehicles, it takes time to replace the existing fleet of cars, and the positive effects will not be realized for some years.

2.7 Advanced Ground Transportation with Biomass Fuel

The previous section analyzed the potential of using batteries combined with traditional engines to reduce the petroleum consumption in the transportation system. However, the scenario used for this analysis can also be extended to determine the combination of PHEVs with biofuels, particularly ethanol made from corn or cellulosic biomass. No similar analysis for using diesel from algae is available at this time.

It is important to note that in October 2010, the U.S. Environmental Protection Agency granted a partial waiver for the use of E15 (15% ethanol and 85% gasoline) for use in light-duty motor vehicles in cars newer than model year 2007. In January 2011, a second waiver was granted that allowed for use of E15 in light-duty vehicles manufactured in 2001–2006. The decision to grant the waivers was the result of testing performed by the Department of Energy (DOE) and information regarding the potential effect of 15 on vehicle emissions [9]. According to the Renewable Fuels Association, the E15 waivers pertain to over 62% of vehicles currently on the roads in the United States. If all passenger cars and pickup trucks were to switch to E15, this would represent 17.5 gal of ethanol use annually [10]. There remain some practical infrastructure barriers, as the current fuel pumps at gasoline retail stations do not support this higher blend. However, as consumers begin to demand more options in biofuels, it is likely that E15 will be made more available in the foreseeable future.

A scenario for a sustainable transportation system based on fuel from biomass has been presented in Ref. [11]. In this analysis, the following four vehicle types combined with various fuel options have been calculated. The preferred mixture in an economy based largely on ethanol (E85) would be 85% ethanol and 15% gasoline that could be used in Flex Fuel automobiles. Currently, the United States is using E10, a mixture of 90% gasoline and 10% ethanol, with ethanol produced from corn. The fuel types used in this analysis are gasoline only, E10 with ethanol made from either corn or cellulosic materials, and E85 with ethanol from either corn or cellulosic materials. The four vehicle combinations are a convention SI engine, an HEV similar to the Toyota Prius, a PHEV20, and a PHEV30. The analysis was based upon an average 2009 performance on the U.S. light vehicle fleet at approximately 20 miles per gallon (mpg), for an HEV at 45 mpg, and for a PHEV20 at 65 mpg, according to Ref. [7]. For ethanol-/gasoline-blended fuels, it was assumed that the gasoline and ethanol are utilized with the same efficiency. That is, the mileage per unit of fuel energy is the same for gasoline and ethanol.

Based on the earlier assumption, the following parameters were calculated:

1. The miles per gallon of fuel, including the gasoline used to make ethanols (mpg).
2. The petroleum required to drive a particular distance for a case vehicle as a percentage of the petroleum required to drive the same distance by a gasoline-fueled SI vehicle.
3. The carbon dioxide emission rate for case vehicles as a percentage of that for SI gasoline only.

Using the earlier assumptions, one can calculate the mpg of fuel as

$$MF_{ij} = \text{miles/gal gas}) \times \left[(\text{gal gas/gal fuel}) + (1 - \text{gal gas/gal fuel}) \right.$$

$$\left. \times \frac{(\text{energy, LHV/gal ethanol})}{(\text{LHV/gal gasoline})} \right]$$

$$= MGO_i \times [FG_j + (1 - FG_j) \times (\text{LHV ratio})] \tag{2.2}$$

where
  MF_{ij} is the mpg of fuel for vehicles type i and fuel type j
  MGO_i is the mpg gasoline-only for vehicle i (see Table 2.4)
  FG_j is the volume fraction of gasoline in fuel type
  $j(1 - FG_j)$ is the volume fraction of ethanol in fuel type j
  LHV ratio is the ratio of lower heating values; LHV ratio = (LHV/gal ethanol)/(LHV/gal gasoline) = 0.6625

The index i indicates the vehicle type as shown in Table 2.4, and the index j denotes the volume fraction of gasoline in an ethanol–gasoline blend as follows: for gasoline only, $FG_1 = 1$; for E10 (i.e., 10 vol% ethanol and 90 vol% gasoline), $FG_2 = 0.90$; and for E85 (85 vol% ethanol, 15 vol% gasoline), $FG_3 = 0.15$. Other ethanol concentrations could be used (Figure 2.7).

The mpg of gasoline in the fuel, including the petroleum-based fuels used in the making of the ethanol in the fuel (by counting the energy of all petroleum-based fuels as gasoline), is given by

$$MG_{ijk} = \frac{MF_{ij}}{[FG_j + (1 - FG_j) \times (0.6625) \times (\text{MJ gasoline used in making 1 gal ethanol})/\text{MJ/gal ethanol}]}$$

$$= \frac{MF_{ij}}{FG_j + (1 - FG_j) \times (0.6625) \times Rk} \tag{2.3}$$

where Rk denotes the gallons of gasoline used to make 1 gal of ethanol. For corn-based ethanol, $k = 1$, $R1 = 0.06$, while for cellulosic-based ethanol, $k = 2$, $R2 = 0.08$, according to Ref. [12].

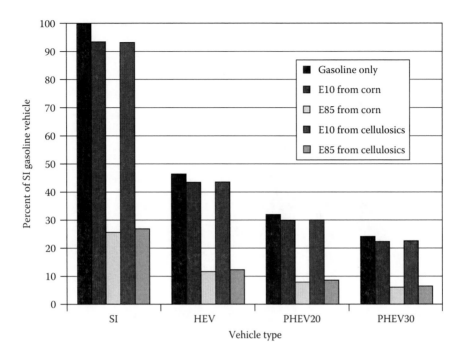

FIGURE 2.7
Petroleum requirement as a percentage of that for SI gasoline vehicle. (Calculated by Kreith, F. and West, R.E., *ASME J. Energy Res. Technol.*, 128(9), 236, September 2006.)

2.7.1 Petroleum Requirement

The petroleum required to produce an ethanol–gasoline blend, including the petroleum used to make the ethanol, is expressed as a percentage of the petroleum required for the same miles traveled by the same vehicle type using gasoline only (Figure 2.7). For a gasoline-only-fueled vehicle of any type, this percentage is 100%.

$$\text{Petroleum requirement (\%)} = 100 \times \frac{(MGO_i)}{(MG_{ijk})} \tag{2.4}$$

2.7.2 Carbon Dioxide Emissions

The CO_2 emissions, including those from making the ethanol and generating the electricity used from the grid by the vehicle, are expressed as a percentage of the emissions produced by the same type of vehicle fueled by gasoline only traveling the same number of miles (Figure 2.8).

$$= 100 \times [FE_m \times (\text{kWh/miles}) \times (\text{g-carbon/kWh})] + (1/MGO_i) \times (1 - FE_m)$$

$$\times [FG_j \times (\text{g-carbon/MJ gasoline}) \times (\text{MJ/gal gasoline})] + (1 - FG_j)$$

$$\times \frac{[(\text{MJ/gal ethanol}) \times (\text{g-carbon/MJ ethanol})]}{[(\text{g-carbon/MJ gasoline used}) \times (\text{MJ/gal gasoline})/(MGO_1)]} \tag{2.5}$$

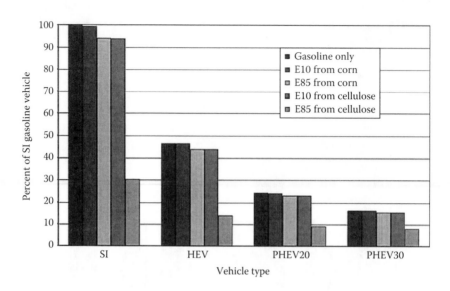

FIGURE 2.8
Carbon dioxide emissions as a percentage of emissions for SI gasoline vehicle.

where FE_m is the fraction of the miles driven by electricity from the grid for a plug-in hybrid vehicle. For $m = 0$, any non-PHEV, $FE_0 = 0$; for a PHEV20, $m = 1$, and $FE_1 = 0.327$ [7]; and for a PHEV30, $m = 2$, and $FE_2 = 0.50$ estimated by analogy with PHEV20. According to Ref. [7], the kWh/mile from the grid = 0.2853, the g-carbon emitted/kWh = 157 (146 average for all electricity generation [13] divided by 0.93, the average transmission efficiency), the g-carbon emitted/MJ gasoline = 94 [7], the MJ/gal gasoline = 121, the MJ/gal ethanol = 80.2, and CE_k the g-carbon emitted/MJ ethanol. For $k = 1$ (corn), $CE_1 = 87$, and for $k = 2$, cellulosics, $CE_2 = 11$.

So

CO_2 production as % of that for a gasoline-only vehicle

$$100 \times \{ FE_m \times (0.2853) \times (157) + (1/MF_{ij}) \times (1 - FE_m)$$
$$= \frac{\times [FG_j \times (94) \times 121 + (1 - FG_j) \times CE_k \times (80.2)] \}}{(94) \times (121)/21} \quad (2.6)$$

The final question to be asked in the utilization of PHEVs is whether or not the existing electricity system of the United States could handle the charging of the batteries in a PHEV-based ground transportation system. This question has recently been answered by an analysis in Ref. [14]. This analysis clearly showed that with a normal commuting scenario, which was based upon statistical information from a major city, if charging occurred during off-peak hours, no additional generational capacity or transmission requirements would be needed to charge a significant portion of the automotive fleet with PHEVs in the system, as illustrated in Figure 2.9. The analysis also showed that the off-peak charging scenario would also add to the profit of the electric utilities.

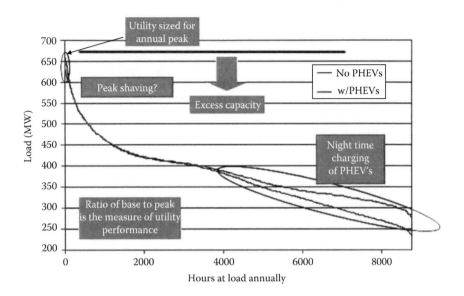

FIGURE 2.9
PHEV impact on utilities' load profile. (Courtesy of NREL, Golden, CO.)

An unexpected result of the analysis was that, although the amount of CO_2 emitted by the electric utility will increase from PHEV charging, if this is compared to the corresponding reduction in tailpipe emission to assess the overall environmental impact, the generation/PHEV transportation system would substantially decrease CO_2 emission even if the current mix of coal, nuclear, and renewable generation facilities were unchanged. This result can be explained, however, because the average efficiency of the electric power system is on the order of 43%, whereas the average efficiency of IC engines is only on the order of 22%. Consequently, the net emission of CO_2 would be reduced by a hybrid system consisting of PHEVs and electric charging during off-peak hours. Moreover, the utility generation profile would be evened out, and this would also contribute to reducing CO_2 emissions, as well as the cost of producing electricity.

In summary, in a PHEV transportation system, if charging of batteries is limited to off-peak hours, this hybrid arrangement can

- Reduce the amount of petroleum consumed by the transportation sector
- Reduce the cost of driving
- Reduce CO_2 emission
- Improve the load profile of electric utilities

Although no quantitative analyses are as yet available, it is believed that the availability of electric storage in the batteries of a fleet of PHEVs could also be used to reduce the peak demand on electric utilities by utilizing the vehicle's batteries as a distributed storage system. Details of such an arrangement would have to be worked out by differential charges, incentives, and taxation arranged between utilities and owners of PHEVs.

2.8 Future All-Electric System

The next step in the development of a viable transportation system could be the all-electric car. All-electric vehicles were mandated in California as part of an effort to reduce air pollution about 25 years ago by the California Air Resources Board. Initially, automakers embraced the idea, but it is likely that the acceptance by Detroit was the result of the mandate that required that at least 2% of all the cars sold in California by any one automaker had to be zero-emission vehicles. The only zero-emission vehicle available at the time was the electric car, and the mandate therefore required selling a certain number of all-electric vehicles. Battery technology at the time was nowhere near ready for commercialization, and in addition, there was no infrastructure available for charging vehicle batteries on the road. As a result, the mandate failed to achieve its objective, and as soon as the mandate was lifted, automakers ceased to make electric vehicles. In the meantime, however, battery technology has evolved to where one could potentially envision an all-electric ground transportation system. Some people propose that there should be enough charging stations built to make it possible to charge batteries anywhere in the country, whereas others propose that there should be, instead of gas stations, battery exchange stations that would simply replace batteries as they reach the end of their charge. At present, batteries take too long to be charged during a trip, and it may be necessary to combine the two ideas to evolve an all-battery transportation system sometime in the near future.

City pollution could be significantly reduced with the use of electric cars due to their zero tailpipe emissions. It is postulated that carbon dioxide emissions could be reduced by up to 40%, depending upon the country's current energy mix. The information from Electric Drive Transportation Association (EDTA) as of January 2015 (Figure 2.10) shows how the market for PHEVs and EVs are growing over time [15].

At this time, it is not possible to assess which electrified powertrain will achieve the bigger market domination over time as the price of fuel, batteries, and other infrastructure all play into that equation. The earlier summary is a snapshot at the time of preparing this text and is a continuous state of flux. The reader is encouraged to follow developments in the current literature included in the bibliography. The EDTA is an excellent source for timely developments regarding this technology.

2.9 Hydrogen for Transportation

Hydrogen was touted as a potential transportation fuel after former President George W. Bush said in his 2001 inaugural address that "a child born today will be driving a pollution-free vehicle... powered by hydrogen." This appeared to be welcome news. However, to analyze whether or not hydrogen is a sustainable technology for transportation, one must take into account all the steps necessary to make the hydrogen from a primary fuel source, get it into the fuel tank, and then power the wheels via a prime mover and a drive train. In other words, one must perform a well-to-wheel analysis, as shown in Figure 2.3. There is a loss in each step, and to obtain the overall efficiency, one must multiple the efficiencies of all the steps. Using natural gas as the primary energy source, a well-to-wheel analysis [16] showed that a hybrid SI car would have a wheel-to-energy

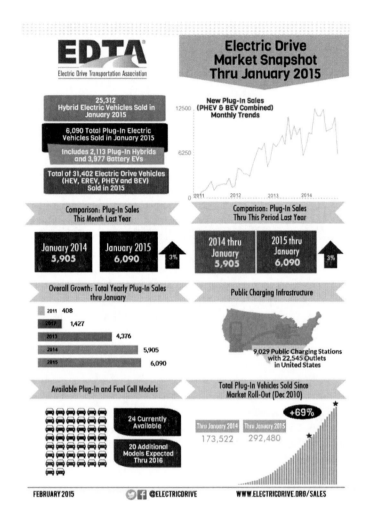

FIGURE 2.10
Growth of market for PHEVs and EVs over time.

efficiency of 32%; a hydrogen-powered fuel cell car with hydrogen made by steam reforming of natural gas would have a wheel-to-energy efficiency of 22%; a hydrogen fuel cell car with hydrogen made by electrolysis with electricity from a natural gas–combined cycle power plant with 55% efficiency would have a well-to-wheel efficiency of 12%, as shown in Table 2.6; and if the hydrogen were produced from photovoltaic cells, the well-to-wheel efficiency of the automobiles would be less than 5%. The estimates in Table 2.6 assume a fuel cell stack efficiency at a peak load of 44.5%, a part load efficiency factor of 1.1, and a transmission efficiency of 90%. For details, see [17].

In addition to the low overall efficiency of a hydrogen transportation system, as well as the high cost compared to other options, it should also be noted that before a hydrogen transportation system could be put into practice, an infrastructure for the distribution and storage of the hydrogen would have to be constructed. An extensive study of the comparative costs of fuel distribution systems conducted at Argonne National Laboratory in 2001 estimated that for a market penetration by the year 2030, 40% of hydrogen vehicles with a

TABLE 2.6

Well-to-Wheel Efficiency of Fuel Cell Vehicle with Hydrogen
Produced by Electrolysis

NG feedstock production efficiency	95%
Conversion efficiency (NG to electricity)	55%
Electrolysis efficiency (electricity to H_2)	63%
Storage and transmission	97%
Compression efficiency	87%
Overall efficiency of fuel production	28%
Total fuel cell well-to-wheel efficiency: $(0.28 \times 445 \times 1.1 \times 0.9)$	12%

mileage of 2.5 times that of average CVs (i.e., about 55 mpg) would minimally cost $320 billion, but could be as high as $600 billion. Based upon these estimates, a national transportation plan based on hydrogen with any currently available technology would be wasteful and inefficient and should not be considered as a pathway to a sustainable energy future [16]. The former U.S. DOE, secretary of Energy and Nobel Laureate, Dr. Steven Chu, concurred with this conclusion and, according to the *New York Times*, cutoff funds for development related to hydrogen fuel cell vehicles in 2009 [19]. Before he left DOE, Chu reversed his support for FCs after several meetings with OEMs, who showed their significant progress and commitment to FCVs. Although still significant infrastructure hurdles exist for hydrogen, the ability to provide a product with a range similar to conventional ICE is very compelling to OEMs, especially when they consider the cost improvement potential of the fuel cell and associated components including new and innovative packaging. Commitments by OEMs to develop hydrogen fuel-cell cars surged between 2012 and 2014. BMW, Toyota, Hyundai, Daimler, Nissan, and Honda all announced plans to commercialize the fuel cell drivetrain, in some cases through collaborative agreements in order to spread early technology risk and accomplish economies of scale.

2.10 Natural Gas as a Transitional Bridging Fuel

Despite its potential as a transportation fuel, except for large buses and trucks, natural gas has heretofore received relatively little attention from the U.S. automobile industry. The only major effort to use natural gas for transportation was the Freedom CAR initiative proposed by President George W. Bush in 2001. This program envisioned replacing gasoline with hydrogen, which at that time was largely produced from natural gas. More than a billion dollars was provided for R&D as well as generous tax incentives. But as shown in Table 2.6, the efficiency of a natural gas/hydrogen vehicle based on a well-to-wheel analysis is less than that of hybrid SI/natural gas vehicle (NGV), and the construction of a hydrogen distribution system would be extremely expensive. Thus, the hydrogen via natural gas effort for transportation was a failure and was terminated in 2010. However, since 2009, new supplies of natural gas have become available in the United States, primarily as a result of fracking technology for natural gas extraction from oil shale deposits. This development has heightened awareness of the potential of natural gas as a bridging fuel for transportation for an eventual zero carbon future. Within this context, MIT has conducted an interdisciplinary investigation that addresses the question, "What is the role of natural gas in a carbon-constrained economy?" [20].

Natural gas is likely to find increased use in the transportation sector with compressed natural gas (CNG) playing an important role, particularly for high-mileage fleets. But the advantage of liquid fuel in transportation indicates that the chemical conversion of the gas into some form of liquid fuel may be a more desirable pathway for the future. It should be noted that a basic infrastructure for distributing natural gas exists, and if CNG were to be used to fuel automobiles, the only major addition at gas stations that have a natural gas outlet would be a compressor to increase the gas pressure above that in the natural gas automobile tank. However, the vast majority of natural gas supplies are delivered to markets by pipeline, and delivery costs typically represent a relatively large fraction of the total cost in the supply chain.

Natural gas vehicles have been in use for trucks and other large vehicles in many parts of the world for some time as shown in Table 2.7. With a starting price of $26,305, a 2013 natural gas Civic costs $8,100 more than the base gasoline model. Big trucks that burn 20,000–40,000 gal a year can easily make up that difference, but it takes far longer for regular consumers, who may use only 500 gal/year. Home fueling stations add $4000–$6000 to the cost.

Range is also a concern. The United States has 1100 natural gas fueling stations, but only about half are open to the public. A natural gas Civic can go around 200 miles on a tank. That's better than an electric car, which might go 100 miles on a charge. But it's less than the 300–350 miles a driver can go on a tank of gasoline in a regular Civic.

All those things could change, but GE is trying to develop a $500 home fueling station, and the federal government could encourage sales with tax credits, as it has done with its $7500 electric vehicle credit. Some states are already giving tax credits to CNG vehicle buyers, including West Virginia—which gives up to $7,500 for smaller vehicles and $20,000 for trucks—and Colorado, which gives up to $6,000.

As availability of natural gas wanes, a number of renewable sources for natural gas or biomethane are available. Biomethane can be produced from any organic matter. Nature produces it naturally in landfills, but it can also be produced in anaerobic digesters or through pyrolysis from sewage, industrial, animal, or crop wastes, or from specific energy crops. The biomethane from landfills used in NGVs reduces greenhouse gases by

TABLE 2.7

Top 11 NGV Countries

Country	No. of NGVs
1. Pakistan	2,850,500
2. Iran	2,070,930
3. Argentina	1,927,007
4. Brazil	1,667,038
5. India	1,100,000
6. Italy	754,659
7. China	550,000
8. Colombia	340,000
9. Thailand	238,583
10. Ukraine	200,019
11. United States	112,000

Source: Gas Vehicle Rep., 10(4), June 2011.

90% according to the California Air Resources Board. After biomethane is produced, it can be injected into natural gas pipeline systems and sold to NGV station operators. A 5% or 10% blend of biomethane with natural gas would add to NGV's greenhouse gas potential. While many other alternative fuels are still in the R&D stage, NGVs are not in that category and are ready to go now.

References

1. Kreith, F. (1999) *Ground Transportation for the 21st Century*. National Conference of State Legislatures, Denver, CO; ASME Press, New York.
2. Kreith, F., West, R.E., and Isler, B. (2002) Legislative and technical perspectives for advanced ground transportation system. *Transportation Quarterly* 96(1), 51–73, Winter 2002.
3. Dismukes, C. et al. (June 2008) Aquatic phototrophs: Efficient alternatives to land-based crops for biofuels. *Current Opinions in Biotechnology* 19, 235–240.
4. Christi, Y. (2007) Biodiesel from microalgae. *Biotechnology Advances* 25, 294–306.
5. Kreith, F. and West, R.E. (2003) Gauging efficiency: Well-to-wheel. *Mechanical Engineering Power*, 20–23.
6. Davis, S. and Diegel, S. (2004) *Transportation Energy Data Book*. Oak Ridge National Laboratory/ U.S. Department of Energy, Oak Ridge, TN.
7. EPRI (2004) Advanced batteries for electric drive vehicles: A technology and cost-effectiveness assessment for battery electric vehicles, power assist hybrid electric vehicles and plug-in hybrid electric vehicles. EPRI Tech Report 1009299, EPRI, Palo Alto, CA.
8. Kreith, F. and West, R.E. (September 2006) A vision for a secure transportation system without hydrogen or oil. *ASME Journal of Energy Resources Technology* 128(9), 236–243.
9. U.S. EPA (2013). E15 (a blend of gasoline and ethanol). www.epa.gov/otaq/regs/fuels/ additive/e15. Accessed April 8, 2014.
10. Renewable Fuels Association. The new fuel: E15. www.ethanolrfa.org/pages/E15. Accessed April 8, 2014.
11. Kreith, F. and West, R.E. (May 2008) A scenario for a secure transportation system based on fuel from biomass. *Journal of Solar Energy Engineering* 130, 1–6.
12. Farrell, A.E., Plevin, R.J., Turner, B.T., Jones, A.D., O'Hare, M., and Kammen, D.M. (2006) Ethanol can contribute to energy and environmental goals. *Science* 311, 506–508.
13. EIA (2007) www.cia.doe.gov/fuelelectric/electricityinfocard2005.
14. Himelich, J.B. and Kreith, F. (2008) Potential benefits of plug-in hybrid electric vehicles for consumers and electric power utilities. In: *Proceedings of ASME 2008 IMEC*, Boston, MA, October 31–November 6, 2008.
15. Electric Drive Transportation Association (EDTA). (2015). www.electricdrive.org. Accessed April 1, 2015.
16. Kreith, F. and West, R.E. (2004) Fallacies of a hydrogen economy: A critical analysis of hydrogen production and utilization. *Journal of Energy Resources Technology* 126, 249–257.
17. Kreith, F., West, R.E., and Isler, B.E. (2002) Efficiency of advanced ground transportation technologies. *Journal of Energy Resources Technology* 24, 173–179.
18. Mince, M. (2001) Infrastructure requirement of advanced technology vehicles. In: *NCSL/TRB Transportation Technology and Policy Symposium*, Argonne National Laboratory, Argonne, IL.
19. Wald, M.L. (2009) *New York Times* News Service, April 8, 2009.
20. Moniz, E. et al. (2011). The future of natural gas: An interdisciplinary MIT study. *MIT Energy Initiative*. http://web.mit.edu/mitei/research/studies/natural-gas-2011.shtml.
21. Bartlett, A.A. (2000) An analysis of U.S. and world oil production patterns using Hubbard-style curves. *Mathematical Geology* 32(1).

Online Resources

http://www.nissanusa.com/leaf-electric-car/index.

http://www.teslamotors.com/models.

http://www.nrel.gov/sustainable_nrel/transportation.html.

http://www.nrel.gov/learning/re_biofuels.html.

http://www.nrel.gov/vehiclesandfuels/energystorage/batteries.html.

http://cta.ornl.gov/vtmarketreport/index.shtml.

http://cta.ornl.gov/data/download31.shtml (a large source of data that can be downloaded for free).

3

Economics Methods*

Walter Short and Rosalie Ruegg

CONTENTS

---

* Modified by Walter Short, retired from the National Renewable Energy Laboratory (NREL), from the original text prepared by Rosalie Ruegg for the 2007 *Handbook of Energy Efficiency and Renewable Energy.*

3.1 Introduction

Economic-evaluation methods facilitate comparisons among energy technology investments. Generally, the same methods can be used to compare investments in energy supply or energy efficiency. All sectors of the energy community need guidelines for making economically efficient energy-related decisions.

This chapter provides an introduction to some basic methods that are helpful in designing and sizing cost-effective systems, and in determining whether it is economically efficient to invest in specific energy efficiency or renewable energy projects. The targeted audience includes analysts, architects, engineers, designers, builders, codes and standards writers, and government policy makers—collectively referred to as the "design community."

The focus is on microeconomic methods for measuring cost-effectiveness of individual projects or groups of projects, with explicit treatment of uncertainty. The chapter does not treat macroeconomic methods and national market-penetration models for measuring economic impacts of energy efficiency and renewable energy investments on the national economy. It provides sufficient guidance for computing the measures of economic performance for relatively simple investment choices, and it provides the fundamentals for dealing with complex investment decisions.

3.2 Making Economically Efficient Choices*

Economic-evaluation methods can be used in a number of ways to increase the *economic efficiency* of energy-related decisions. There are methods that can be used to obtain the largest possible savings in energy costs for a given energy budget; there are methods that can be used to achieve a targeted reduction in energy costs for the lowest possible efficiency/renewable energy investment; and there are methods that can be used to determine how much it pays to spend on energy efficiency and renewable energy to lower total lifetime costs, including both *investment costs* and energy costs.

The first two ways of using economic-evaluation methods (i.e., to obtain the largest savings for a fixed budget and to obtain a targeted savings for the lowest budget) have more limited applications than the third, which aims at minimizing total costs or maximizing *net benefits* (*NB*) (net savings (*NS*)) from expenditure on energy efficiency and renewables. As an example of the first, a plant owner may budget a specific sum of money for the purpose of retrofitting the plant for energy efficiency. As an example of the second, designers may be required by state or federal building standards and/or codes to reduce the design energy loads of new buildings below some specified level. As an example of the third, engineers may be required by their clients to include, in a production plant, those energy efficiency and renewable energy features that will pay off in terms of lower overall production costs over the long run.

Note that economic efficiency is not necessarily the same as engineering thermal efficiency. For example, one furnace may be more "efficient" than another in the engineering technical sense, if it delivers more units of heat for a given quantity of fuel than another. Yet, it may not be economically efficient if the first cost of the higher output

---

* This section is based on a treatment of these concepts provided by Marshall and Ruegg (1980a).

furnace outweighs its fuel savings. The focus in this chapter is on economic efficiency, not engineering efficiency.

Economic efficiency is conceptually illustrated in Figures 3.1 through 3.3 with an investment in energy efficiency. Figure 3.1 shows the level of energy conservation, Q_c, that maximizes NB from energy conservation—that is, the level that is most profitable over the long run. Note that it corresponds to the level of energy conservation at which the curves are most distant from one another.

Figure 3.2 shows how "marginal analysis" can be used to find the same level of conservation, Q_c, that will yield the largest NB. It depicts changes in the total benefits and cost curves (i.e., the derivatives of the curves in Figure 3.1) as the level of energy conservation is increased. The point of intersection of the marginal curves coincides with the most profitable level of energy conservation indicated in Figure 3.1. This is the point at which the cost of adding one more unit of conservation is just equal to the corresponding benefits in terms of energy savings (i.e., the point at which "marginal costs" and "marginal benefits" are equal). To the left of the point of intersection, the additional

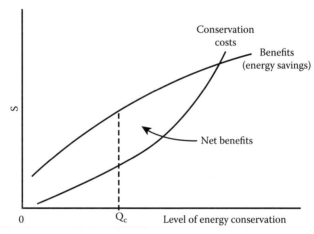

FIGURE 3.1
Maximizing net benefits.

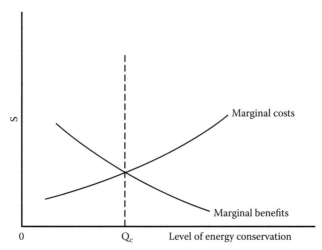

FIGURE 3.2
Equating marginal benefits and marginal costs.

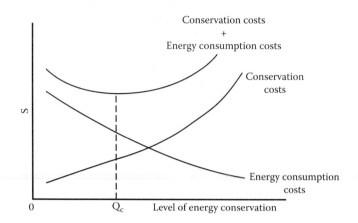

FIGURE 3.3
Minimizing LCC.

benefits from increasing the level of conservation by another unit are greater than the additional costs, and it pays to invest more. To the right of the point of intersection, the costs of an addition to the level of conservation exceed the benefits—and the level of total NB begins to fall, as shown in Figure 3.1. Figure 3.3 shows that the most economically efficient level of energy conservation, Q_c, is that for which the total cost curve is at a minimum.

The most economically efficient level of conservation is the same, Q_c, in Figures 3.1 through 3.3. Three different approaches to finding Q_c are illustrated: finding the maximum difference between benefits and costs; finding the point where marginal benefits equal marginal costs; and finding the lowest life-cycle costs. The graphical methods of Figures 3.1 through 3.3 are captured by the quantitative methods described in the section that follows.

3.3 Economic Evaluation Methods*

There are a number of closely related, commonly used methods for evaluating economic performance. These include the *life-cycle cost* (LCC) method, levelized cost of energy (LCOE) method, net present value (NPV) or NB (net present worth) method, benefit/cost (or savings-to-investment) ratio (SIR) method, internal rate-of-return (IRR) method, overall rate-of-return (ORR) method, and discounted payback (DPB) method. All of these methods are used when the important effects can be measured in dollars. If incommensurable effects are critical to the decision, it is important that they also be taken into account. But, because only quantified effects are included in the calculations for these economic methods, unquantified effects must be treated outside the models. Brief treatments of the methods are provided; some additional methods are identified but not treated. For more comprehensive treatments, see Ruegg and Marshall (1990).

---

* These methods are treated in detail in Ruegg and Marshall (1990).

3.3.1 Life-Cycle Cost (LCC) Method

The life-cycle costing method sums, for each investment alternative, the costs of acquisition, maintenance, repair, replacement, energy, and any other monetary costs (less than any income amounts, such as salvage value) that are affected by the investment decision. The *time value of money* must be taken into account for all amounts, and the amounts must be considered over the relevant period. All amounts are usually measured either in *present value* or annual value dollars. This is discussed later in Sections 3.5.2 and 3.5.3. At a minimum, for comparison, the investment alternatives should include a "base-case" alternative of not making the energy efficiency or renewable investment, and at least one case of an investment in a specific efficiency or renewable system. Numerous alternatives may be compared. The alternative with the lowest LCC that meets the investor's objective and constraints is the preferred investment. This least-cost solution is analogous to the least cost presented in Figure 3.3.

The following is a formula for finding the LCCs of each alternative:

$$LCC_{A1} = I_{A1} + E_{A1} + M_{A1} + R_{A1} - S_{A1} \qquad (3.1)$$

where
  LCC_{A1} = life-cycle cost of alternative A1
  I_{A1} = present-value investment costs of alternative A1
  E_{A1} = present-value energy costs associated with alternative A1
  M_{A1} = present-value nonfuel operating and maintenance cost of A1
  R_{A1} = present-value repair and replacement costs of A1
  S_{A1} = present-value resale (or salvage) value less disposal cost associated with alternative A1

The LCC method is particularly useful for decisions that are made primarily on the basis of cost-effectiveness, such as whether a given energy efficiency or renewable energy investment will lower total cost (e.g., the sum of investment and operating costs). It can be used to compare alternative designs or sizes of systems, as long as the systems provide the same service. The method, if used correctly, can be used to find the overall cost-minimizing combination of energy efficiency investments and energy supply investments within a given facility. However, in general, it cannot be used to find the best investment, because totally different investments do not provide the same service.

3.3.2 Levelized Cost of Energy (LCOE) Method

The LCOE is similar to the LCC method, in that it considers all the costs associated with an investment alternative and takes into account the time value of money for the analysis period. However, it is generally used to compare two alternative energy supply technologies or systems, for example, two electricity production technologies that may or may not provide exactly the same service ,that is, the same level of energy production. It differs from the LCC in that it usually considers taxes, but like LCC, frequently ignores financing costs.

The LCOE is the value that must be received for each unit of energy produced to ensure that all costs and a reasonable profit are made. Profit is ensured by discounting future revenues at a discount rate that equals the rate of return that might be gained on other investments of comparable risk, that is, the opportunity cost of capital. This can be represented in the following equation:

$$\sum_{t=1}^{t=N} \frac{LCOE * Q_t}{(1+d')^t} = \sum_{t=0}^{t=N} \frac{C_t}{(1+d)^t} \tag{3.2}$$

where
  N = the analysis period
  Q_t = the amount of energy production in period t
  C_t = the cost incurred in period t
  d′ = the discount rate or opportunity cost of capital; if d′ is a real discount rate (excludes inflation) then the LCOE will be in real (constant) dollar terms, while if d′ is a normal discount rate, the LCOE will be in nominal (current) dollar terms
  d = the discount rate used to bring future costs back to their present value. If those costs are expressed in real dollars, then the discount rate d should be a real discount rate; while, if they are in nominal dollars, the discount rate should be a nominal discount rate

3.3.3 Net Present Value (NPV) or Net Benefits (NB) Method

The NPV method finds the excess of benefits over costs, where all amounts are discounted for their time value. (If costs exceed benefits, net losses result.)

The NPV method is also often called the "net present worth" or "NS" method. When this method is used for evaluating a cost-reducing investment, the cost savings are the benefits, and it is often called the "NS" method.

Following is a formula for finding the NPV from an investment, such as an investment in energy efficiency or renewable energy systems:

$$NPV_{A1:A2} = \sum_{t=0}^{N} \frac{B_t - C_t}{(1+d)^t} \tag{3.3}$$

where
  $NPV_{A1:A2}$ = NB, that is, present value benefits (savings) net of present value costs for alternative A1 as compared with alternative A2
  B_t = benefits in year t, which may be defined to include energy savings associated with using alternative A1 instead of alternative A2
  C_t = costs in year t associated with alternative A1 as compared with a mutually exclusive alternative A2
  d = discount rate

The NPV (NB) method is useful for deciding whether to make a given investment and for designing and sizing systems. It is not appropriate for comparing investments that provide different services.

3.3.4 Benefit-to-Cost Ratio (BCR) or Savings-to-Investment Ratio (SIR) Method

This method divides benefits by costs or, equivalently, savings by investment. When used to evaluate energy efficiency and renewable energy systems, benefits are in terms of energy cost savings. The numerator of the SIR is usually constructed as energy savings, and net of maintenance and repair costs; and the denominator as the sum of investment costs and the present value of replacement costs less salvage value (capital cost items). However, depending on the objective, sometimes only initial investment costs are placed in the denominator and the other costs are subtracted in the numerator—or sometimes only the investor's equity capital is placed in the denominator. Like the three preceding methods, this method is based on discounted cash flows.

Unlike the three preceding methods that provided a performance measure in dollars, this method gives the measure as a dimensionless number. The higher the ratio, the more the dollar savings realized per dollar of investment. In particular, a value greater than 1 is generally required for an investment to be considered economically efficient.

Following is a commonly used formula for computing the ratio of savings-to-investment costs:

$$\text{SIR}_{A1:A2} = \frac{\sum_{t=0}^{N}\left(CS_t(1+d)^{-t}\right)}{\sum_{t=0}^{N}\left(I_t(1+d)^{-t}\right)} \tag{3.4}$$

where
$\text{SIR}_{A1:A2}$ = savings-to-investment ratio for alternative A1 relative to mutually exclusive alternative A2
CS_t = cost savings (excluding those investment costs in the denominator) plus any positive benefits of alternative A1 as compared with mutually exclusive alternative A2
I_t = additional investment costs for alternative A1 relative to A2

Note that the particular formulation of the ratio with respect to the placement of items in the numerator or denominator can affect the outcome. One should use a formulation appropriate to the decision maker's objectives.

The ratio method can be used to determine whether or not to accept or reject a given investment on economic grounds. It also can be used for design and size decisions and other choices among mutually exclusive alternatives, if applied incrementally (i.e., the investment and savings are the difference between the two mutually exclusive alternatives). A primary application of the ratio method is to set funding priorities among projects competing for a limited budget. When it is used in this way—and when project costs are "lumpy" (making it impossible to fully allocate the budget by taking projects in order according to the size of their ratios)—SIR should be supplemented with the evaluation of alternative sets of projects using the NPV or NB method.

3.3.5 Internal Rate-of-Return (IRR) Method

The IRR method solves for the discount rate for which dollar savings are just equal to dollar costs over the analysis period; that is, the rate for which the NPV is zero. This discount rate is the rate of return on the investment. It is compared to the investor's minimum acceptable rate of return to determine whether the investment is desirable. Unlike the preceding three techniques, the IRR does not call for the inclusion of a prespecified discount rate in the computation, but, rather, solves for a discount rate.

The rate of return is typically calculated by a process of trial and error, by which various compound rates of interest are used to discount cash flows until a rate is found for which the NPV of the investment is zero. The approach is the following: compute NPV using Equation 3.3, except substitute a trial interest rate for the discount rate, d, in the equation. A positive NPV means that the IRR is greater than the trial rate; a negative NPV means that the IRR is less than the trial rate. Based on the information, try another rate. By a series of iterations, find the rate at which NPV equals zero.

Computer algorithms, graphical techniques, and—for simple cases—discount-factor tabular approaches are often used to facilitate IRR solutions (Ruegg and Marshall, 1990, pp. 71–72). Expressing economic performance as a rate of return can be desirable for ease in comparing the returns on a variety of investment opportunities, because returns are often expressed in terms of annual rates of return. The IRR method is useful for accepting or rejecting individual investments or for allocating a budget. For designing or sizing projects, the IRR method, like the SIR, must be applied incrementally. It is not recommended for selecting between mutually exclusive investments with significantly different lifetimes (e.g., a project with a high annual return of 35% for 20 years is a much better investment than a project with the same 35% annual return for only 2 years).

IRR is a widely used method, but it is often misused, largely due to shortcomings that include the possibility of

- No solution (the sum of all nondiscounted returns within the analysis period are less than the investment costs)
- Multiple solution values (some costs occur later than some of the returns)
- Failure to give a measure of overall return associated with the project over the analysis period (returns occurring before the end of the analysis are implicitly assumed to be reinvested at the same rate of return as the calculated IRR. This may or may not be possible).

3.3.6 Overall Rate-of-Return (ORR) Method

The ORR method corrects for the last two shortcomings expressed earlier for the IRR. Like the IRR, the ORR expresses economic performance in terms of an annual rate of return over the analysis period. But unlike the IRR, the ORR requires, as input, an explicit reinvestment rate on interim receipts and produces a unique solution value.[*] The explicit reinvestment rate makes it possible to express net cash flows (excluding investment costs) in terms of their future value at the end of the analysis period. The ORR is then easily computed with a closed-form solution as shown in Equation 3.5.

---

[*] As shown in Equation 3.5, the reinvestment rate is also used to bring all investments back to their present value. Alternatively, investments after time zero can be discounted by the overall growth rate. In this case, a unique solution is not guaranteed, and the ORR must be found iteratively (Stermole and Stermole, 2000).

$$\text{ORR}_{A1:A2} = \left[\frac{\left[\sum_{t=0}^{N} (B_t - C_t)(1+r)^{N-t} \right]}{\sum_{t=0}^{N} \left[\frac{I_t}{(1+r)^t} \right]} \right]^{1/N} - 1 \tag{3.5}$$

where

ORR$_{A1:A2}$ = overall rate of return on a given investment alternative A1 relative to a mutually exclusive alternative A2 over a designated study period

B_t = benefits from a given alternative relative to a mutually exclusive alternative A2 over time period t

C_t = costs (excluding that part of investment costs on which the return is to be maximized) associated with a given alternative relative to a mutually exclusive alternative A2 over time t

r = the reinvestment rate at which net returns can be reinvested, usually set equal to the discount rate

N = the length of the study period

I_t = investment costs in time t on which the return is to be maximized

The ORR is recommended as a substitute for the IRR, because it avoids some of the limitations and problems of the IRR. It can be used for deciding whether or not to fund a given project, for designing or sizing projects (if it is used incrementally), and for budget-allocation decisions.

3.3.7 Discounted Payback (DPB) Method

This evaluation method measures the elapsed time between the time of an initial investment and the point in time at which accumulated discounted savings or benefits—net of other accumulated discounted costs—are sufficient to offset the initial investment, taking into account the time value of money. (If costs and savings are not discounted, the technique is called "simple payback.") For the investor who requires a rapid return of investment funds, the shorter the length of time until the investment pays off, the more desirable is the investment.

To determine the *DPB* period, find the minimum value of Y (year in which payback occurs) such that the following equality is satisfied.

$$\sum_{t=1}^{Y} \frac{B_t - C_t'}{(1+d)^t} = I_0 \tag{3.6}$$

where

B_t = benefits associated in period t with one alternative as compared with a mutually exclusive alternative

C_t' = costs in period t (not including initial investment costs) associated with an alternative as compared with a mutually exclusive alternative in period t

I_0 = initial investment costs of an alternative as compared with a mutually exclusive alternative, where the initial investment cost comprises total investment costs

DPB is often—correctly—used as a supplementary measure when project life is uncertain. It is used to identify feasible projects when the investor's time horizon is constrained. It is used as a supplementary measure in the face of uncertainty to indicate how long capital is at risk. It is a rough guide for accept/reject decisions. It is also overused and misused. Because it indicates the time at which the investment just breaks even, it is not a reliable guide for choosing the most profitable investment alternative, as savings or benefits after the payback time could be significant.

3.3.8 Other Economic-Evaluation Methods

A variety of other methods have been used to evaluate the economic performance of energy systems, but these tend to be hybrids of those presented here. One of these is the required revenue method, which computes a measure of the before-tax revenue in present or annual value dollars required to cover the costs on an after-tax basis of an energy system (Ruegg and Short, 1988, pp. 22–23). Mathematical programming methods have also been used to evaluate the optimal size or design of projects, as well as other mathematical and statistical techniques.

3.4 Risk Assessment

Many of the inputs to the evaluation methods mentioned earlier will be highly uncertain at the time an investment decision must be made. To make the most informed decision possible, an investor should employ these methods within a framework that explicitly accounts for risk and uncertainty.

Risk assessment provides decision makers with information about the "risk exposure" inherent in a given decision—that is, the probability that the outcome will be different from the "best-guess" estimate. Risk assessment is also concerned with the "risk attitude" of the decision maker, which describes his/her willingness to take a chance on an investment of uncertain outcome. Risk assessment techniques are typically used in conjunction with the evaluation methods outlined earlier; and not as stand-alone evaluation techniques.

The risk assessment techniques range from simple and partial to complex and comprehensive. Though none takes the risk out of making decisions, the techniques—if used correctly—can help the decision maker make more informed choices in the face of uncertainty.

This chapter provides an overview of the following probability-based risk assessment techniques:

- Expected value (EV) analysis
- Mean-variance criterion (MVC) and coefficient of variation (CV)
- Risk-adjusted discount rate (RADR) technique
- Certainty equivalent (CE) technique
- Monte Carlo simulation
- Decision analysis
- Real options analysis (ROA)
- Sensitivity analysis

There are other techniques that are used to assess the risks and uncertainty (e.g., CAP_M and break-even analysis), but those are not treated here.

3.4.1 Expected Value (EV) Analysis

EV analysis provides a simple way of taking into account uncertainty about input values, but it does not provide an explicit measure of risk in the outcome. It is helpful in explaining and illustrating risk attitudes.

How to calculate EV: An "expected value" is the sum of the products of the dollar value of alternative outcomes, a_i ($i = 1,\ldots,$ n), and their probabilities of occurrence, p_i. The EV of the decision is calculated as follows:

$$EV = a_1 p_1 + a_2 p_2 + \cdots + a_n p_n \tag{3.7}$$

Example of EV analysis: The following simplified example illustrates the combining of EV analysis and NPV analysis to support a purchase decision.

Assume that a not-for-profit organization must decide whether to buy a given piece of energy-saving equipment. Assume that the unit purchase price of the equipment is $100,000, the yearly operating cost is $5,000 (obtained by a fixed-price contract), and both costs are known with certainty. The annual energy cost savings, on the other hand, are uncertain, but can be estimated in probabilistic terms as shown in Table 3.1 in the columns headed a_1, p_1, a_2, and p_2. The present-value calculations are also given in Table 3.1.

If the equipment decision was based only on NPV, calculated with the "best-guess" energy savings (column a_1), the equipment purchase would be found to be uneconomic with a NPV of $ – 483. But if the possibility of greater energy savings is taken into account by using the EV of savings rather than the best guess, the conclusion is that, over repeated applications, the equipment is expected to be *cost-effective*. The expected NPV of the energy-saving equipment is $25,000 per unit.

Advantages and disadvantages of the EV technique: An advantage of the technique is that it predicts a value that tends to be closer to the actual value than a simple "best-guess" estimate over repeated instances of the same event, provided, of course, that the input probabilities can be estimated with some accuracy.

A disadvantage of the EV technique is that it expresses the outcome as a single-value measure, such that there is no explicit measure of risk. Another is that the estimated outcome

TABLE 3.1

Expected Value (EV) Example

| Year | Equipment Purchase $1000 | Operating Costs $1000 | Energy Savings | | | | PV[a] Factor | PV $1000 |
			a_1 $1000	p_1	a_2 $1000	p_2		
0	−100	—	—	—	—	—	1	−100
1		−5	25	0.8	50	0.2[b]	0.926	23.1
2		−5	30	0.8	60	0.2	0.857	26.6
3		−5	30	0.7	60	0.3	0.794	27.0
4		−5	30	0.6	60	0.4	0.735	27.2
5		−5	30	0.8	60	0.2	0.681	21.1
								25.0

Note: Expected NPV.
[a] Present-value calculations are based on a discount rate of 8%.
[b] Probabilities sum to 1.0 in a given year.

is predicated on many replications of the event, with the EV, in effect, a weighted average of the outcome over many like events. But the EV is unlikely to occur for a single instance of an event. This is analogous to a single coin toss: the outcome will be either heads or tails, not the probabilistic-based weighted average of both.

EV and risk attitude: EVs are useful in explaining risk attitude. Risk attitude may be thought of as a decision maker's preference between taking a chance on an uncertain money payout of known probability versus accepting a sure money amount. Suppose, for example, a person were given a choice between accepting the outcome of a fair coin toss where heads means winning $10,000 and tails means losing $5000 and accepting a certain cash amount of $2000. EV analysis can be used to evaluate and compare the choices. In this case, the EV of the coin toss is $2500, which is $500 more than the certain money amount. The "risk-neutral" decision maker will prefer the coin toss because of its higher EV. The decision maker who prefers the $2000 certain amount is demonstrating a "risk-averse" attitude. On the other hand, if the certain amount were raised to $3000 and the first decision maker still preferred the coin toss, he or she would be demonstrating a "risk-taking" attitude. Such trade-offs can be used to derive a "utility function" that represents a decision maker's risk attitude.

The risk attitude of a given decision maker is typically a function of the amount at risk. Many people who are risk averse when faced with the possibility of significant loss, become risk neutral—or even risk taking, when potential losses are small. Because decision makers vary substantially in their risk attitudes, there is a need to assess not only risk exposure (i.e., the degree of risk inherent in the decision) but also the risk attitude of the decision maker.

3.4.2 Mean-Variance Criterion (MVC) and Coefficient of Variation (CV)

These techniques can be useful in choosing among risky alternatives, if the mean outcomes and standard deviations (variation from the mean) can be calculated.

Consider a choice between two projects—one with higher mean NB and a lower standard deviation than the other. This situation is illustrated in Figure 3.4. In this case, the project whose probability distribution is labeled B can be said to have stochastic dominance over the project labeled A. Project B is preferable to Project A, both on grounds that its output is likely to be higher and that it entails less risk of loss. But what if Project A, the alternative with higher risk, has the higher mean NB, as illustrated in Figure 3.5? If this were the case, the MVC would provide inconclusive results.

When there is no stochastic dominance of one project over the other(s), it is helpful to compute the CV to determine the relative risk of the alternative projects. The CV indicates

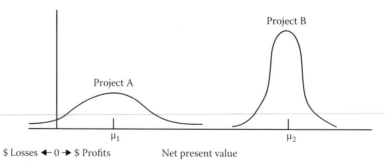

FIGURE 3.4
Stochastic dominance as demonstrated by mean-variance criterion.

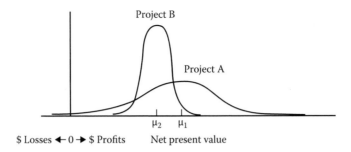

FIGURE 3.5
Inconclusive results from mean-variance criterion.

which alternative has the lower risk per unit of project output. Risk-averse decision makers will prefer the alternative with the lower CV, other things being equal. The CV is calculated as follows:

$$CV = \frac{\sigma}{\mu} \tag{3.8}$$

where
 CV = coefficient of variation
 σ = standard deviation
 μ = mean

The principal advantage of these techniques is that they provide quick, easy-to-calculate indications of the returns and risk exposure of one project relative to another. The principal disadvantage is that the MVC does not provide a clear indication of preference when the alternative with the higher mean output has the higher risk, or vice versa.

3.4.3 Risk-Adjusted Discount Rate (RADR) Technique

The RADR technique takes account of risk through the discount rate. If a project's benefit stream is riskier than that of the average project in the decision maker's portfolio, a higher-than-normal discount rate is used; if the benefit stream is less risky, a lower-than-normal discount rate is used. If costs are the source of the higher-than-average uncertainty, a lower-than-normal discount rate is used and vice versa. The greater the variability in benefits or costs, the greater the adjustment in the discount rate.

The RADR is calculated as follows:

$$RADR = RFR + NRA + XRA \tag{3.9}$$

where
 RADR = risk-adjusted discount rate
 RFR = risk-free discount rate, generally set equal to the treasury bill rate
 NRA = "normal" risk adjustment to account for the average level of risk encountered in the decision maker's operations
 XRA = extra risk adjustment to account for risk greater or less than normal risk

TABLE 3.2

RADR Example

Year	Costs ($M)	Revenue ($M)	PV Costs[a] ($M)	PV Revenue[a] ($M)	NPV ($M)
0	80	—	80	—	−80
1	5	20	4	17	13
2	5	20	4	14	10
3	5	20	4	12	8
4	5	20	3	10	7
5	5	20	3	9	6
6	5	20	3	7	4
7	5	20	2	6	4
Total NPV					−28

[a] Costs are discounted with a discount rate of 12%; revenue with a discount rate of 18%.

An example of using the RADR technique is the following: A company is considering an investment in a new type of alternative energy system with high payoff potential and high risk on the benefits side. The projected cost and revenue streams and the discounted present values are shown in Table 3.2. The treasury bill rate, taken as the risk-free rate, is 8%. The company uses a normal risk adjustment of 5% to account for the average level of risk encountered in its operations. This investment is judged to be twice as risky as the company's average investment, so an additional risk adjustment of 5% is added to the risk-adjusted discount rate. Hence, the RADR is 18%. With this RADR, the NPV of the investment is estimated to be a loss of $28 million. On the basis of this uncertainty analysis, the company would be advised not to accept the project.

Advantages of the RADR technique are that it provides a way to account for both *risk exposure* and *risk attitude*. Moreover, RADR does not require any additional steps for calculating NPV once a value of the RADR is established. The disadvantage is that it provides only an approximate adjustment. The value of the RADR is typically a rough estimate based on sorting investments into risk categories and adding a "fudge factor" to account for the decision maker's risk attitude. It generally is not a fine-tuned measure of the inherent risk associated with variation in cash flows. Further, it typically is biased toward investments with short payoffs because it applies a constant RADR over the entire analysis period, even though risk may vary over time.

3.4.4 Certainty Equivalent (CE) Technique

The CE technique adjusts investment cash flows by a factor that will convert the measure of economic worth to a "CE" amount—the amount a decision maker will find equally acceptable to a given investment with an uncertain outcome. Central to the technique is the derivation of the certainty equivalent factor (CEF), which is used to adjust net cash flows for uncertainty.

Risk exposure can be built into the CEF by establishing categories of risky investments for the decision maker's organization and linking the CEF to the CV of the returns—greater variation translating into smaller CEF values. The procedure is as follows:

1. Divide the organization's portfolio of projects into risk categories. Examples of investment risk categories for a private utility company might be the following: low-risk investments—expansion of existing energy systems and equipment

replacement; moderate-risk investments—adoption of new, conventional energy systems; and high-risk investments—investment in new alternative energy systems.

2. Estimate the CVs (see Section 3.4.2) for each investment-risk category (e.g., on the basis of historical risk-return data).

3. Assign CEFs by year, according to the coefficients of variation, with the highest-risk projects being given the lowest CEFs. If the objectives are to reflect only risk exposure, set the CEFs such that a risk-neutral decision maker will be indifferent between receiving the estimated certain amount and the uncertain investment. If the objective is to reflect risk attitude as well as risk exposure, set the CEFs such that the decision maker with his or her own risk preference will be indifferent.

To apply the technique, proceed with the following steps:

4. Select the measure of economic performance to be used—such as the measure of NPV (i.e., NB).

5. Estimate the net cash flows and decide in which investment-risk category the project in question fits.

6. Multiply the yearly net cash flow amounts by the appropriate CEFs.

7. Discount the adjusted yearly net cash flow amounts with an RFR (an RFR is used because the risk adjustment is accomplished by the CEFs).

8. Proceed with the remainder of the analysis in the conventional way.

In summary, the CE NPV is calculated as follows:

$$NPV_{CE} = \sum_{t=0}^{N} \left[\frac{CEF_t(B_t - C_t)}{(1 + RFD)^t} \right], \tag{3.10}$$

where
NPV_{CE} = NPV adjusted for uncertainty by the CE technique
B_t = estimated benefits in time period t
C_t = estimated costs in time period t
RFD = risk-free discount rate

Table 3.3 illustrates the use of this technique for adjusting NPV calculations for an investment in a new, high-risk alternative energy system. The CEF is set at 0.76 and is assumed to be constant with respect to time.

A principal advantage of the CE Technique is that it can be used to account for both risk exposure and risk attitude. Another is that it separates the adjustment of risk from discounting and makes it possible to make more precise risk adjustments over time. A major disadvantage is that the estimation of CEF is only approximate.

3.4.5 Monte Carlo Simulation

Monte Carlo simulation entails the iterative calculation of the measure of economic worth from probability functions of the input variables. The results are expressed as a probability

TABLE 3.3

CE Example (Investment-Risk Category;
High-Risk—New-Alternative Energy System)

Yearly Net Cash Flow ($M)		CV	CEF	RFD Discount Factors[a]	NPV ($M)
1	−100	0.22	0.76	0.94	−71
2	−100	0.22	0.76	0.89	−68
3	20	0.22	0.76	0.84	13
4	30	0.22	0.76	0.79	18
5	45	0.22	0.76	0.75	26
6	65	0.22	0.76	0.7	35
7	65	0.22	0.76	0.67	33
8	65	0.22	0.76	0.63	31
9	50	0.22	0.76	0.59	22
10	50	0.22	0.76	0.56	21
Total NPV					60

[a] The RFD is assumed equal to 6%.

density function and as a cumulative distribution function. The technique, thereby, enables explicit measures of risk exposure to be calculated. One of the economic-evaluation methods treated earlier is used to calculate economic worth; a computer is employed to sample repeatedly—hundreds of times—from the probability distributions and make the calculations. Monte Carlo simulation can be performed by the following steps:

1. Express variable inputs as probability functions. Where there are interdependencies among input values, multiple probability density functions, tied to one another, may be needed.

2. For each input for which there is a probability function, draw randomly an input value; for each input for which there is only a single value, take that value for calculations.

3. Use the input values to calculate the economic measure of worth and record the results.

4. If inputs are interdependent, such that input X is a function of input Y, first draw the value of Y, then draw randomly from the X values that correspond to the value of Y.

5. Repeat the process many times until the number of results is sufficient to construct a probability density function and a cumulative distribution function.

6. Construct the probability density function and cumulative distribution function for the economic measure of worth, and perform statistical analysis of the variability.

The strong advantage of the technique is that it expresses the results in probabilistic terms, thereby providing explicit assessment of risk exposure. A disadvantage is that it does not explicitly treat risk attitude; however, by providing a clear measure of risk exposure, it facilitates the implicit incorporation of risk attitude in the decision. The necessity of expressing inputs in probabilistic terms and the extensive calculations are also often considered disadvantages.

3.4.6 Decision Analysis

Decision analysis is a versatile technique that enables both risk exposure and risk attitude to be taken into account in the economic assessment. It diagrams possible choices, costs, benefits, and probabilities for a given decision problem in "decision trees," which are useful in understanding the possible choices and outcomes.

Although it is not possible to capture the richness of this technique in a brief overview, a simple decision tree, shown in Figure 3.6, is discussed to give a sense of how the technique is used. The decision problem is whether to lease or build a facility. The decision must be made now, based on uncertain data. The decision tree helps to structure and analyze the problem. The tree is constructed left to right and analyzed right to left. The tree starts with a box representing a decision juncture or node—in this case, whether to lease or build a facility. The line segments branching from the box represent the two alternative paths: the upper one the lease decision and the lower one the build decision. Each has a cost associated with it that is based on the expected cost to be incurred along the path. In this example, the minimum expected cost of $6.26 million is associated with the option to build a facility.

An advantage of this technique is that it helps to understand the problem and to compare alternative solutions. Another advantage is that, in addition to treating risk exposure, it can also accommodate risk attitude by converting benefits and costs to utility values (not addressed here). A disadvantage is that the technique, as typically applied, does not provide an explicit measure of the variability of the outcome.

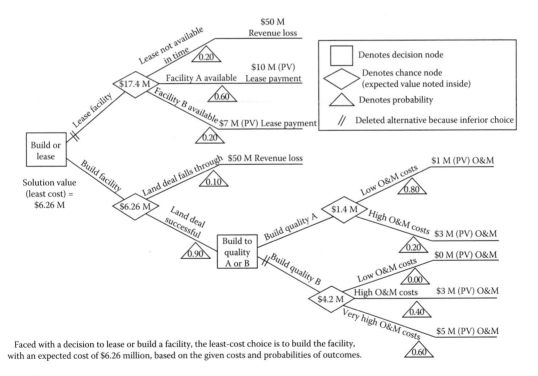

Faced with a decision to lease or build a facility, the least-cost choice is to build the facility, with an expected cost of $6.26 million, based on the given costs and probabilities of outcomes.

FIGURE 3.6
Decision tree: build versus lease.

3.4.7 Real Options Analysis (ROA)

ROA is an adaptation of financial options valuation techniques* to real asset investment decisions. ROA is a method used to analyze decisions in which the decision maker has one or more options regarding the timing or sequencing of an investment. It explicitly assumes that the investment is partially or completely *irreversible*, that there exists leeway or *flexibility* about the timing of the investment, and that it is subject to *uncertainty* over future payoffs. Real options can involve options (and combinations) to: defer, sequence, contract, shut down temporarily, switch uses, abandon, or expand the investment. This is in contrast to the NPV method that implies the decision is a "now or never" choice.

The value of an investment with an option is said to equal the value of the investment using the traditional NPV method (that implicitly assumes no flexibility or option) plus the value of the option. The analysis begins by construction of a decision tree with the option decision embedded in it. There are two basic methods to solve for the option value: the risk-adjusted replicating portfolio (RARP) approach and the risk-neutral probability (RNP) approach. The RARP discounts the expected project cash flows at a RADR, while the RNP approach discounts CE cash flows at a risk-free rate. In other words, the RARP approach takes the cash flows essentially as is, and adjusts the discount rate per time period to reflect that fact that the risk changes as one moves through the decision tree (e.g., risk declines with time as more information becomes available). In the RNP approach, the cash flows themselves are essentially adjusted for risk and discounted at a risk-free rate.

Copeland and Antikarov provide an overall four-step approach for ROA[†]:

1. Step 1—Compute a base-case traditional NPV (e.g., without flexibility).
2. Step 2—Model the uncertainty using (binominal) event trees (still without flexibility; e.g., without options)—although uncertainty is incorporated, the "expected" value of Step 2 should equal that calculated in Step 1.
3. Step 3—Create a decision tree incorporating decision nodes for options, as well as other (nondecision and nonoption decisions) nodes.
4. Step 4—Conduct an ROA by valuing the payoffs, working backward in time, node by node, using the RARP or RNP approach to calculate the ROA value of the investment.

3.4.8 Sensitivity Analysis

Sensitivity analysis is a technique for taking into account uncertainty that does not require estimates of probabilities. It tests the sensitivity of economic performance to alternative values of key factors about which there is uncertainty. Although sensitivity analysis does not provide a single answer in economic terms, it does show decision makers how the

---

* Financial options valuation is credited to Fisher Black and Myron Scholes who demonstrated mathematically that the value of a European call option—an option, but not the obligation, to purchase a financial asset for a given price (i.e., the exercise or strike price) on a particular date (i.e., the expiry date) in the future—depends on the current price of the stock, the volatility of the stock's price, the expiry date, the exercise price, and the risk-free interest rate. (See Black and Scholes, 1973.)

† See Dixit and Pindyck (1994), which is considered the "bible" of real options, and Copeland and Antikarov (2001) which offers more practical spreadsheet methods.
Ibid, pp. 220 and 239–240.

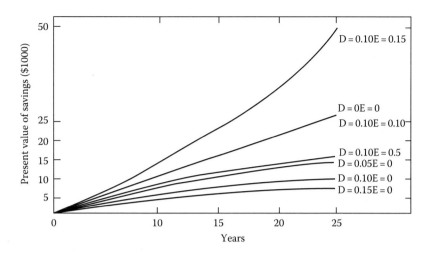

FIGURE 3.7
Sensitivity of present-value energy savings to time horizons, discount rates, and energy price escalation rates.

economic viability of a renewable energy or efficiency project changes as fuel prices, discount rates, time horizons, and other critical factors vary.

Figure 3.7 illustrates the sensitivity of fuel savings realized by a solar energy heating system to three critical factors: time horizons (0–25 years), discount rates (D equals 0%, 5%, 10%, and 15%), and energy escalation rates (E equals 0%, 5%, 10%, and 15%). The present value of savings is based on yearly fuel savings valued initially at $1000.

Note that, other things being equal, the present value of savings increase with time—but less with higher discount rates and more with higher escalation rates. The huge impact of fuel price escalation is most apparent when comparing the top line of the graph (D = 0.10, E = 0.15) with the line next to the bottom (D = 0.10, E = 0). The present value of savings at the end of 25 years is approximately $50,000 with a fuel escalation rate of 15%, and only about $8000 with no escalation, other things being equal. Whereas the quantity of energy saved is the same, the dollar value varies widely, depending on the escalation rate.

This example graphically illustrates a situation frequently encountered in the economic justification of energy efficiency and renewable energy projects: The major savings in energy costs, and thus the bulk of the benefits, accrue in the later years of the project and are highly sensitive to both the assumed rate of fuel-cost escalation and the discount rate. If the two rates are set equal, they will be offsetting as shown by the straight line labeled D = 0 E = 0 and D = 0.10 E = 0.10.

3.5 Building Blocks of Evaluation

Beyond the formula for the basic evaluation methods and risk assessment techniques, the practitioner needs to know some of the "nuts-and-bolts" of carrying out an economic analysis. He or she needs to know how to structure the evaluation process; how to choose a method of evaluation; how to estimate dollar costs and benefits; how to perform discounting operations; how to select an analysis period; how to choose a discount rate; how to adjust for inflation; how to take into account taxes and financing; how to treat residual values; and how to reflect assumptions and constraints, among other things. This section provides brief guidelines for these topics.

3.5.1 Structuring the Evaluation Process and Selecting a Method of Evaluation

A good starting point for the evaluation process is to define the problem and the objective. Identify any constraints to the solution and possible alternatives. Consider if the best solution is obvious, or if economic analysis and risk assessment are needed to help make the decision. Select an appropriate method of evaluation and a risk assessment technique. Compile the necessary data and determine what assumptions are to be made. Apply appropriate formula(s) to compute a measure of economic performance under risk. Compare alternatives and make the decision, taking into account any incommensurable effects that are not included in the dollar benefits and costs. Take into account the risk attitude of the decision maker, if it is relevant.

Although the six evaluation methods given earlier are similar, they are also sufficiently different, in that they are not always equally suitable for evaluating all types of energy investment decisions. For some types of decisions, the choice of method is more critical than for others. Figure 3.8 categorizes different investment types and the most suitable evaluation methods for each. If only a single investment is being considered, the "accept/reject" decision can often be made by any one of several techniques, provided the correct criterion is used.

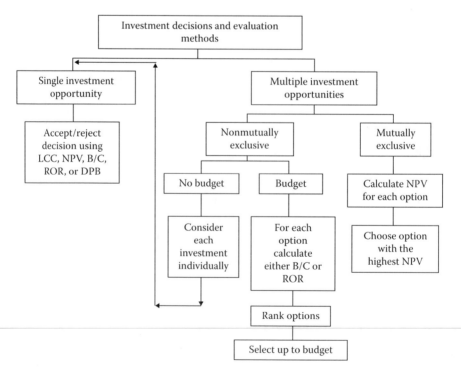

FIGURE 3.8
Investment decisions and evaluation methods.

Accept/reject criteria:
LCC technique—LCC must be lower as a result of the energy efficiency or renewable energy investment than without it.

- NPV (NB) technique—NPV must be positive as a result of the investment.
- B/C (SIR) technique—B/C (SIR) must be greater than 1.
- IRR technique—the IRR must be greater than the investor's minimum acceptable rate of return.
- DPB technique—the number of years to achieve DPB must be less than the project life or the investor's time horizon, and there are no cash flows after payback is achieved that would reverse payback.

If multiple investment opportunities are available, but only one investment can be made (i.e., they are mutually exclusive), any of the methods (except DPB) will usually work, provided they are used correctly. However, the NPV method is usually recommended for this purpose, because it is less likely to be misapplied. The NPV of each investment is calculated and the investment with the highest present value is the most economic. This is true even if the investments require significantly different initial investments, have significantly different times at which the returns occur, or have different useful lifetimes. Examples of mutually exclusive investments include different system sizes (e.g., three different photovoltaic array sizes are being considered for a single rooftop), different system configurations (e.g., different turbines are being considered for the same wind farm), and so forth.

If the investments are not mutually exclusive, then (as shown in Figure 3.8) one must consider whether there is an overall budget limitation that would restrict the number of economic investments that might be undertaken. If there is no budget (i.e., no limitation on the investment funds available), than there is really no comparison to be performed and the investor simply makes an "accept/reject" decision for each investment individually as described earlier.

If funds are not available to undertake all of the investments (i.e., there is a budget), then the easiest approach is to rank the alternatives, with the best having the highest benefit-to-cost ratio or rate of return. (The investment with the highest NPV will not necessarily be the one with the highest rank, because present value does not show return per unit investment.) Once ranked, those investments at the top of the priority list are selected until the budget is exhausted.

In the case where a fast turnaround on investment funds is required, DPB is recommended. The other methods, although more comprehensive and accurate for measuring an investment's lifetime profitability, do not indicate the time required for recouping the investment funds.

3.5.2 Discounting

Some or all investment costs in energy efficiency or renewable energy systems are incurred near the beginning of the project and are treated as "first costs." The benefits, on the other hand, typically accrue over the life span of the project in the form of yearly energy saved or produced. To compare benefits and costs that accrue at different points in time, it is necessary to put all cash flows on a time-equivalent basis. The method for converting cash flows to a time-equivalent basis is often called "discounting."

The value of money is time-dependent for two reasons: First, inflation or deflation can change the buying power of the dollar; and second, money can be invested over time to yield a return over and above inflation. For these two reasons, a given dollar amount today will be worth more than that same dollar amount a year later. For example, suppose a person were able to earn a maximum of 10% interest per annum risk-free. He or she would require $1.10 a year from now to be willing to forego having $1 today. If the person were indifferent between $1 today and $1.10 a year from now, then the 10% rate of interest would indicate that person's time preference for money. The higher the time preference, the higher the rate of interest required to make future cash flows equal to a given value today. The rate of interest for which an investor feels adequately compensated for trading money now for money in the future is the appropriate rate to use for converting present sums to future equivalent sums and future sums to present equivalent sums (i.e., the rate for discounting cash flows for that particular investor). This rate is often called the "discount rate."

To evaluate correctly the economic efficiency of an energy efficiency or renewable energy investment, it is necessary to convert the various expenditures and savings that accrue over time to a lump-sum, time-equivalent value in some base year (usually the present), or to annual values. The remainder of this section illustrates how to discount various types of cash flows.

Discounting is illustrated by Figure 3.9 in a problem of installing, maintaining, and operating a heat pump, as compared to an alternative heating/cooling system. The life-cycle cost calculations are shown for two reference times. The first is the present, and it is therefore called a present value. The second is based on a yearly time scale and is called an annual value. These two reference points are the most common in economic evaluations of investments. When the evaluation methods are derived properly, each time basis will give the same relative ranking of investment priorities.

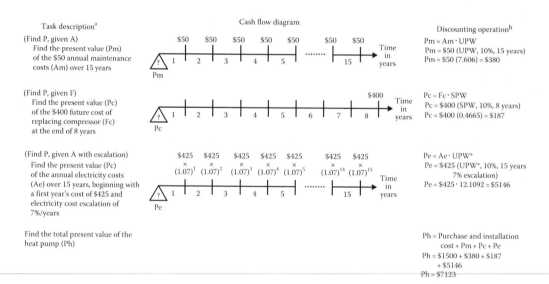

FIGURE 3.9
Determining present-value LCCs: heat pump example. *Note:* [a]P, present value; A, annual value; F, future value. [b]UPW, uniform present worth factor; SPW, single present worth factor; UPW*, uniform present worth factor with energy escalation. Purchase and installation costs are $1500 incurred initially. (From Ruegg, R.T. and Marshall, H.E., *Building Economics: Theory and Practice*, Chapman & Hall, New York, 1990.)

The assumptions for the heat pump problem—which are given only for the sake of illustration and not to suggest actual prices—are as follows:

1. The residential heat pump (not including the ducting) costs $1500 to purchase and
   install.
2. The heat pump has a useful life of 15 years.
3. The system has annual maintenance costs of $50 every year during its useful life,
   fixed by contractual agreement.
4. A compressor replacement is required in the eighth year at a cost of $400.
5. The yearly electricity cost for heating and cooling is $425, evaluated at the outset,
   and increased at a rate of 7% per annum due to rising electricity prices.
6. The discount rate (a nominal rate that includes an inflation adjustment) is 10%.
7. No salvage value is expected at the end of 15 years.

The LCCs in the sample problem are derived only for the heat pump and not for alternative heating/cooling systems. Hence, no attempt is made to compare alternative systems
in this discounting example. To do so would require similar calculations of life-cycle costs
for other types of heating/cooling systems. Total costs of a heat pump system include
costs of purchase and installation, maintenance, replacements, and electricity for operation. Using the present as the base-time reference point, we need to convert each of these
costs to the present before summing them. If we assume that the purchase and installation costs occur at the base reference point (the present), the $1500 is already in present
value terms.

Figure 3.9 illustrates how to convert the other cash flows to present values. The first task
is to convert the stream of annual maintenance costs to present value. The maintenance
costs, as shown in the cash flow diagram of Figure 3.9, are $50 per year, measured in cur-
rent dollars (i.e., dollars of the years in which they occur). The triangle indicates the value
to be found. Here we follow the practice of compounding interest at the end of each year.
The present refers to the beginning of year one.

The discounting operation for calculating the present value of maintenance costs
(last column of Figure 3.9) is to multiply the annual maintenance costs times the uniform present worth (UPW) factor. The UPW is a multiplicative factor computed from
the formula given in Table 3.4, or taken from a look-up table of factors that have been
published in many economics textbooks. UPW factors make it easy to calculate the
present values of a uniform series of annual values. For a discount rate of 10% and a
time period of 15 years, the UPW factor is 7.606. Multiplying this factor by $50 gives a
present value maintenance cost equal to $380. Note that the $380 present value of $50
per year incurred in each of 15 years is much less than simply adding $50 for 15 years
(i.e., $750). Discounting is required to achieve correct statements of costs and benefits
over time.

The second step is to convert the one-time future cost of compressor replacement, $400,
to its present value. The operation for calculating the present value of compressor replacement is to multiply the *future value* of the compressor replacement times the single-
payment present worth (SPW) factor, which can be calculated from the formula in Table 3.4,
or taken from a discount factor look-up table. For a discount rate of 10% and a time period
of 15 years, the SPW factor is 0.4665. Multiplying this factor by $400 gives a present-value
cost of the compressor replacement of $187, as shown in the last column of Figure 3.9.

TABLE 3.4

Discount Formulas

Standard Nomenclature	Use When	Standard Notation	Algebraic Form
Single compound amount	Given P; to find F	(SCA, d%, N)	$F = P(1+d)^N$
Single present worth	Given F; to find P	(SPW, d%, N)	$P = F\dfrac{1}{(1+d)^N}$
Uniform compound amount	Given A; to find F	(UCA, d%, N)	$F = A\dfrac{(1+d)^N - 1}{d}$
Uniform sinking fund	Given F; to find A	(USF, d%, N)	$A = F\dfrac{d}{(1+d)^N - 1}$
Uniform capital recovery	Given P; to find A	(UCR, d%, N)	$A = P\dfrac{d(1+d)^N}{(1+d)^N - 1}$
Uniform present worth	Given A; to find P	(UPW, d%, N)	$P = A\dfrac{(1+d)^N - 1}{d(1+d)^N}$
Uniform present worth modified	Given A escalating at a rate e; to find P	(UPW*, d%, e, N)	$P = A\dfrac{(1+e)\left[1 - \left(\dfrac{1+e}{1+d}\right)^N\right]}{d-e}$

Note: P, a present sum of money; F, a future sum of money, equivalent to P at the end of N periods of time at a discount rate of d; N, number of interest periods; A, an end-of-period payment (or receipt) in a uniform series of payments (or receipts) over N periods at *discount rate d*, usually annually; e, a rate of escalation in A in each of N periods.

Again, note that discounting makes a significant difference in the measure of costs. Failing to discount the $400 would result in an overestimate of cost, in this case of $213.

The third step is to convert the annual electricity costs for heating and cooling to present value. A year's electricity costs, evaluated at the time of installation of the heat pump, are assumed to be $425. Electricity prices, for purposes of illustration, are assumed to increase at a rate of 7% per annum. This is reflected in Table 3.4 by multiplying $425 times $(1.07)^t$ where $t = 1, 2, \ldots, 15$. The electricity cost at the end of the fourth year, for example, is $425(1.07)^4 = 55.

The discounting operation for finding the present value of all electricity costs (shown in Figure 3.9) is to multiply the initial, yearly electricity costs times the appropriate UPW* factor. (An asterisk following UPW denotes that a term for price escalation is included.) The UPW or UPW* discount formulas in Table 3.4 can also be used to obtain present values from annual costs or multiplicative discount factors from look-up tables can be used. For a period of 15 years, a discount rate of 10%, and an escalation rate of 7%, the UPW* factor is 12.1092. Multiplying the factor by $425 gives a present value of electricity costs of $5146. Note, once again, that failing to discount (i.e., simply adding annual electricity expenses in current prices) would overestimate costs by $1229 ($6376–$5146). Discounting with a UPW factor that does not incorporate energy price escalation would underestimate costs by $1913 ($5146–$3233).

The final operation described in Figure 3.9 is to sum purchase and installation cost and the present values of maintenance, compressor replacement, and electricity costs. Total LCCs of the heat pump in present value terms are $7213. This is one of the amounts that a designer would need for comparing the cost-effectiveness of heat pumps to alternative heating/cooling systems.

Only one discounting operation is required for converting the present value costs of the heat pump to annual value terms. The total present value amount is converted to the total annual value simply by multiplying it by the uniform capital recovery (UCR) factor—in this case the UCR for 10% and 15 years. The UCR factor, calculated with the UCR formula given in Table 3.4, is 0.13147. Multiplying this factor by the total present value of $7213 gives the cost of the heat pump as $948 in annual value terms. The two figures—$7213 and $948 per year—are time-equivalent values, made consistent through the discounting.

Figure 3.9 provides a model for the designer who must calculate present values from all kinds of benefit or cost streams. Most distributions of values occurring in future years can be handled with the SPW, the UPW, or the UPW* factors.

3.5.3 Discount Rate

Of the various factors affecting the NB of energy efficiency and renewable energy investments, the discount rate is one of the most dramatic. A project that appears economic at one discount rate will often appear uneconomic at another rate. For example, a project that yields NS at a 6% discount rate might yield net losses if evaluated with a 7% rate.

As the discount rate is increased, the present value of any future stream of costs or benefits is going to become smaller. High discount rates tend to favor projects with quick payoffs over projects with benefits deferred further in the future.

The discount rate should be set equal to the rate of return available on the next-best investment opportunity of similar risk to the project in question—that is, it should indicate the opportunity cost of the investor.

The discount rate may be formulated as a "real rate" exclusive of general price inflation or as a "nominal rate" inclusive of inflation. The former should be used to discount cash flows that are stated in *constant dollars*. The latter should be used to discount cash flows stated in *current dollars*.

3.5.4 Inflation

Inflation is a rise in the general price level. Because future price changes are unknown, it is frequently assumed that prices will increase at the rate of inflation. Under this assumption, it is generally easier to conduct all economic evaluations in constant dollars and to discount those values using "real" discount rates. For example, converting the constant dollar annual maintenance costs in Figure 3.9 to a present value can be easily done by multiplying by a UPW factor (calculated using a real discount rate) because the maintenance costs do not change over time. However, some cash flows are more easily expressed in current dollars—for example, equal loan payments, tax depreciation, etc. These can be converted to present values using a nominal discount rate.

3.5.5 Analysis Period

The analysis period is the length of time over which costs and benefits are considered in an economic evaluation. The analysis period need not be the same as either the "useful life" or the "economic life," two common concepts of investment life. The useful life is the period over which the investment has some value; that is, the investment continues to conserve or provide energy during this period. Economic life is the period during which the investment in question is the least-cost way of meeting the requirement. Often, economic life is shorter than useful life.

The selection of an analysis period will depend on the objectives and perspective of the decision maker. A speculative investor who plans to develop a project for immediate sale, for example, may view the relevant time horizon as that short period of ownership from planning and acquisition of property to the first sale of the project. Although the useful life of a solar domestic hot water heating system, for example, might be 20 years, a speculative home builder might operate on the basis of a 2-year time horizon, if the property is expected to change hands within that period. Only if the speculator expects to gain the benefit of those energy savings through a higher selling price for the building, will the higher first cost of the solar energy investment likely be economic.

If an analyst is performing an economic analysis for a particular client, that client's time horizon should serve as the analysis period. If an analyst is performing an analysis in support of public investment or a policy decision, the life of the system or building is typically the appropriate analysis period.

When considering multiple investment options, it is best with some evaluation methods (such as LCC, IRR, and ORR) to use the same analysis period. With others like NPV and BCR, different analysis periods can be used. If an investment's useful life is shorter than the analysis period, it may be necessary to consider reinvesting in that option at the end of its useful life. If an investment's useful life is longer than the analysis period, a salvage value may need to be estimated.

3.5.6 Taxes and Subsidies

Taxes and subsidies should be taken into account in economic evaluations, because they may affect the economic viability of an investment, the return to the investor, and the optimal size of the investment. Taxes, which may have positive and negative effects, include— but are not limited to—income taxes, sales taxes, property taxes, excise taxes, capital gain taxes, depreciation recapture taxes, tax deductions, and tax credits.

Subsidies are inducements for a particular type of behavior or action. They include grants—cash subsidies of specified amounts; government cost sharing; loan-interest reductions; and tax-related subsidies. Income tax credits for efficiency or renewable energy expenditures provide a subsidy by allowing specific deductions from the investor's tax liability. Property tax exemptions eliminate the property taxes that would otherwise add to annual costs. Income tax deductions for energy efficiency or renewable energy expenses reduce annual tax costs. The imposition of higher taxes on nonrenewable energy sources raises their prices and encourages efficiency and renewable energy investments.

It is important to distinguish between a before-tax cash flow and an after-tax cash flow. For example, fuel costs are a before-tax cash flow (they can be expensed), while a production tax credit for electricity from wind is an after-tax cash flow.

3.5.7 Financing

Financing of an energy investment can alter the economic viability of that investment. This is especially true for energy efficiency and renewable energy investments that generally have large initial investment costs with returns spread out over time. Ignoring financing costs when comparing these investments against conventional sources of energy can bias the evaluation against the energy efficiency and renewable energy investments.

Financing is generally described in terms of the amount financed, the loan period, and the interest rate. Unless specified otherwise, a uniform payment schedule is usually assumed. Generally, financing improves the economic effectiveness of an investment if the after-tax nominal interest rate is less than the investor's nominal discount rate.

Financing essentially reduces the initial outlay in favor of additional future outlays over time—usually equal payments for a fixed number of years. These cash flows can be treated like any other: The equity portion of the capital cost occurs at the start of the first year, and the loan payments occur monthly or annually. The only other major consideration is the tax deductibility of the interest portion of the loan payments.

3.5.8 Residual Values

Residual values may arise from salvage (net of disposal costs) at the end of the life of systems and components, from reuse values when the purpose is changed, and from remaining value when assets are sold prior to the end of their lives. The present value of residuals can generally be expected to decrease, other things equal, as (1) the discount rate rises, (2) the equipment or building deteriorates, and (3) the time horizon lengthens.

To estimate the residual value of energy efficiency or renewable energy systems and components, it is helpful to consider the amount that can be added to the selling price of a project or building because of those systems. It might be assumed that a building buyer will be willing to pay an additional amount equal to the capitalized value of energy savings over the remaining life of the efficiency or renewable investment. If the analysis period is the same as the useful life, there will be no residual value.

3.6 Economic Analysis Software for Renewable Energy Investments

Over the last three to four decades, a large number of models for the economic analysis of renewable energy systems have been developed. In the early years, the emphasis was on simpler models for the analysis of solar hot water and space heating. In the last decade, the emphasis has shifted to models of renewable electric technologies as those technologies have become more and more cost competitive. At the same time due to the complexities of the electric system, the models have become more and more sophisticated with respect to both system performance and system economics and financing. The need for reliable power and the variability of wind and solar are dealt with in many models by performance projections down to the hourly level. Similarly, the complex ownership and regulation of power plants and electric utilities has led to increasingly sophisticated financing and ownership structures in today's models. We will examine these intricacies as we briefly review a handful of the more prominent models used in the United States with an emphasis on the economic measures used, the technologies treated, and the different areas of emphasis of the different models.

System advisor model (available at https://sam.nrel.gov/) SAM is probably the most sophisticated of the tools available today for the analysis of renewable energy technologies in the electric sector. Developed by the National Renewable Energy Laboratory (NREL), "SAM makes performance predictions and cost of energy estimates for grid-connected power projects based on installation and operating costs and system design parameters that you specify as inputs to the model." (Gilman and Dobos 2012). It calculates the performance for each of the 8760 hours of the year which can be viewed at the hourly level or through more aggregated measures like capacity factors and seasonal or annual output. The technologies it can evaluate include

- Photovoltaic systems (flat plate and concentrating)
- Parabolic trough concentrating solar power systems
- Power tower concentrating solar power systems
- Linear Fresnel concentrating solar power systems
- Dish-Stirling concentrating solar power systems
- Conventional fossil-fuel thermal systems
- Solar water heating for residential or commercial buildings*
- Large and small wind power projects
- Geothermal power and coproduction
- Biomass power

SAM can evaluate the economics of these systems from the perspective of different owners and developer perspectives to include

- Residential rooftop
- Commercial rooftop
- Utility scale (power purchase agreement)
  - Single owner
  - Leveraged partnership flip
  - All equity partnership flip
  - Sale leaseback

SAM calculates the following measures based on the cash flows for the ownership, financing, and other descriptors input by the user:

- Payback period (buildings only)
- Revenue with and without renewable energy system
- LCOE
- NPV
- Power purchase agreement price (electricity sales price)
- IRR

In making these financial calculations, SAM can account for a wide range of incentives including

- Investment-based incentives
- Capacity-based incentives
- Production-based incentives

---

* This is the only non-electric-production technology that can be evaluated with SAM.

- Investment tax credits
- Production tax credits
- Depreciation (MACRS, straight-line, custom)

Cost of Renewable Energy Spreadsheet Tool (CREST is available at https://financere.nrel.gov/finance/content/crest-cost-energy-models) CREST was developed for NREL by Sustainable Energy Advantage. It is distinguished from SAM primarily in that it is easier to use, is spreadsheet based, and has fewer technical performance, financing, and economic-metric details and options. It calculates the first year cost of energy or the LCOE for photovoltaics, concentrating solar power, wind, and geothermal electricity technologies, as well as anaerobic digestion technologies. It can account for different cost-based incentives and several ownership structures.

HOMER (available at: http://homerenergy.com/index.html) HOMER can be used to design and analyze hybrid power systems, that can include storage, conventional generators and combined heat and power systems along with photovoltaics, wind, hydropower, and biomass. HOMER was also originally developed at NREL and is now supported by HOMER Energy. It is distinguished from NREL's SAM and CREST models described earlier primarily in that it can be used for grid and off-grid applications of hybrid and distributed energy systems. To analyze these systems, HOMER's performance calculations are made on an hourly basis for every hour of a year and show the changing mix over time of contributing generators from the hybrid system.

RETScreen (available from the Canadian government at: http://www.retscreen.net/ang/home.php): RETScreen is actually two separate programs—RETScreen 4, an Excel spreadsheet program (available at: http://www.retscreen.net/ang/version4.php), and RETScreen Plus, a Windows-based performance measurement and verification program. We focus here on the economic analysis part, RETScreen 4. It differs from the NREL suite of models described earlier primarily in that it addresses both electric and nonelectric renewable energy technologies as well as efficiency and cogeneration projects. It includes a large database of international resource and weather data for analyzing systems in almost all global locations, although it makes annual performance approximations, not hourly estimates. It calculates the standard set of economic metrics, for example, IRR, NPV, payback, through cash flow analysis that considers financing and tax provisions.

3.7 Summary

There are multiple methods for evaluating economic performance and multiple techniques of risk analysis that can be selected and combined to improve decisions in energy efficiency and renewable energy investments. Economic performance can be stated in a variety of ways, depending on the problem and preferences of the decision maker: as NPV, as LCCs, as the cost of energy, as a rate of return, as years to payback, or as a ratio. To reflect the reality that most decisions are made under conditions of uncertainty, risk assessment techniques can be used to reflect the risk exposure of the project and the risk attitude of the decision maker. Rather than expressing results in single, deterministic terms, they can be expressed in probabilistic terms, thereby revealing the likelihood that the outcome will differ from the best-guess answer. These methods and techniques can be used to decide whether or not to invest in a given energy efficiency or renewable energy system; to determine which system design or size is economically efficient; to find the combination of components and systems that are expected to be cost-effective; to estimate how long before a project will break even; and to decide which energy-related investments are likely to provide the highest rate of return to the investor. The methods support the goal of achieving economic efficiency—which may differ from engineering technical efficiency. There are many models available today that can assist in evaluating the economic and financial viability of an investment in renewable energy.

3.8 Defining Terms

Analysis period—Length of time over which costs and benefits are considered in an economic evaluation.

Benefit/cost (B/C) or saving-to-investment (SIR) ratio—A method of measuring the economic performance of alternatives by dividing present-value benefits (savings) by present-value costs.

Constant dollars—Values expressed in terms of the general purchasing power of the dollar in a base year. Constant dollars do not reflect price inflation or deflation.

Cost-effective investment—The least-cost alternative for achieving a given level of performance.

Current dollars—Values expressed in terms of actual prices of each year (i.e., current dollars reflect price inflation or deflation).

Discount rate—Based on the opportunity cost of capital, this minimum acceptable rate of return is used to convert benefits and costs occurring at different times to their equivalent values at a common time.

Discounted payback period—The time required for the discounted annual net benefits derived from an investment to pay back the initial investment.

Discounting—A technique for converting cash flows that occur over time to equivalent amounts at a common point in time using the opportunity cost for capital.

Economic efficiency optimization—Maximizing net benefits or minimizing costs for a given level of benefits (i.e., "getting the most for your money").

Economic life—That period of time over which an investment is considered to be the least-cost alternative for meeting a particular objective.

Future value (worth)—The value of a dollar amount at some point in the future, taking into account the opportunity cost of capital.

Internal rate of return—The discount rate that equates total discounted benefits with total discounted costs.

Investment costs—The sum of the planning, design, and construction costs necessary to obtain or develop an asset.

Levelized cost of energy—The before-tax revenue required per unit of energy to cover all costs plus a profit/return on investment equal to the discount rate used to levelize the costs.

Life-cycle cost—The total of all relevant costs associated with an asset or project over the analysis period.

Net benefits—Benefits minus costs.

Present value (worth)—Past, present, or future cash flows all expressed as a lump sum amount as of the present time, taking into account the time value of money.

Real options analysis—Method used to analyze investment decisions in which the decision maker has one or more options regarding the timing or sequencing of investment.

Risk assessment—As applied to economic decisions, the body of theory and practice that helps decision makers assess their risk exposures and risk attitudes in order to increase the probability that they will make economic choices that are best for them.

Risk attitude—The willingness of decision makers to take chances on investments with uncertain outcomes. Risk attitudes may be classified as risk averse, risk neutral, and risk taking.

Risk exposure—The probability that a project's economic outcome will be less favorable than what is considered economically desirable.

Sensitivity analysis—A non-probability-based technique for reflecting uncertainty that entails testing the outcome of an investment by altering one or more system parameters from the initially assumed values.

Time value of money—The amount that people are willing to pay for having money today rather than some time in the future.

Uncertainty—As used in the context of this chapter, a lack of knowledge about the values of inputs required for an economic analysis.

References

Black, F. and Scholes, M. 1973. The pricing of options and corporate liabilities. *Journal of Political Economy* 81: 637–659.

Copeland, T. and Antikarov, V. 2001. *Real Options: A Practitioner's Guide*. Texere, New York.

Dixit, A.K. and Pindyck, R.S. 1994. *Investment under Uncertainty*. Princeton University Press, Princeton, NJ.

Gilman, P. and Dobos, A. 2012. System advisor model, SAM 2011.12.2: General description, NREL/TP-6A20-53437. National Renewable Energy Laboratory, Golden, CO, February 2012.

Marshall, H.E. and Ruegg, RT. 1980a. Principles of economics applied to investments in energy conservation and solar energy systems. In *Economics of Solar Energy and Conservation Systems*, eds. F. Kreith and R. West, pp. 123–173. CRC Press, Boca Raton, FL.

Ruegg, R.T. and Marshall, H.E. 1990. *Building Economics: Theory and Practice*. Chapman & Hall, New York.

Ruegg, R.T. and Short, W. 1988. Economic methods. In *Economic Analysis of Solar Thermal Energy Systems*, eds. R. West and F. Kreith, pp. 18–83. The MIT Press, Cambridge, MA.

Stermole, F.J. and Stermole, J.M. 2000. *Economic Evaluation and Investment Decision Methods*. Investment Evaluations Corporation, Lakewood, CO.

4

Analysis Methods for Building Energy Auditing

Moncef Krarti

CONTENTS

4.1 Introduction

Since the oil embargo of 1973, significant improvements have been made in the energy efficiency of new buildings. However, the vast majority of the existing building stock is more than 20 years old and does not meet current energy efficiency construction standards (EIA, 2008). Therefore, energy retrofits of existing buildings will be required for decades to come if the overall energy efficiency of the building stock is to meet the standards.

Investing to improve the energy efficiency of buildings provides an immediate and relatively predictable positive cash flow resulting from lower energy bills. In addition to the conventional financing options available to owners and building operators (such as loans and leases), other methods are available to finance energy retrofit projects for buildings. One of these methods is performance contracting in which payment for a retrofit project is contingent upon its successful outcome. Typically, an energy services company (ESCO) assumes all the risks for a retrofit project by performing the engineering analysis and obtaining the initial capital to purchase and install equipment needed for energy efficiency improvements. Energy auditing is an important step used by energy service companies to insure the success of their performance contracting projects.

Moreover, several large industrial and commercial buildings have established internal energy management programs based on energy audits to reduce waste in energy use or to comply with the specifications of some regulations and standards. Other building owners and operators take advantage of available financial incentives typically offered by utilities or state agencies to perform energy audits and implement energy conservation measures (ECMs).

In the 1970s, building energy retrofits consisted of simple measures such as shutting off lights, turning down heating temperatures, turning up air-conditioning temperatures, and reducing the hot water temperatures. Today, building energy management includes a comprehensive evaluation of almost all the energy systems within a facility. Therefore, the energy auditor should be aware of key energy issues such as the subtleties of electrical utility rate structures and of the latest building energy efficiency technologies and their applications.

This chapter describes a general but systematic procedure for energy auditing suitable for both commercial buildings and industrial facilities. Some of the commonly recommended ECMs are briefly discussed. A case study for an office building is presented to illustrate the various tasks involved in an energy audit. Finally, an overview is provided to outline the existing methods for measurement and verification of energy savings incurred by the implementation of ECMs.

4.2 Types of Energy Audits

The term *energy audit* is widely used and may have different meanings depending on the energy service companies. Energy auditing of buildings can range from a short walk-through of the facility to a detailed analysis with hourly computer simulation. Generally, four types of energy audits can be distinguished as briefly described in the following (Krarti, 2010).

4.2.1 Walk-Through Audit

This audit consists of a short on-site visit of the facility to identify areas where simple and inexpensive actions can provide immediate energy use and/or operating cost savings. Some engineers refer to these types of actions as operating and maintenance (O&M) measures. Examples of O&M measures include setting back heating set point temperatures, replacing broken windows, insulating exposed hot water or steam pipes, and adjusting boiler fuel–air ratio.

4.2.2 Utility Cost Analysis

The main purpose of this type of audit is to carefully analyze the operating costs of the facility. Typically, the utility data over several years are evaluated to identify the patterns of energy use, peak demand, weather effects, and potential for energy savings. To perform this analysis, it is recommended that the energy auditor conducts a walk-through survey to get acquainted with the facility and its energy systems.

It is important that the energy auditor understands clearly the utility rate structure that applies to the facility for several reasons:

- To check the utility charges and insure that no mistakes were made in calculating the monthly bills. Indeed, the utility rate structures for commercial and industrial facilities can be quite complex with ratchet charges and power factor penalties.

- To determine the most dominant charges in the utility bills. For instance, peak demand charges can be significant portion of the utility bill especially when ratchet rates are applied. Peak shaving measures can be then recommended to reduce these demand charges.

- To identify whether or not the facility can benefit from using other utility rate structures to purchase cheaper fuel and reduce its operating costs. This analysis can provide a significant reduction in the utility bills especially with implementation of the electrical deregulation and the advent of real-time pricing rate structures.

Moreover, the energy auditor can determine whether or not the facility is prime for energy retrofit projects by analyzing the utility data. Indeed, the energy use of the facility can be normalized and compared to indices (for instance, the energy use per unit of floor area—for commercial buildings—or per unit of a product, for industrial facilities).

4.2.3 Standard Energy Audit

The standard audit provides a comprehensive energy analysis for the energy systems of the facility. In addition to the activities described for the walk-through audit and for the utility cost analysis described earlier, the standard energy audit includes the development of a baseline for the energy use of the facility and the evaluation of the energy savings and the cost-effectiveness of appropriately selected ECMs. The step-by-step approach of the standard energy audit is similar to that of the detailed energy audit that is described later on in the following section.

Typically, simplified tools are used in the standard energy audit to develop baseline energy models and to predict the energy savings of ECMs. Among these tools are the degree-day methods and linear regression models (Fels, 1986). In addition, a simple payback analysis is generally performed to determine the cost-effectiveness of ECMs.

4.2.4 Detailed Energy Audit

This audit is the most comprehensive but also time-consuming energy audit type. Specifically, the detailed energy audit includes the use of instruments to measure energy use for the whole building and/or for some energy systems within the building (for instance, by end uses, lighting systems, office equipment, fans, and chillers). In addition, sophisticated computer simulation programs are typically considered for detailed energy audits to evaluate and recommend energy retrofits for the facility.

The techniques available to perform measurements for an energy audit are diverse. During on-site visit, handheld and clamp-on instruments can be used to determine the variation of some building parameters such as the indoor air temperature, the luminance level, and the electrical energy use. When long-term measurements are needed, sensors are typically used and connected to a data-acquisition system so measured data can be stored and be remotely accessible. Recently, nonintrusive load monitoring (NILM) techniques have been proposed (Shaw et al., 1998). The NILM technique can determine the real-time energy use of the significant electrical loads in a facility using only a single set of sensors at the facility service entrance. The minimal effort associated with using the NILM technique when compared to the traditional submetering approach (which requires separate set of sensors to monitor energy consumption for each end use) makes the NILM a very attractive and inexpensive load monitoring technique for energy service companies and facility owners.

The computer simulation programs used in the detailed energy audit can provide typically the energy use distribution by load type (i.e., energy use for lighting, fans, chillers, and boilers). They are often based on dynamic thermal performance of the building energy systems and require typically high level of engineering expertise and training. These simulation programs range from those based on the bin method (Knebel, 1983) to those that provide hourly building thermal and electrical loads such as DOE-2 (LBL, 1980). The reader is referred to Krarti (2010) for more detailed discussion of the energy analysis tools that can be used to estimate energy and cost savings attributed to ECMs.

In the detailed energy audit, more rigorous economical evaluation of the ECMs is generally performed. Specifically, the cost-effectiveness of energy retrofits may be determined based on the life-cycle cost (LCC) analysis rather than the simple payback period analysis. LCC analysis takes into account a number of economic parameters such as interest, inflation, and tax rates. Krarti (2010) describes some of the basic analysis tools that are often used to evaluate energy efficiency projects.

4.3 General Procedure for a Detailed Energy Audit

To perform an energy audit, several tasks are typically carried out depending on the type of the audit and the size and function of the audited building. Some of the tasks may have to be repeated, reduced in scope, or even eliminated based on the findings of other tasks. Therefore, the execution of an energy audit is often not a linear process and is rather iterative. However, a general procedure can be outlined for most buildings.

4.3.1 Step 1: Building and Utility Data Analysis

The main purpose of this step is to evaluate the characteristics of the energy systems and the patterns of energy use for the building. The building characteristics can be collected from the architectural/mechanical/electrical drawings and/or from discussions with building operators. The energy use patterns can be obtained from a compilation of utility bills over several years. Analysis of the historical variation of the utility bills allows the energy auditor to determine if there are any seasonal and weather effects on the building energy use. Some of the tasks that can be performed in this step are presented here with the key results expected from each task and are noted:

- Collect at least 3 years of utility data (*to identify a historical energy use pattern*).
- Identify the fuel types used (electricity, natural gas, oil, etc.) (*to determine the fuel type that accounts for the largest energy use*).

- Determine the patterns of fuel use by fuel type (*to identify the peak demand for energy use by fuel type*).
- Understand utility rate structure (energy and demand rates) (*to evaluate if the building is penalized for peak demand and if cheaper fuel can be purchased*).
- Analyze the effect of weather on fuel consumption (*to pinpoint any variations of energy use related to extreme weather conditions*).
- Perform utility energy use analysis by building type and size (building signature can be determined including energy use per unit area) (*to compare against typical indices*).

4.3.2 Step 2: Walk-Through Survey

From this step, potential energy savings measures should be identified. The results of this step are important since they determine if the building warrants any further energy auditing work. The following are some of the tasks involved in this step:

- Identify the customer concerns and needs.
- Check the current O&M procedures.
- Determine the existing operating conditions of major energy use equipment (lighting, HVAC systems, motors, etc.).
- Estimate the occupancy, equipment, and lighting (energy use density and hours of operation).

4.3.3 Step 3: Baseline for Building Energy Use

The main purpose of this step is to develop a base-case model that represents the existing energy use and operating conditions for the building. This model is to be used as a reference to estimate the energy savings incurred from appropriately selected ECMs. There are major tasks to be performed during this step:

- Obtain and review architectural, mechanical, electrical, and control drawings.
- Inspect, test, and evaluate building equipment for efficiency, performance, and reliability.
- Obtain all occupancy and operating schedules for equipment (including lighting and HVAC systems).
- Develop a baseline model for building energy use.
- Calibrate the baseline model using the utility data and/or metered data.

4.3.4 Step 4: Evaluation of Energy Savings Measures

In this step, a list of cost-effective ECMs is determined using both energy savings and economical analysis. To achieve this goal, the following tasks are recommended:

- Prepare a comprehensive list of ECMs (using the information collected in the walk-through survey).

- Determine the energy savings due to the various ECMs pertinent to the building using the baseline energy use simulation model developed in phase 3.
- Estimate the initial costs required to implement the ECMs.
- Evaluate the cost-effectiveness of each ECM using an economical analysis method (simple payback or LCC analysis).

Tables 4.1 and 4.2 provide summaries of the energy audit procedure recommended, respectively, for commercial buildings and for industrial facilities. Energy audits for thermal and electrical systems are separated since they are typically subject to different utility rates.

TABLE 4.1

Energy Audit Summary for Residential and Commercial Buildings

Phase	Thermal Systems	Electrical Systems
Utility analysis	• Thermal energy use profile (building signature). • Thermal energy use per unit area (or per student for schools or per bed for hospitals). • Thermal energy use distribution (heating, DHW, process, etc.). • Fuel types used. • Weather effect on thermal energy use. • Utility rate structure.	• Electrical energy use profile (building signature). • Electrical energy use per unit area (or per student for schools or per bed for hospitals). • Electrical energy use distribution (cooling, lighting, equipment, fans, etc.). • Weather effect on electrical energy use. • Utility rate structure (energy charges, demand charges, power factor penalty, etc.).
On-site survey	• Construction materials (thermal resistance type and thickness). • HVAC system type. • DHW system. • Hot water/steam use for heating. • Hot water/steam for cooling. • Hot water/steam for DHW. • Hot water/steam for specific applications (hospitals, swimming pools, etc.).	• HVAC system type. • Lighting type and density. • Equipment type and density. • Energy use for heating. • Energy use for cooling. • Energy use for lighting. • Energy use for equipment. • Energy use for air handling. • Energy use for water distribution.
Energy use baseline	• Review architectural, mechanical, and control drawings. • Develop a base-case model (using any baselining method ranging from very simple to more detailed tools). • Calibrate the base-case model (using utility data or metered data).	• Review architectural, mechanical, electrical, and control drawings. • Develop a base-case model (using any baselining method ranging from very simple to more detailed tools). • Calibrate the base-case model (using utility data or metered data).
ECMs	• Heat recovery system (heat exchangers). • Efficient heating system (boilers). • Temperature setback. • EMCS. • HVAC system retrofit. • DHW use reduction. • Cogeneration.	• Energy-efficient lighting. • Energy-efficient equipment (computers). • Energy-efficient motors. • HVAC system retrofit. • EMCS. • Temperature setup. • Energy-efficient cooling system (chiller). • Peak demand shaving. • TES system. • Cogeneration. • Power factor improvement. • Reduction of harmonics.

TABLE 4.2

Energy Audit Summary for Industrial Facilities

Phase	Thermal Systems	Electrical Systems
Utility analysis	• Thermal energy use profile (building signature). • Thermal energy use per unit of a product. • Thermal energy use distribution (heating, process, etc.). • Fuel types used. • Analysis of the thermal energy input for specific processes used in the production line (such as drying). • Utility rate structure.	• Electrical energy use profile (building signature). • Electrical energy use per unit of a product. • Electrical energy use distribution (cooling, lighting, equipment, process, etc.). • Analysis of the electrical energy input for specific processes used in the production line (such as drying). • Utility rate structure (energy charges, demand charges, power factor penalty, etc.).
On-site survey	• List of equipment that use thermal energy. • Perform heat balance of the thermal energy. • Monitor the thermal energy use of all or part of the equipment. • Determine the by-products of thermal energy use (such as emissions and solid waste).	• List of equipment that use electrical energy. • Perform heat balance of the electrical energy. • Monitor the electrical energy use of all or part of the equipment. • Determine the by-products of electrical energy use (such as pollutants).
Energy use baseline	• Review mechanical drawings and production flow charts. • Develop a base-case model (using any baselining method). • Calibrate the base-case model (using utility data or metered data).	• Review electrical drawings and production flow charts. • Develop a base-case model (using any baselining method). • Calibrate the base-case model (using utility data or metered data).
ECMs	• Heat recovery system. • Efficient heating and drying system. • EMCS. • HVAC system retrofit. • Hot water and steam use reduction. • Cogeneration (possibly with solid waste from the production line).	• Energy-efficient motors. • Variable speed drives. • Air compressors. • Energy-efficient lighting. • HVAC system retrofit. • EMCS. • Cogeneration (possibly with solid waste from the production line). • Peak demand shaving. • Power factor improvement. • Reduction of harmonics.

4.4 Common Energy Conservation Measures

In this section, some ECMs commonly recommended for commercial and industrial facilities are briefly discussed. It should be noted that the list of ECMs presented in this section does not pretend to be exhaustive nor comprehensive. It is provided merely to indicate some of the options that the energy auditor can consider when performing an energy analysis of a commercial or an industrial facility. More discussion of energy efficiency measures for various building energy systems is provided in later chapters of this book. However, it is strongly advised that the energy auditor keeps abreast of any new technologies that can improve the building energy efficiency. Moreover, the energy auditor should recommend the ECMs only based on a sound economical analysis for each ECM.

4.4.1 Building Envelope

For some buildings, the envelope (i.e., walls, roofs, floors, windows, and doors) can have an important impact on the energy used to condition the facility. The energy auditor should determine the actual characteristics of the building envelope. During the survey, a descriptive sheet for the building envelope should be established to include information such as materials of construction (for instance, the level of insulation in walls, floors, and roofs) and the area and the number of building envelope assemblies (for instance, the type and the number of panes for the windows). In addition, comments on the repair needs and recent replacement should be noted during the survey.

The following are some of the commonly recommended ECMs to improve the thermal performance of building envelope:

(a) *Addition of thermal insulation*: For building surfaces without any thermal insulation, this measure can be cost-effective.

(b) *Replacement of windows*: When windows represent a significant portion of the exposed building surfaces, using more energy-efficient windows (high *R*-value, low-emissivity glazing, air tight, etc.) can be beneficial in both reducing the energy use and improving the indoor comfort level.

(c) *Reduction of air leakage*: When infiltration load is significant, leakage area of the building envelope can be reduced by simple and inexpensive weather-stripping techniques.

The energy audit of the envelope is especially important for residential buildings. Indeed, the energy use from residential buildings is dominated by weather since heat gain and/or loss from direct conduction of heat or from air infiltration/exfiltration through building surfaces accounts for a major portion (50%–80%) of the energy consumption. For commercial buildings, improvements to building envelope are often not cost-effective due to the fact that modifications to the building envelope (replacing windows, adding thermal insulation in walls) are typically considerably expensive. However, it is recommended to systematically audit the envelope components not only to determine the potential for energy savings but also to insure the integrity of its overall condition. For instance, thermal bridges—if present—can lead to heat transfer increase and to moisture condensation. The moisture condensation is often more damaging and costly than the increase in heat transfer since it can affect the structural integrity of the building envelope.

4.4.2 Electrical Systems

For most commercial buildings and a large number of industrial facilities, the electrical energy cost constitutes the dominant part of the utility bill. Lighting, office equipment, and motors are the electrical systems that consume the major part of energy in commercial and industrial buildings.

(a) *Lighting*: Lighting for a typical office building represents on average 40% of the total electrical energy use. There are a variety of simple and inexpensive measures to improve the efficiency of lighting systems. These measures include the use of energy-efficient lighting lamps and ballasts, the addition of reflective devices, delamping (when the luminance levels are above the recommended levels by

TABLE 4.3

Typical Efficiencies of Motors

Motor Size (HP)	Standard Efficiency (%)	Premium Efficiency (%)
1	73.0	85.5
2	75.0	86.5
3	77.0	86.5
5	80.0	89.5
7.5	82.0	89.5
10	85.0	91.7
15	86.0	92.4
20	87.5	93.0
30	88.0	93.6
40	88.5	93.6
50	89.5	94.1

the standards), and the use of daylighting controls. Most lighting measures are especially cost-effective for office buildings for which payback periods are less than 1 year.

(b) *Office equipment*: Office equipment constitutes the fastest-growing part of the electrical loads especially in commercial buildings. Office equipment includes computers, fax machines, printers, and copiers. Today, there are several manufacturers that provide energy-efficient office equipment (such those that comply with the U.S. EPA Energy Star specifications). For instance, energy-efficient computers automatically switch to a low-power *sleep* mode or off mode when not in use.

(c) *Motors*: The energy cost to operate electric motors can be a significant part of the operating budget of any commercial and industrial building. Measures to reduce the energy cost of using motors include reducing operating time (turning off unnecessary equipment), optimizing motor systems, using controls to match motor output with demand, using variable speed drives for air and water distribution, and installing energy-efficient motors. Table 4.3 provides typical efficiencies for several motor sizes.

In addition to the reduction in the total facility electrical energy use, retrofits of the electrical systems decrease space cooling loads and therefore further reduce the electrical energy use in the building. These cooling energy reductions as well as possible increases in thermal energy use (for space heating) should be accounted for when evaluating the cost-effectiveness of improvements in lighting and office equipment.

4.4.3 Daylighting Controls

Several studies indicated that daylighting can offer a cost-effective alternative to electrical lighting for commercial and institutional buildings. Through sensors and controllers, daylighting can reduce and even eliminate the use of electrical lighting required to provide sufficient illuminance levels inside office spaces. Recently, a simplified calculation method has been developed by Krarti et al. (2005) to estimate the reduction in the total lighting energy use due to daylighting with dimming controls for office buildings. The method has been shown to apply for office buildings in the United States as well as in Egypt (El Mohimem et al., 2005). The simplified calculation method is easy to use and can be used as a predesign tool to assess the potential of daylighting in saving electricity use associated with artificial lighting for office buildings.

To determine the percent savings, f_d, in annual use of artificial lighting due to implementing daylighting using daylighting controls in office buildings, Krarti et al. (2005) found that the following equation can be used:

$$f_d = b[1 - \exp(-a\tau_w A_w / A_p)]\frac{A_p}{A_f} \tag{4.1}$$

where

A_w/A_p is the window to perimeter floor area. This parameter provides a good indicator of the window size relative to the daylit floor area

A_p/A_f is the perimeter to total floor area. This parameter indicates the extent of the daylit area relative to the total building floor area. Thus, when $A_p/A_f = 1$, the whole building can benefit from daylighting

a and b are the coefficients that depend only on the building location and are given by Table 4.4 for various sites throughout the world

τ_w is the visible transmittance of the glazing

TABLE 4.4

Coefficients a and b of Equation 4.1 for Various Locations throughout the World

Location	a	b	Location	a	b
Atlanta	19.63	74.34	Casper	19.24	72.66
Chicago	18.39	71.66	Portland	17.79	70.93
Denver	19.36	72.86	Montreal	18.79	69.83
Phoenix	22.31	74.75	Quebec	19.07	70.61
New York City	18.73	66.96	Vancouver	16.93	68.69
Washington, DC	18.69	70.75	Regina	20.00	70.54
Boston	18.69	67.14	Toronto	19.30	70.48
Miami	25.13	74.82	Winnipeg	19.56	70.85
San Francisco	20.58	73.95	Shanghai	19.40	67.29
Seattle	16.60	69.23	Kuala Lumpur	20.15	72.37
Los Angeles	21.96	74.15	Singapore	23.27	73.68
Madison	18.79	70.03	Cairo	26.98	74.23
Houston	21.64	74.68	Alexandria	36.88	74.74
Fort Worth	19.70	72.91	Tunis	25.17	74.08
Bangor	17.86	70.73	Sao Paulo	29.36	71.19
Dodge City	18.77	72.62	Mexico 91	28.62	73.63
Nashville	20.02	70.35	Melbourne	19.96	67.72
Oklahoma City	20.20	74.43	Roma	16.03	72.44
Columbus	18.60	72.28	Frankfurt	15.22	69.69
Bismarck	17.91	71.50	Kuwait	21.98	65.31
Minneapolis	18.16	71.98	Riyadh	21.17	72.69
Omaha	18.94	72.30			

4.4.4 HVAC Systems

The energy use due to HVAC systems can represent 40% of the total energy consumed by a typical commercial building. The energy auditor should obtain the characteristics of major HVAC equipment to determine the condition of the equipment, their operating schedule, their quality of maintenance, and their control procedures. A large number of measures can be considered to improve the energy performance of both primary and secondary HVAC systems. Some of these measures are listed:

(a) *Setting up/back thermostat temperatures*: When appropriate, setting back heating temperatures can be recommended during unoccupied periods. Similarly, setting up cooling temperatures can be considered.

(b) *Retrofit of constant air volume systems*: For commercial buildings, variable air volume (VAV) systems should be considered when the existing HVAC systems rely on constant volume fans to condition part or the entire building.

(c) *Installation of heat recovery systems*: Heat can be recovered from some HVAC equipment. For instance, heat exchangers can be installed to recover heat from air handling unit (AHU) exhaust air streams and from boiler stacks.

(d) *Retrofit of central heating plants*: The efficiency of a boiler can be drastically improved by adjusting the fuel–air ratio for proper combustion. In addition, installation of new energy-efficient boilers can be economically justified when old boilers are to be replaced.

(e) *Retrofit of central cooling plants*: Currently, there are several chillers that are energy efficient and easy to control and operate and are suitable for retrofit projects.

It should be noted that there is a strong interaction between various components of heating and cooling system. Therefore, a whole-system analysis approach should be followed when retrofitting a building HVAC system. Optimizing the energy use of a central cooling plant (which may include chillers, pumps, and cooling towers) is one example of using a whole-system approach to reduce the energy use for heating and cooling buildings.

4.4.5 Compressed Air Systems

Compressed air has become an indispensable tool for most manufacturing facilities. Its uses range from air-powered hand tools and actuators to sophisticated pneumatic robotics. Unfortunately, staggering amounts of compressed air are currently wasted in a large number of facilities. It is estimated that only a fraction of 20%–25% of input electrical energy is delivered as useful compressed air energy. Leaks are reported to account for 10%–50% of the waste, while misapplication accounts for 5%–40% of loss in compressed air (Howe and Scales, 1998).

To improve the efficiency of compressed air systems, the auditor can consider several issues including whether or not compressed air is the right tool for the job (for instance, electric motors are more energy efficient than air-driven rotary devices), how compressed air is applied (for instance, lower pressures can be used to supply pneumatic tools), how it is delivered and controlled (for instance, the compressed air needs to be turned off when the process is not running), and how compressed air system is managed (for each machine or process, the cost of compressed air needs to be known to identify energy and cost savings opportunities).

4.4.6 Energy Management Controls

With the constant decrease in the cost of computer technology, automated control of a wide range of energy systems within commercial and industrial buildings is becoming increasingly popular and cost-effective. An energy management and control system (EMCS) can be designed to control and reduce the building energy consumption within a facility by continuously monitoring the energy use of various equipment and making appropriate adjustments. For instance, an EMCS can automatically monitor and adjust indoor ambient temperatures, set fan speeds, open and close AHU dampers, and control lighting systems.

If an EMCS is already installed in the building, it is important to recommend a system tune-up to insure that the controls are properly operating. For instance, the sensors should be calibrated regularly in accordance with manufacturers' specifications. Poorly calibrated sensors may cause increase in heating and cooling loads and may reduce occupant comfort.

4.4.7 Indoor Water Management

Water and energy savings can be achieved in buildings by using water-savings fixtures instead of the conventional fixtures for toilets, faucets, showerheads, dishwashers, and clothes washers. Savings can also be achieved by eliminating leaks in pipes and fixtures.

Table 4.5 provides typical water use of conventional and water-efficient fixtures for various end uses. In addition, Table 4.5 indicates the hot water use by each fixture as a fraction of the total water. With water-efficient fixtures, savings of 50% of water use can be achieved for toilets, showers, and faucets.

4.4.8 Advanced Technologies

The energy auditor may consider the potential of implementing and integrating new technologies within the facility. It is therefore important that the energy auditor understands these new technologies and knows how to apply them. The following are among the new technologies that can be considered for commercial and industrial buildings:

(a) *Building envelope technologies*: Recently, several materials and systems have been proposed to improve the energy efficiency of building envelope and especially windows including

  • Spectrally selective glasses that can optimize solar gains and shading effects

TABLE 4.5

Usage Characteristics of Water-Using Fixtures

End Use	Conventional Fixtures	Water-Efficient Fixtures	Usage Pattern	Hot Water (%)
Toilets	3.5 gal/flush	1.6 gal/flush	4 flushes/day	0
Showers	5.0 gal/min	2.5 gal/min	5 min/shower	60
Faucets	4.0 gal/min	2.0 gal/min	2.5 min/day	50
Dishwashers	14.0 gal/load	8.5 gal/load	0.17 loads/day	100
Clothes washers	55.0 gal/load	42.0 gal/load	0.3 loads/day	25
Leaks	10% of total use	2% of total use	N/A	50

- Chromogenic glazings that change its properties automatically depending on temperature and/or light level conditions (similar to sunglasses that become dark in sunlight)
- Building integrated photovoltaic (PV) panels that can generate electricity while absorbing solar radiation and reducing heat gain through building envelope (typically roofs)

(b) *Light pipe technologies*: While the use of daylighting is straightforward for perimeter zones that are near windows, it is not usually feasible for interior spaces, particularly those without any skylights. Recent but still emerging technologies allow to *pipe* light from roof or wall-mounted collectors to interior spaces that are not close to windows or skylights.

(c) *HVAC systems and controls*: Several strategies can be considered for energy retrofits:

- Heat recovery technologies such as rotary heat wheels and heat pipes can recover 50%–80% of the energy used to heat or cool ventilation air supplied to the building.
- Desiccant-based cooling systems are now available and can be used in buildings with large dehumidification loads during long periods (such as hospitals, swimming pools, and supermarket fresh produce areas).
- Geothermal heat pumps can provide an opportunity to take advantage of the heat stored underground to condition building spaces.
- Thermal energy storage (TES) systems offer a mean of using less expensive off-peak power to produce cooling or heating to condition the building during on-peak periods. Several optimal control strategies have been developed in recent years to maximize the cost savings of using TES systems.

(d) *Cogeneration*: This is not really a new technology. However, recent improvements in its combined thermal and electrical efficiency made cogeneration cost-effective in several applications including institutional buildings such as hospitals and universities.

4.5 Net-Zero Energy Retrofits

While the concept of the zero net energy (ZNE) buildings has been mostly applied to new construction, it can be considered for retrofit projects including residential buildings. Typically, ZNE is defined in terms of either site energy or resource energy. Site energy consists of energy produced and consumed at the building site. Source or primary energy includes site energy as well as the energy used to generate, transmit, and distribute this site energy. Therefore, source energy provides a better indicator of the energy use of buildings and their impact on the environment and the society and thus is better suited for ZNE building analysis. An analysis based on source energy effectively allows different fuel types, such as electricity and natural gas commonly used in buildings, to be encompassed together.

It is generally understood that ZNE buildings produce as much energy as they consume on-site annually. These buildings typically include aggressive energy efficiency measures and active solar water heating systems. Moreover, ZNE buildings employ grid-tied, net-metered renewable energy generation technologies, generally PV systems, to produce electricity. Effectively, ZNE buildings use the grid as *battery storage* to reduce required generation system capacity.

Unlike the case for new construction, analysis methods and case studies of net zero energy (NZE) retrofits of existing buildings are limited. A more detailed description of NZE retrofitting approaches, analysis techniques, and some case studies is provided by Krarti (2012). In this section, a case study for NZE retrofitting of existing homes in Mexico is provided. Specifically and in a study by Griego et al. (2012), NZE retrofits of Mexican residential building have been carried out. In the study, various combinations of energy efficiency are applied to arrive at an optimum set of recommendations for existing residential and new construction residential buildings. The optimum point is the minimum annualized energy-related costs and the corresponding annual source energy savings. Two separate optimizations are performed for an existing-unconditioned home and an existing-conditioned home.

The baseline annualized energy-related costs for the conditioned and unconditioned homes are US$797 and US$557, respectively. These costs are obtained before any of the energy efficiency and thermal comfort measures are applied. Therefore, it is determined that the cost of improved thermal comfort for a typical home in Salamanca is roughly US$240/year. The optimum point for the unconditioned case occurs at 17.3% annual energy savings and a corresponding minimum cost of US$433 as shown in Figure 4.1. The conditioned case on the other hand has a greater opportunity for energy savings and achieves a minimum cost of US$542 at 35.0% energy savings as indicated in Figure 4.2. The optimum point for the conditioned case includes implementing methods to reduce miscellaneous plug loads, R-1.4 m^{2-} K/W roof assembly, low-flow showerheads and sinks, an electric stove, 100% compact fluorescent lamps, and R-0.35 m^{2-} K/W trunk–branch domestic hot water (DHW) pipe distribution. The optimum point for the unconditioned case includes all of the same measures as the conditioned model with the exception of the added roof insulation.

The comparison between the unconditioned and the conditioned optimization results outlined in Figure 4.3 reveals that the optimum point for the conditioned case (US$542) is roughly the cost neutral point for the unconditioned case (US$557). It is also useful to compare energy end uses to gauge measures with highest potential of the energy savings. Figures 4.4 and 4.5 include a summary of the total annual source energy consumption by

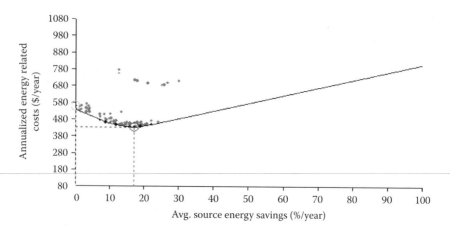

FIGURE 4.1
Retrofit optimization path for an unconditioned home in Salamanca, Mexico.

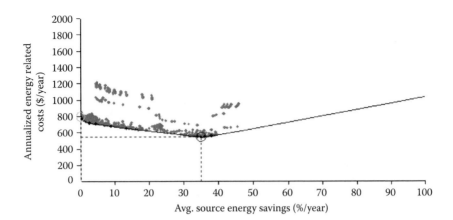

FIGURE 4.2
Retrofit optimization path for an air-conditioned home in Salamanca, Mexico.

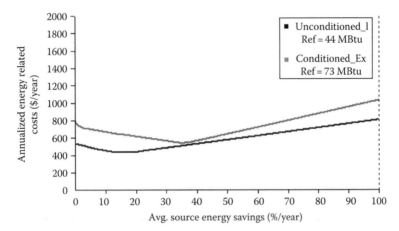

FIGURE 4.3
Comparison of retrofit optimization paths for unconditioned and conditioned homes.

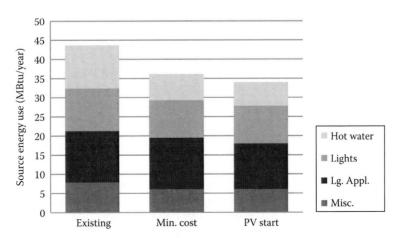

FIGURE 4.4
Annual end-use source energy for NZE retrofitting of an unconditioned home in Salamanca, Mexico.

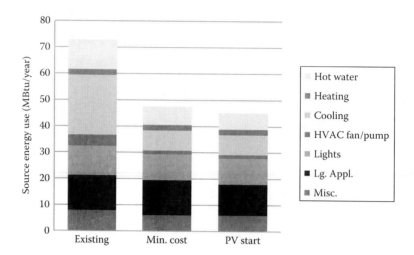

FIGURE 4.5
Annual end-use source energy for NZE retrofitting of an air-conditioned home in Salamanca, Mexico.

end use for the unconditioned and conditioned home models, respectively. The minimum cost option and the PV start options are compared with the baseline model. In the unconditioned case, the largest energy savings are obtained for the hot water, miscellaneous equipment, and lighting. The conditioned case has similar energy savings as the unconditioned case for miscellaneous equipment and domestic hot water; however, the greatest area for energy savings is for cooling, which is primarily attributed to the use of roof insulation.

The two renewable energy technologies evaluated in this study are solar domestic hot water systems and PV systems. Although solar DHW has a large potential for energy savings, the high implementation cost makes it unfeasible. Figures 4.1 and 4.2 show points hovering significantly above the optimization curve; those points are associated to combinations of measures that include solar domestic hot water systems. Note that the cost for labor and materials is assumed to be comparable to those in the United States.

The results for the PV system are shown by the sloped line leading to 100% energy savings. The size of PV to arrive at a ZNE solution for the unconditioned home model is a 3 kW system, and the conditioned home model is a 4 kW system. Both systems are south facing and installed in inclined panels to match the latitude in Salamanca. The slope of the line toward ZNE is relatively shallow where the annualized energy cost for 100% annual energy savings is US$920 for the unconditioned and US$1185 for the conditioned cases. Similar to the solar DHW system, U.S. costs are assumed for PV material and labor cost. PV technology may only be desirable with the appropriate subsidies for implementation costs.

The Predicted Mean Vote (PMV) thermal comfort analysis is used as verification for the optimization results by evaluating the improved indoor thermal comfort after implementing the recommended energy efficiency measures. First, the PMV ratings above and below the acceptable comfort range (i.e., PMV values between –1 and 1) are determined, the baseline and optimal cases for both conditioned and unconditioned building models. The unconditioned building model shows roughly 1550 h above 1 PMV and 150 h below 1 PMV annually. This is in contrast to the conditioned building models, where thermal comfort is maintained throughout the year, as expected.

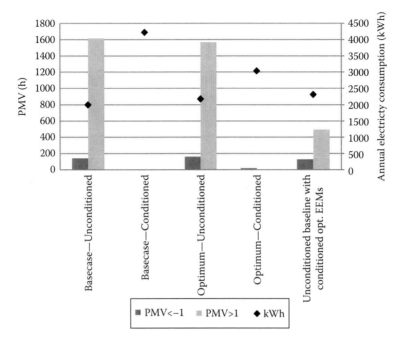

FIGURE 4.6
Comparison of various existing home configurations with and without NZE retrofits including thermal comfort analysis.

The thermal comfort analysis is also applied at the optimum building models. The annual energy consumption in the conditioned new construction and existing building models decreases relative to the baseline when thermal insulation is added to the building and, as predicted, the annual PMV ratings remain relatively constant. However, when the optimum set of ECMs from the conditioned case are applied to the unconditioned baseline home model, the number of hours outside of the thermal comfort zone decreases significantly as indicated in Figure 4.6.

In the conditioned baseline retrofit case, the cost of installing an electric heat pump is estimated at US$4394 for a 3.5 tons unit. However, when roof insulation is added to the unconditioned retrofit building, the number of hours outside of the thermal comfort zone decreases by almost 60% for a much lower initial cost of roughly US$426.

4.6 Verification Methods of Energy Savings

Energy conservation retrofits are deemed cost-effective based on predictions of energy and cost savings. However, several studies have found that large discrepancies exist between actual and predicted energy savings. Due to the significant increase in the activities of ESCOs, the need became evident for standardized methods for measurement and verification of energy savings. This interest has led to the development of the *North American Energy Measurement and Verification Protocol* published in 1996 and later expanded and revised under the *International Performance Measurement and Verification Protocol* (IPMVP, 1997, 2002, 2007).

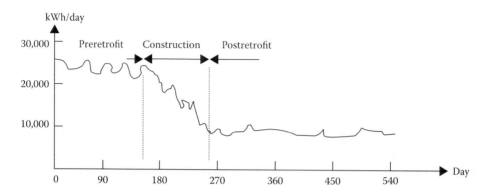

FIGURE 4.7
Daily variation of a building energy consumption showing preretrofit, construction, and postretrofit periods.

In order to estimate the energy savings incurred by an energy project, it is important to first identify the implementation period of the project, that is, the construction phase where the facility is subject to operational or physical changes due to the retrofit (Guiterman and Krarti, 2011). Figure 4.7 illustrates an example of the variation of the electrical energy use in a facility that has been retrofitted from constant volume to a VAV HVAC system. The time-series plot of the facility energy use clearly indicated the duration of the construction period, the end of the preretrofit period, and the start of the postretrofit period. The duration of the construction period depends on the nature of the retrofit project and can range from few hours to several months.

In principle, the measurement of the retrofit energy savings can be obtained by simply comparing the energy use during pre- and postretrofit periods. Unfortunately, the change in energy use between the pre- and postretrofit periods is not only due to the retrofit itself but also to other factors such as changes in weather conditions, levels of occupancy, and HVAC operating procedures. It is important to account for all these changes to determine accurately the retrofit energy savings.

Several methods have been proposed to measure and verify energy savings of implemented ECMs in commercial and industrial buildings. Some of these techniques are briefly described:

Regression models: The early regression models used to measure savings adapted the variable-base degree-day method. Among these early regression models, the Princeton Scorekeeping Method is the method that uses measured monthly energy consumption data and daily average temperatures to calibrate a linear regression model and determine the best values for non-weather-dependent consumption, the temperature at which the energy consumption began to increase due to heating or cooling (the change-point or base temperature), and the rate at which the energy consumption increased. Several studies have indicated that the simple linear regression model is suitable for estimating energy savings for residential buildings. However, subsequent work has shown that the PRISM model does not provide accurate estimates for energy savings for most commercial buildings (Ruch and Claridge, 1992). Single-variable (temperature) regression models require the use of at least four-parameter segmented-linear or change-point regressions to be suitable for commercial buildings. Figure 4.8 illustrates the basic functional forms commonly used for ambient-temperature linear

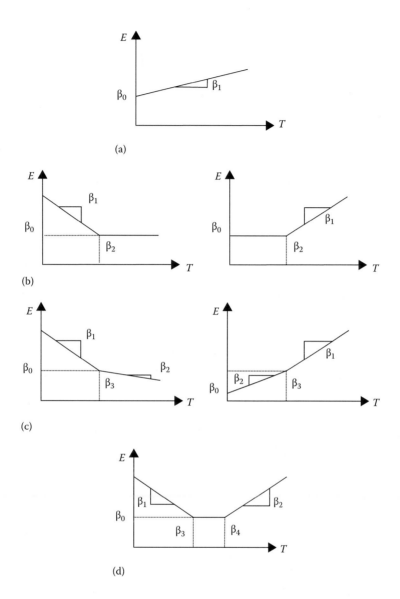

FIGURE 4.8
Basic forms of single-variable regression models using (a) 2 parameters (2P), (b) 3 parameters (3P), (c) 4 parameters (4P), and (d) 5 parameters (5P).

regression models. There are regression models that are also called change-point or segmented-linear models. Table 4.6 summarizes the mathematical expressions of four change-point models and their applications. In general, the change-point regression models are more suitable for predicting heating rather than cooling energy use. Indeed, these regression models assume steady-state conditions and are insensitive to the building dynamic effects, solar effects, and nonlinear HVAC system controls such as on–off schedules.

Katipamula et al. (1994) proposed multiple linear regression models to include as independent variables internal gain, solar radiation, wind, and humidity ratio, in addition to

TABLE 4.6

Mathematical Expressions and Applications of Change-Point Regression Models

Model Type	Mathematical Expression	Applications
Two-parameter (2-P) model	$E = \beta_0 + \beta_1 \cdot T$	Buildings with constant air volume systems and simple controls
Three-parameter (3-P) model	Heating $E = \beta_0 + \beta_1 \cdot (\beta_2 - T)^+$ Cooling $E = \beta_0 + \beta_1 \cdot (T - \beta_2)^+$	Buildings with envelope-driven heating or cooling loads (most residential buildings follow this model)
Four-parameter (4-P) model	Heating $E = \beta_0 + \beta_1 \cdot (\beta_3 - T)^+ - \beta_2 \cdot (T - \beta_3)^+$ Cooling $E = \beta_0 - \beta_1 \cdot (\beta_3 - T)^+ + \beta_2 \cdot (T - \beta_3)^+$	Buildings with VAV systems and/or with high latent loads; also, buildings with nonlinear control features (such as economizer cycles and hot deck reset schedules)
Five-parameter (5-P) model	$E = \beta_0 + \beta_1 \cdot (\beta_3 - T)^+ + \beta_2 \cdot (T - \beta_4)^+$	Buildings with systems that use the same energy source for both heating and cooling (i.e., heat pumps, electric heating and cooling systems)

the outdoor temperature. For the buildings considered in their analysis, Katipamula et al. found that wind and solar radiation have small effects on the energy consumption. They also found that internal gains have generally modest impact on energy consumption. Katipamula et al. (1998) discussed in more details the advantages and the limitations of multivariate regression modeling.

Time-variant models: There are several techniques that are proposed to include the effect of time variation of several independent variables on estimating the energy savings due to retrofits of building energy systems. Among these techniques are the artificial neural networks (Krarti et al., 1998), Fourier series (Dhar et al., 1998), and NILM (Shaw et al., 1998). These techniques are typically involved and require high level of expertise and training.

4.7 Summary

Retrofitting of buildings encompasses a wide variety of tasks and requires expertise in a number of areas to determine the best ECMs suitable for an existing facility. This chapter provided a description of a general but a systematic approach to perform energy audits. If followed carefully, the approach helps facilitate the process of analyzing a seemingly endless array of alternatives and complex interrelationships between building and energy system components. In particular, the chapter discussed net-zero energy retrofitting of buildings using detailed simulation analysis and optimization techniques to ensure that over a period of 1 year, the retrofitted building would effectively consume as much source energy as it produces.

References

Dhar, A., Reddy, T.A., and Claridge, D.E. 1998. Modeling hourly energy use in commercial buildings with fourier series functional forms. *ASME Solar Energy Engineering Journal*, 120(3), 217.

EIA. 2008. *2005 Residential Energy Consumption Survey*. Washington, DC: U.S. Energy Information Administration.

El Mohimem, M.A., Hanna, G., and Krarti, M. 2005. Analysis of daylighting benefits for office building in Egypt. *ASME Journal of Solar Energy Engineering*, 127(3), 366–370.

Fels, M. 1986. Special issue devoted to measuring energy savings: The scorekeeping approach. *Energy and Buildings*, 9(2), 127–136.

Griego, D., Krarti, M., and Hernandez-Guerrero, A. 2012. Optimization of energy efficiency and thermal comfort for residential buildings in Salamanca, Mexico. *Energy and Buildings*, 50, 550–549.

Guiterman, T. and Krarti, M. 2011. Analysis of measurement and verification methods for energy retrofits applied to residential buildings. *ASHRAE Transactions*, 117(2), 382–394.

Howe, B. and Scales, B. 1998. Beyond leaks: Demand-side strategies for improving compressed air efficiency. *Energy Engineering*, 95, 31.

IPMVP. 1997. *International Performance Monitoring and Verification Protocol*. U.S. Department of Energy DOE/EE-0157. Washington, DC: U.S. Government Printing Office.

IPMVP. 2002. *International Performance Monitoring and Verification Protocol, Concepts and Options for Determining Energy and Water Savings*, Volume 1. U.S. Department of Energy DOE/GO-102002-1554. Washington, DC: U.S. Government Printing Office.

IPMVP. 2007. *International Performance Monitoring and Verification Protocol, Concepts and Options for Determining Energy and Water Savings*. Washington, DC: U.S. Government Printing Office. http://www.evo-world.org (accessed February 5, 2014).

Katipamula, S., Reddy, T.A., and Claridge, D.E. 1994. Development and application of regression models to predict cooling energy use in large commercial buildings. *Proceedings of the ASME/JSES/JSES International Solar Energy Conference*, San Francisco, CA, p. 307.

Katipamula, S., Reddy, T.A., and Claridge, D.E. 1998. Multivariate regression modeling. *ASME Solar Energy Engineering Journal*, 120(3), 177.

Knebel, D.E. 1983. *Simplified Energy Analysis Using the Modified Bin Method*. Atlanta, GA: American Society of Heating, Refrigeration, and Air-Conditioning Engineers.

Krarti, M. 2010. *Energy Audit of Building Systems: An Engineering Approach*, 2nd Ed. Boca Raton, FL: CRC Press.

Krarti, M. 2012. *Weatherization and Energy Efficiency Improvement for Existing Homes: An Engineering Approach*, 1st Ed. Boca Raton, FL: CRC Press.

Krarti, M., Erickson, P., and Hillman, T. 2005. A simplified method to estimate energy savings of artificial lighting use from daylighting. *Building an Environment*, 40, 747–754.

Krarti, M., Kreider, J.F., Cohen, D., and Curtiss, P. 1998. Estimation of energy savings for building retrofits using neural networks. *ASME Journal of Solar Energy Engineering*, 120(3), 211.

LBL. 1980. *DOE-2 User Guide*, Version 2.1, LBL report No. LBL-8689 Rev. 2. Berkeley, CA: Lawrence Berkeley Laboratory.

Ruch, D. and Claridge, D.E. 1992. A four-parameter change-point model for predicting energy consumption in commercial buildings. *ASME Journal of Solar Energy Engineering*, 104, 177.

Shaw, S.R., Abler, C.B., Lepard, R.F., Luo, D., Leeb, S.B., and Norford L.K. 1998. Instrumentation for high performance non-intrusive electrical load monitoring. *ASME Journal of Solar Energy Engineering*, 120(3), 224.

5

Electrical Energy Management in Buildings

Craig B. Smith and Kelly E. Parmenter

CONTENTS

5.1 Principal Electricity Uses in Buildings

5.1.1 Introduction: The Importance of Energy Efficiency in Buildings

A typical building is designed for a 40-year economic life. This implies that the existing inventory of buildings—with all their good and bad features—is turned over very slowly. Today we know it is cost-effective to design a high degree of energy efficiency into new buildings, because the savings on operating and maintenance costs will repay the initial investment many times over. Many technological advances have occurred in the last few decades, resulting in striking reductions in the energy usage required to operate buildings safely and comfortably. Added benefits of these developments are reductions in air pollution and greenhouse gas emissions, which have occurred as a result of generating less electricity.

There are hundreds of building types, and buildings can be categorized in many ways—by use, type of construction, size, or thermal characteristics, to name a few. For simplicity, two designations will be used here: residential and commercial. Industrial facilities are not included here, but are discussed in Chapter 10.

The residential category includes features common to single family dwellings, condominiums and townhouses, and multifamily apartments. In 2012, there were approximately 115 million occupied housing units in the United States (U.S. Census Bureau 2012). The commercial category includes a major emphasis on office buildings, as well as a less detailed discussion of features common to retail stores, hospitals, restaurants, and laundries. The extension to other types is either obvious or can be pursued by referring to the literature. There are roughly five million commercial buildings in the United States, estimated to total 83 billion square feet in 2012 (EIA 2013a). Three-quarters of these buildings are 25 or more years old, and the average building age is about 50 years (SMR Research Corporation 2011). Most of this space is contained in buildings larger than 10,000 ft².

Total energy consumption in the two sectors has evolved over time and since the previous three editions of this book (see Table 5.1):

There has been a remarkable shift in the energy sources used by the residential and commercial sectors in the decades since 1975. Natural gas, which increased rapidly in these sectors prior to 1975, flattened out and has remained more or less constant. The use of petroleum has decreased. The big change has been the dramatic increase in electricity sales to the residential and commercial sectors, more than doubling from 1980 to 2011 and increasing by 2.6 times since 1975.

TABLE 5.1

Total Energy Consumption Trends, Quadrillion BTU

	Year						
	1955	**1965**	**1975**	**1985**	**1995**	**2005**	**2011**
Residential	7.3	10.6	14.8	16.0	18.5	21.6	21.6
Commercial	3.9	5.8	9.5	11.5	14.7	17.9	18.0
Total	11.2	16.5	24.3	27.5	33.2	39.5	39.6

Source: EIA, Annual energy review 2011. DOE/EIA-0384(2011), U.S. Energy Information Administration, Washington, DC, 2012, http://www.eia.gov/totalenergy/data/annual/pdf/aer.pdf.

The approach taken in this chapter is to list two categories of specific strategies that are cost-effective methods for using electricity efficiently. The first category includes those measures that can be implemented at low capital cost using existing facilities and equipment in an essentially unmodified state. The second category includes technologies that require retrofitting, modification of existing equipment, or new equipment or processes. Generally, moderate to substantial capital investments are also required.

5.1.2 Electricity Use in Residential and Commercial Buildings

Figures 5.1 and 5.2 illustrate estimated breakdowns of purchased electricity by major end use for the residential and commercial sectors, respectively. The data are from the Energy Information Administration's (EIA's) most recent estimates (EIA 2013a,b).

The single most significant residential end use of electricity is space cooling (19%), followed by lighting (13%), water heating (9%), refrigeration (8%), televisions and related equipment (6%), and space heating (6%). The combination of other uses such as clothes washers and dryers, personal computers, dishwashers, fans, pumps, etc. is also substantial, accounting for nearly 40% of residential electricity use.

In the commercial sector, space conditioning—that is, the combination of heating, ventilating, and air conditioning (HVAC)—is the top user of electricity, accounting for 28%. When HVAC equipment are considered as separated end uses, lighting rises to the top as the largest single end use of electricity in the commercial sector at 21%. Refrigeration (8%), non–personal computer (non-PC) office equipment (5%), and PC office equipment (5%) are also significant end uses of electricity. In addition, other end uses such as water heating, transformers, medical equipment, elevators and escalators, fume hoods, etc. account for the remaining third of electricity use in commercial buildings.

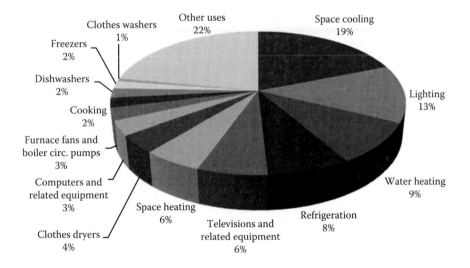

FIGURE 5.1
Breakdown of residential electricity end use, 2011, 1424 billion kWh. (From EIA, Annual energy outlook 2013 with projections to 2040. Residential sector key indicators and consumption. DOE/EIA-0383(2013), U.S. Energy Information Administration, Washington, DC, 2013b, http://www.eia.gov/oiaf/aeo/tablebrowser/.)

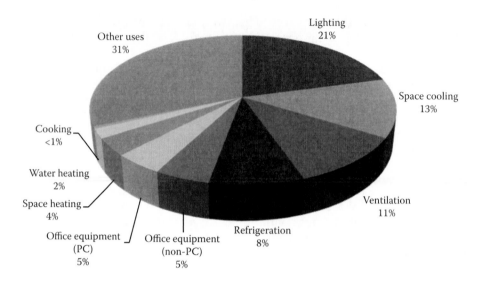

FIGURE 5.2
Breakdown of commercial electricity end use, 2011, 1319 billion kWh. (From EIA, Annual energy outlook 2013 with projections to 2040. Commercial sector key indicators and consumption. DOE/EIA-0383(2013), U.S. Energy Information Administration, Washington, DC, 2013a, http://www.eia.gov/oiaf/aeo/tablebrowser/.)

5.1.2.1 Residential Electricity Use

Space conditioning (including both space cooling and heating and associated fans and pumps) is the most significant end use of electricity in residential buildings, accounting for more than one-quarter of total electricity purchases today. EIA projects space conditioning's share of total residential electricity use will continue to be substantial with the possibility of increases in demand in the absence of greater efficiency improvements. Electricity is used in space heating and cooling to drive fans and compressors, to provide a direct source of heat (resistance heating), to provide an indirect source of heat or *cool* (heat pumps),* and for controls.

At 13% in 2011, residential lighting electricity use is up from 9% in 2001, due in part to a smaller share of electricity being consumed by other end uses. However, EIA expects lighting electricity use to decline considerably over the next decade both in terms of absolute quantity of electricity delivered to homes to power lights (35% reduction between 2011 and 2023) and in terms of lighting's relative share of residential end use (down to 8% in 2023). These projected reductions are due to the Energy Independence and Security Act (EISA) of 2007, which as of January 1, 2013, has been phasing in standards to replace incandescent lamps with more efficient compact fluorescent (CFL) and light-emitting diode (LED) lamps.

The share of residential electricity use by water heating has stayed relatively constant at 9% during the last decade. EIA projects the total electricity delivered to water heating systems will increase by 15% over the next decade and water heating's relative share of electricity use will rise to over 10% by 2023. Electricity use for this purpose currently occurs in electric storage tank water heaters, tankless electric water heaters,

---

* Heat pumps are discussed in detail in Chapter 9 of this handbook.

and heat pump water heaters. Solar water heating is another alternative that is used on a limited basis.

Refrigerators are another important energy end use in the residential sector, accounting for 8% of residential electricity consumption in 2011, down from 14% in 2001. EIA projects refrigeration's share of residential electricity use as well as the absolute quantity of electricity delivered to refrigerators will maintain roughly constant between now and 2023. For the last half century, virtually every home in the United States has had a refrigerator. Therefore, refrigerators have fully penetrated the residential sector for some time. However, significant changes related to energy use have occurred during this period as new standards have been implemented. For one, the average size of refrigerators has more than doubled from less than 10 ft^3 in 1947 to over 20 ft^3 today. Meanwhile, the efficiency of refrigerators has increased dramatically. The net result is that current refrigerators use about one-quarter as much energy (~450 kWh/year in 2011) as smaller units did 40 years ago (~1800 kWh/year in 1972), according to data from the Association of Home Appliance Manufacturers. Furthermore, when new efficiency standards take effect in Fall 2014, new refrigerators will use about one-fifth of the energy use of those in the early 1970s.

TVs, computers, and related equipment collectively account for over 9% of residential electricity use. These types of electronic systems have proliferated in homes during the last few decades. Clothes washers and dryers, cooking, dishwashers, and freezers account for another 11%, while *other* uses (including small electric devices, heating elements, and motors not included earlier) make up the balance (22%) of electricity use in the residential sector.

5.1.2.2 Commercial Electricity Use

For the commercial sector as a whole, HVAC dominates electricity use, with space cooling, space heating, and ventilation systems accounting for over one-quarter of electricity use. This trend is also true for most types of commercial buildings where space conditioning is used. There are exceptions of course—in energy-intensive facilities such as laundries, the process energy will be most important. Electricity is used in space conditioning to run fans, pumps, chillers, cooling towers, and heat pumps. Other uses include electric resistance heating (e.g., in terminal reheat systems) or electric boilers.

Commercial lighting is generally next in importance to HVAC for total use of electricity, except in those facilities with energy-intensive processes. Interior lighting is predominantly fluorescent, with an increasing portion of metal halide, a small fraction of incandescent, and a growing use of LED lamps. High-efficiency fluorescent lamps, electronic ballasts, CFL lamps, and improved lighting controls are now the norm. Incandescent lamps still see use in retail for display lighting as well as in older buildings or for decorative or esthetic applications. Lighting—estimated by the EIA to represent 21% of commercial sector electricity use in 2011—shows a reduction over the last two decades, dropping from 27.7% in 1992. The EIA predicts lighting's share of commercial building electricity use will continue to drop, reaching an estimated 15% by 2040 (EIA 2013a).

Refrigeration is an important use of energy in supermarkets and several other types of commercial facilities. As in residential applications, commercial refrigeration's share of electricity use has decreased in the past few decades due to efficiency gains. For example, in 2011, refrigeration accounted for 8% of commercial electricity use, down from 10% in 1992; EIA predictions estimate its share will drop to 7% over the next decade (EIA 2013a).

Commercial electricity use by office equipment has changed substantially over the last 20 years; its share grew from 7% in 1992 to 18% in 1999 and has now decreased to 10% in 2011. Much of the increase at the turn of the millennium was due to a greater use of computers. Computers have now greatly penetrated the commercial sector, and efficiency is constantly improving. As a result, EIA estimates the absolute quantity of energy used by computers will stay relatively constant over the next decade.

Water heating is another energy use in commercial buildings, but here circulating systems (using a heater, storage tank, and pump) are more common. Many possibilities exist for using heat recovery as a source of hot water. The share of electricity use for water heating in commercial buildings has grown from about 1% in 1999 to 2% in 2011. The EIA predicts the absolute quantity of electricity use for water heating will stay relatively constant over the next couple of decades (EIA 2013a).

In commercial buildings, the balance of the electricity use is for elevators, escalators, and miscellaneous items.

5.2 Setting Up an Energy Management Program

The general procedure for establishing an energy management program in buildings involves five steps:

- Review historical energy use.
- Perform energy audits.
- Identify energy management opportunities (EMOs).
- Implement changes to save energy.
- Monitor the energy management program, set goals, and review progress.

Each step will be described briefly. These steps have been designed for homeowners, apartment owners, and commercial building owners or operators to carry out. Many electric utilities also provide technical and financial assistance for various types of energy efficiency studies and improvements, so building owners and operators are encouraged to seek support from their local utility.

5.2.1 Review of Historical Energy Use

Utility records can be compiled to establish electricity use for a recent 12-month period. These should be graphed (see Figure 5.3) so that annual variations and trends can be evaluated. By placing several years (e.g., last year, this year, and next year projected) on the graph, past trends can be reviewed and future electricity use can be compared with goals. Alternatively, several energy graphs can be compared, or energy use versus production determined (meals served, for a restaurant, kilograms of laundry washed, for a laundry, etc.).

Previously, the norm was to review monthly data to look for energy use trends; however, daily, hourly, 15 min, and even 1 min energy data is now much more available due to the increasing deployment of interval meters. Studying trends in interval data allows building owners to have a greater understanding of exactly when electricity is used in addition to how much is used in a given billing cycle. This additional knowledge can

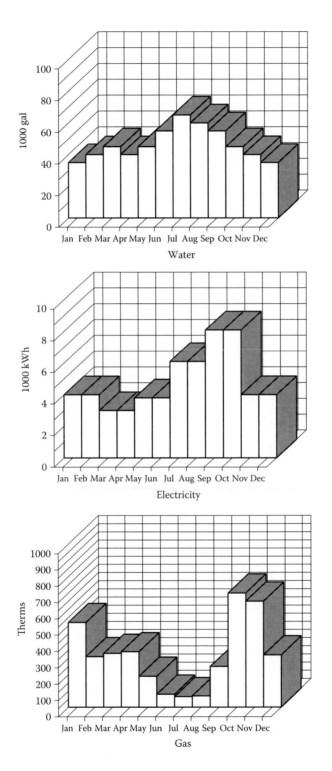

FIGURE 5.3
Sample graph: Historical energy use in an office building.

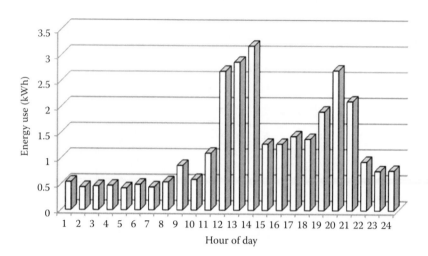

FIGURE 5.4
Example of single family home daily electric load profile, summer.

help building owners or operators manage loads during peak demand periods. Figure 5.4 illustrates a daily load profile generated with hourly interval data for a single family home in California. The graph has two daily peaks: the midday peak is when the home's pool pump is on and the evening peak is when the family turns on lights, cooks dinner, uses the spa, and watches television.

5.2.2 Perform Energy Audits

Figures 5.5 and 5.6 are data sheets used in performing an energy audit of a building. The Building Energy Survey Form (Figure 5.5) provides a gross indication of how energy is used in the building in meeting the particular purpose for which it was designed. This form would not be applicable to single family residences, but it could be used with apartments. It is primarily intended for commercial buildings.

Figure 5.6 is a form used to gather information concerning energy used by each piece of equipment in the building. When totaled, the audit results can be compared with the historical energy use records plotted on Figures 5.3 and 5.4. The energy audit results show a detailed breakdown and permit identification of major energy using equipment items.

Another way to perform energy audits is to use a laptop and a commercially available database or spreadsheet program to record the data and make the calculations. If the workload is extensive, the program can include *look-up* tables of frequently used electrical loads, utility rates, and other essential information to automate the process. We have used teams of engineers with laptops to rapidly survey and collect the energy data from large commercial facilities. See Figure 5.7 for example results from a lighting audit conducted several years ago.

In some cases, monitoring of key equipment may be warranted to improve the accuracy of the audit results. Monitoring provides insight into the actual time-based loads of equipment. It can also be used to simply indicate when loads are turned on or off. It is particularly valuable for highly variable or weather-dependent loads, including HVAC equipment. Monitoring can be accomplished with temporary data loggers that are placed on equipment for a representative period of time (e.g., 2–4 weeks). Permanent monitoring devices integrated with building management system can also be used to characterize

---

Building Description

- Name_____ Age_____ years Construction material_____
- Location_____
- No. of floors_____ Gross floor area_____ m² (ft²) Net floor area_____ m² (ft²)
- Percentage of surface area that is glazed_____ % Double or single pane (circle)
- Type of air conditioning system (describe)_____
- Type of heating system (describe)_____
- Cooling degree days_____ Heating degree days_____
- Percentage breakdown of lighting equipment: Fluorescent_____ % HID_____ %
- Incandescent_____ % Other_____ % Lighting controls?_____
- Other unique attributes (describe)_____

Building Mission

- What is facility used for_____
- Full time occupancy (employees)_____ persons
- Transient occupancy (visitors or public)_____ persons
- Hours of operations per year_____
- Unit of production per year_____ Unit is_____

Installed Capacity

- Installed capacity for lighting_____ kW
- Total installed capacity of electric drives greater than 7.5 kW (10 hp)
  (motors, pumps, fans, elevators, chillers, etc.)_____ hp × 0.746 =_____ kW
- Total steam requirements_____ lbs/day or_____ kg/day
- Total gas requirements_____ ft³/day or BTU/h or_____ m³/day
- Total other fuel requirements_____

Annual Energy End Use

Energy Form × Conversion	kBTU/year Metric Units			Conversion MJ/year	
• Electricity _____ kWh/year × 3.41	=_____	_____ kWh/year	×	3.6	=_____
• Steam _____ lb/year × 1.00	=_____	_____ kg/year	×	2.32	=_____
• Natural gas _____ cf/year × 1.03	=_____	_____ m3/year	×	38.4	=_____
• Oil _____ gls/year × $\begin{Bmatrix} \#2\ 139 \\ \#6\ 150 \end{Bmatrix}$	=_____	_____ l/year	×	$\begin{Bmatrix} \#2\ 38.9 \\ \#6\ 41.8 \end{Bmatrix}$	=_____
• Coal _____ tons/year × 24,000	=_____	_____ kg/year	×	28.0	=_____
• Other _____ ×	=_____	_____	×		=_____
Totals	_____				_____

Energy Use Performance Factors (EUPFs) for Building

- EUPF 1 = MJ/year(kBTU/year) ÷ Net floor area   = _____ MJ/m²year(kBTU/ft² year)
- EUPF 2 = MJ/year(kBTU/year) ÷ Average annual occupancy = _____ MJ/person · year
                                                              (kBTU/person · year)
- EUPF 3 = MJ/year(kBTU/year) ÷ Annual units of production = _____ MJ/unit · year
                                                              (kBTU/unit · year)

FIGURE 5.5
Building energy survey form.

energy use patterns on a continuous basis. Major advantages of continuous monitoring include the ability to spot energy use anomalies early and to measure performance of energy efficiency improvements against savings targets.

Energy audits can vary from simple 1-day walkthroughs to comprehensive multiday studies. Often significant opportunities are captured in a simple audit, but more detailed

Conversion Factors

Facility Name_____	Date_____ By_____ Sheet___ Of_____	Multiply by to get
Location_____	Period of survey: 1 day 1 week 1 month 1 year	kWh 3.6 MJ
Symbols: K = 10^3 M = 10^6	Notes_____	Btu/h 0.000293 kW
		hp 0.746 kW

Item No.	Equipment Description	Power			Est. Hrs Use Per Period	kWh	Conv. Factor	Total Energy Use Per Period (MJ)
		Name Plate Rating (Btu/h, kW, hp, etc.)	Conv. Factor to kW	Est.% Load (100%, 50%, etc.)				

FIGURE 5.6
Energy audit data sheet.

studies may be necessary to justify large capital projects. (See Chapter 4 for further information on energy audits.)

5.2.3 Identify Energy Management Opportunities

An overall estimate should be made of how effectively the facility uses its energy resources. This is difficult to do in many cases, because so many operations are unique. An idea can be obtained, however, by comparing similar buildings located in similar climates. Several online benchmarking tools exist, including EPA's ENERGY STAR Portfolio Manager. Table 5.2 shows representative values of energy use intensity (EUI), which is a measure of energy use per square foot of building area. The table includes median values for several types of commercial buildings and illustrates the range in performance factors that exists across these different building types.

Next, areas or equipment that use the greatest amounts of electricity should be examined. Each item should be reviewed and these questions asked:

- Is this actually needed?
- How can the same equipment be used more efficiently?
- How can the same purpose be accomplished with less energy?
- Can the equipment be modified to use less energy?
- Would new, more efficient equipment be cost-effective?

Lighting Energy Savings Summary

Existing annual kWh:	181,828 kWh	Existing kW draw:	36.37 kW
Proposed annual kWh:	95,234 kWh	Proposed kW draw:	18.52 kW
Annual kWh savings:	86,594 kWh	kW savings:	17.85 kW
% kWh savings:	47.6%	% kW savings:	49.1%

Prepare by:
Annual Energy $$ saved: $12,094
Estimated PRE-REBATE cost: $13,592

Lighting Inventory, Recommendations, and Savings

Item #	Location	Existing				Recommended			Savings			Estimated	
		Weekly Hours	Qty	Fixtures	W/Fix	Qty	Fixture	W/Fix	kW	Annual kWh	Annual Energy ($)	Unit Cost ($)	Total Cost ($)
1	Presidents office	60	8	75 Watt INC Spotlight	75	8	18 Watt CFL/SI/Ref.	18	0.5	1368	219	22	176
2	Presidents office	60	6	2-F40T12(40W)/STD	96	6	2-F32T8(32W)/ELEC	61	0.2	630	101	48	288
3	V.P. office	60	8	75 WATT INC Spotlight	75	8	18 Watt CFL/SI/Ref.	18	0.5	1368	219	22	176
4	V.P. office	60	4	2-F40T12(40W)/STD	96	4	2-F32T8(32W)/ELEC	61	0.1	420	67	48	192
5	Night lighting	168	4	2-F40T12(40W)/STD	96	4	2-F32T8(32W)/ELEC	61	0.1	1176	145	48	192
6	Women's restroom mirror	60	12	25 Watt INC	25		None						
7	Women's restroom	60	6	2-F40T12(34W)/U/STD	94	6	2-F40T12(34W)/U/ELEC	60	0.2	612	98	48	288
8	Men's restroom	60	6	2-F40T12(34W)/U/STD	94	6	2-F40T12(34W)/U/ELEC	60	0.2	612	98	48	288
9	Main office area	60	56	3-F40T12(40W)/2-Class 11	136	56	3-F32T8(32W)/1-ELEC	90	2.6	7728	1240	48	2688
10	Storage room	25	1	100 Watt INC	100	1	28 Watt PL CFL/SI	30	0.1	88	19	32	32
11	Parking garage	168	26	500 Watt Quartz	350	26	175 Watt MH	205	3.8	31,668	3906	200	5200
12	Parking garage	168	8	2-F96T12(75W)/STD	173	8	20F96T8(50W)/ELEC	104	0.6	4637	572	60	480
13	Physical plant	80	28	2-F96T12(215W)/WHO/STD	450	28	2-F96T12(95W)/HO/ELEC	166	8.0	31,808	4741	110	3080
14	Physical plant	80	16	100 Watt INC	100	16	28 Watt PL CFL/SI	30	1.0	4480	668	32	512
Total			162						17.9	86,594	$12,094	N/A	$13,592

FIGURE 5.7
Sample energy audit results XYZ corporation.

TABLE 5.2

Median EUI Values for Representative Building Types

Building Type	Source EUI[a] (kBTU/ft^2)	Site EUI[b] (kBTU/ft^2)
Fast food	1170	418
Food sales	570	193
Restaurant/cafeteria	434	207
Strip and enclosed mall	247	94
College/university	244	104
Lodging	163	72
Outpatient and health care	163	62
Retail (nonmall)	139	53
Public assembly	89	42

Source: EPA, CBECS national median source energy use and performance comparisons by building type, 2011, http://www.energystar.gov/ia/business/tools_resources/new_bldg_design/2003_ CBECSPerformanceTargetsTable.pdf. (Accessed August 13, 2013).

[a] Source EUI represents the quantity of raw fuel used to operate a building; it captures energy losses associated with the generation, transmission, and distribution of electricity.

[b] Site EUI represents the quantity of energy used directly on site.

5.2.4 Implement Changes

Once certain actions to save energy have been identified, an economic analysis will be necessary to establish the economic benefits and to determine if the cost of the action is justified. (Refer to Chapter 3 for guidance.) Those changes that satisfy the economic criteria of the building owner (or occupant) will then be implemented. Economic criteria might include a minimum return on investment (e.g., 25%), a minimum payback period (e.g., 2 years), or a minimum benefit–cost ratio (e.g., 2.0).

5.2.5 Monitor the Program, Establish Goals

This is the final—and perhaps most important—step in the program. A continuing monitoring program is necessary to ensure that energy savings do not gradually disappear as personnel return to their old ways of operation, equipment gets out of calibration, needed maintenance is neglected, etc. Also, setting goals (they should be realistic) provides energy management personnel with targets against which they can gauge their performance and the success of their programs.

5.2.6 Summary of Energy Management Programs

The foregoing has been outlined in two tables to provide a step-by-step procedure for electrical energy management in buildings. Table 5.3 is directed at the homeowner or apartment manager, while Table 5.4 has been prepared for the commercial building owner or operator. Industrial facilities are treated separately; refer to Chapter 10. As mentioned to previously, prior to undertaking an independent energy management program, ask your electric utility if you qualify for technical support or financial assistance.

One problem in performing the energy audit is determining the energy used by each item of equipment. In many cases, published data are available—as in Table 5.5 for residential appliances. In other cases, engineering judgments must be made, the manufacturer consulted, or instrumentation provided to actually measure energy use.

TABLE 5.3

Energy Management Plan for the Homeowner or Apartment Manager

Review Historical Data

1. Collect utility bills for a recent 12-month period.
2. Add up the bills and calculate total kWh, total $, average kWh (divide total by 12), average $, and note the months with the lowest and highest kWh. If you have access to interval data, investigate daily load profiles and note the hours of peak usage.
3. Calculate a seasonal variation factor (svf) by dividing the kWh for the greatest month by the kWh for the lowest month.

Perform Energy Audits

4. Identify all electrical loads greater than 1 kW (1000 W). Refer to Table 5.5 for assistance. Most electrical appliances have labels indicating the wattage. If not, use the relation $W = V \times A$.
5. Estimate the number of hours per month each appliance is used.
6. Estimate the percentage of full load (pfl) by each device under normal use. For a lamp, it is 100%; for water heaters and refrigerators, which cycle on and off, about 30%, for an electric range, about 25% (only rarely are *all* burners *and* the oven used), etc.
7. For each device, calculate kWh by multiplying: kW × hours/month × pfl = kWh/month.
8. Add up all kWh calculated by this method. The total should be smaller than the average monthly kWh calculated in Step 2.
9. Note: if the svf is greater than 1.5, the load shows strong seasonal variation, for example, summer air conditioning, winter heating, etc. If this is the case, make two sets of calculation, one for the lowest month (when the fewest loads are operating) and one for the highest month.
10. Make a table listing the wattage of each lamp and the estimated numbers of hours of use per month for each lamp. Multiply Watts times hours for each, sum, and divide by 1000. This gives kWh for the lighting loads. Add this to the total shown.
11. Add the refrigerator, television, and all other appliances or tools that use 5 kWh/month or more.
12. By this process, you should now have identified 80%–90% of electricity using loads. Other small appliances that are used infrequently can be ignored. The test is to now compare with the average month (high or low month if svf is greater than 1.5). If your total is too high, you have overestimated the pfl or the hours of use.
13. Now rank each appliance in descending order of kWh used per month. Your list should read approximately like this:

    First: Air conditioning (hot climates)
    Second: Lighting
    Third: Water heating
    Fourth: Refrigeration
    Fifth: Televisions and related equipment
    Sixth: Space heating (electric)
    Seventh to last: All others

Identify Opportunities and Apply Energy Management Principles

14. Attack the highest priority loads first. There are three general things that can be done to save energy and/or lower energy bills: (1) reduce kW (more efficient lamps and appliances); (2) reduce hours of use (turn lights off, etc.); (3) shift operation of loads to off-peak periods (if your electricity rates are higher during peak hours). Refer to the text for detailed suggestions.
15. *Educate* and get support from family members or other occupants of the space.

Monitor Program, Calculate Savings

16. After the energy management program has been initiated, examine subsequent utility bills to determine if you are succeeding.
17. Calculate energy and cost savings by comparing utility data.
18. Continue to set goals and try to meet them.

TABLE 5.4

Energy Management Plan for Commercial Building Operator

Review Historical Data

1. Collect utility bills for a recent 12-month period.
2. Add up the bills and calculate total kWh, total $, average kWh (divide total by 12), average $, and note the months with the lowest and highest kWh. If you have access to interval data, investigate daily load profiles and note hours of peak usage.
3. Calculate an svf by dividing the kWh for the greatest month by the kWh for the lowest month.
4. Prepare a graph of historical energy use (see Figures 5.3 and 5.4).

Perform Energy Audits

5. Evaluate major loads. In commercial buildings, loads can be divided into four categories:
   a. HVAC (fans, pumps, chillers, heaters, cooling towers)
   b. Lighting
   c. Office equipment and appliances (elevators, computers, cash registers, copy machines, hot water heaters, etc.)
   d. Process equipment (as in laundries, restaurants, bakeries, shops, etc.)
   Items a–c are common to all commercial operations and will be discussed here. Item d overlaps with industry and the reader should also refer to Chapter 10. Generally items a, b, and d account for the greatest use of electricity and should be examined in that order.
6. In carrying out the energy audit, focus on major loads. Items that together comprise less than 1% of the total connected load in kW can often be ignored with little sacrifice in accuracy.
7. Use the methodology described earlier and in Chapter 4 for conducting the audit.
8. Compare audit results with historical energy use. If 80%–90% of the total (according to the historical records) has been identified, this is generally adequate.

Formulate the Energy Management Plan

9. Secure management commitment. The need for this varies with the size and complexity of the operation. However, any formal program will cost something, in terms of salary for the energy coordinator as well as (possibly) an investment in building modifications and new equipment. At this stage, it is very important to project current energy usage and costs ahead for the next 3–5 years, make a preliminary estimate of potential savings (typically 10%–50% per year), and establish the potential payback or return on investment in the program.
10. Develop a list of EMOs and estimate the cost of each EMO and its payback. Methods for economic analysis are given in Chapter 3. For ideas and approaches useful for identifying EMOs, refer to the text.
11. Communicate the plan to employees, department heads, equipment operators, etc. Spell out who will do what, why there is a need, what are the potential benefits and savings. Make the point (if appropriate) that *the energy you save may save your job.* If employees are informed, understand the purpose, and realize that the plan applies to everyone, including the president, cooperation is increased.
12. Set goals for department managers, building engineers, equipment operators, etc., and provide monthly reports so that they can measure their performance.
13. Enlist the assistance of all personnel in (1) better *housekeeping and operations* (e.g., turning off lights, keeping doors closed) and (2) locating obvious wastes of electricity (e.g., equipment operating needlessly, better methods of doing jobs).

Implement Plan

14. Implementation should be done in two parts. First, carry out operational and housekeeping improvements with a goal of, say, 10% reduction in electricity use at essentially no cost and no reduction in quality of service or quantity of production. Second, carry out those modifications (retrofitting of buildings, new equipment, process changes) that have been shown to be economically attractive.
15. As changes are made, it is important to continue to monitor electricity usage to determine if goals are being realized. Additional energy audits may be justified.

Evaluate Progress, Management Report

16. Compare actual performance to the goals established in Item 12. Make corrections for weather variations, increases or decreases in production or number of employees, addition of new buildings, etc.
17. Provide a summary report of energy quantities and dollars saved.
18. Prepare new plans for the future.

TABLE 5.5

Typical Residential Energy Usage for Common Appliances

Electric Appliances	Power (W)
Home entertainment	
Radio	10–20
Stereo	50–400
Speakers	50
Television	
Tube, 25″–27″	90–120
LCD, >40″	150–220
DLP, >40″	200–240
Plasma, >40″	400–480
CD player	10–20
DVD player	20–25
Gaming stations	20–210
Personal computer	
Laptop	20–50
Desktop CPU	30–120
Desktop monitor	30–150
Computer printer	100
Aquarium	50–1210
Food preparation	
Blender	300
Coffee maker	400–1,200
Dishwasher	1,200–2,400
Frying pan	1,200
Hot plate	1,200
Mixer	127
Microwave oven	750–1,100
Range	
Oven bake unit	2,300–3,200
Broil unit	3,600
Self-cleaning feature	4,000
Toaster	750–1,400
Trash compactor	400
Waffle iron	1200
Waste dispenser	450
Refrigerator/freezer	
Top freezer, <10 years old	440–600 kWh/year
Side by side, <10 years old	600–1200 kWh/year
Laundry	
Electric clothes dryer	1,800–5,000
Clothes washer	350–500
Iron (hand)	1,000–1,800
Water heater	2,500–5,500

(Continued)

TABLE 5.5 (*Continued*)

Typical Residential Energy Usage for Common Appliances

Electric Appliances	Power (W)
Housewares	
Clock	2
Floor polisher	305
Sewing machine	75–100
Vacuum cleaner	630–1,400
Comfort conditioning	
Air conditioner (room)	600–2,000
Central air conditioner	2,000–5,000
Electric blanket	60–180
Heating pad	65
Dehumidifier	257–785
Fan	
Attic	370
Ceiling	50–175
Furnace	300–1,000
Portable	55–250
Heater (portable, baseboard)	750–1,500
Electric furnace with fan	10,500
Heat pump	2,900–10,000
Humidifier	177
Health and beauty	
Hair dryer	1,000–1,875
Curling iron	50
Heat lamp (infrared)	250
Shaver	15
Tooth brush	1
Swimming pool and spa	
Sweep pump (3/4 hp)	560
Filter pump (1–1/2 hp)	1,120
Filter pump (2 hp)	1,500
Spa electric heater	1,500–5,500

5.3 Electricity-Saving Techniques by Category of End Use

This section discusses strategies for saving energy in the major end uses found in residential and commercial buildings. Projects that can be implemented in a short time at zero or low capital cost are presented along with retrofit and new design strategies. The ordering of topics corresponds approximately to their importance in terms of building energy use. Smaller end uses and processes specific to select building types are excluded.

5.3.1 Residential HVAC

Residential HVAC units using electricity are generally air conditioning systems, electric resistance heaters, heat pumps, and ventilation fans. Cooling systems range from window air conditioning units to central air conditioning systems. Evaporative coolers are also used in some climates. Heater types range from electric furnaces, small radiant heaters, duct heaters, and strip or baseboard heaters, to embedded floor or ceiling heating systems. Heat pump systems can be used for both heating and cooling and are a highly efficient alternative.

Principal operational and maintenance strategies for existing equipment include the following:

- System maintenance and cleanup
- Thermostat calibration and set-back
- Night cool down
- Improved controls and operating procedures
- Heated or cooled volume reduction
- Reduction of infiltration and exfiltration losses

System maintenance is an obvious but often neglected energy-saving tool. Dirty heat transfer surfaces decrease in efficiency. Clogged filters increase pressure drops and pumping power. Inoperable or malfunctioning dampers can waste energy and prevent proper operation of the system.

In residential systems, heating and cooling are generally controlled by a central or local thermostat. Thermostats should be set to about 26°C (78°F) or higher for cooling and 20°C (68°F) or lower for heating when occupants are home and awake. During periods when occupants are away or asleep, the temperature set points should be adjusted to reduce heating and cooling energy use. As a first step, check the calibration of the thermostat, since these low cost devices can be inaccurate by as much as ±5°C. Programmable thermostats are now widely available. These can be programmed by the user to set-back or set-forward temperature automatically depending on the time of day and day of week. By eliminating the need for manual control, they ensure that the settings will indeed be changed, whereas manual resetting of thermostats depends on occupant diligence. Some utilities have set up demand response programs in which they can communicate with smart thermostats, also referred to as programmable communicating thermostats (PCTs). Participating customers can program their PCTs to control the temperature in response to pricing signals or other demand response event notifications from the utility, thereby reducing peak demand and lowering energy costs. In some programs, the utility uses the PCT to control the customer's equipment directly during high peak periods. A general rule of thumb is that for every 1°C of thermostat set-back (heating) or set-forward (cooling) during an 8 h period, there is a 1% savings in annual heating or cooling energy costs. (The energy savings are generally lower in more severe climates.)

Sometimes simple changes in controls or operating procedures will save energy. In cooling, use night air for summer cool-down. When the outside air temperature is cool, turn off the air conditioner and circulate straight outside air or simply open windows. If fan units have more than one speed, use the lowest speed that provides satisfactory operation. Also, check the balance of the system and the operation of dampers and vents to insure that heating and cooling is provided in the correct quantities where needed.

Reducing the volume of the heated or cooled space can yield energy savings. This can be accomplished by closing vents, doors, or other appropriate means. Usually, it is not necessary to heat or cool an entire residence; the spare bedroom is rarely used, halls can be closed off, etc.

A major cause of energy wastage is air entering or leaving a home. Unintentional air transfer toward the inside is referred to as *infiltration*, and unintentional air transfer toward the outside is referred to as *exfiltration*. However, *infiltration* is often used to imply air leakage both into and out of a home, and this is the terminology used in this chapter. In a poorly *sealed* residence, infiltration of cold or hot air will increase heating or cooling energy use. A typical home loses 25%–40% of its HVAC energy through infiltration. Infiltration also affects concentrations of indoor pollutants and can cause uncomfortable drafts. Air can infiltrate through numerous cracks and spaces created during building construction, such as those associated with electrical outlets, pipes, ducts, windows, doors, and gaps between ceilings, walls, floors, and so on. Infiltration results from temperature and pressure differences between the inside and outside of a home caused by wind, natural convection, and other forces. Major sources of air leakage are attic bypasses (paths within walls that connect conditioned spaces with the attic), fireplaces without dampers, leaky ductwork, window and door frames, and holes drilled in framing members for plumbing, electrical, and HVAC equipment. According to the U.S. Department of Energy, the most significant source for infiltration is the combination of walls, ceilings, and floors, which comprises 31% of the total infiltration in a typical home. Ducts (15%), fireplaces (14%), plumbing penetrations (13%), doors (11%), and windows (10%) are also substantial contributors to infiltration. Of lesser consequence are fans and vents (4%) and electrical outlets (2%).

To combat infiltration, builders of energy-efficient homes use house wraps, caulking, foam insulation, tapes, and other seals. Sealing ducts in the home is also important to prevent the escape of heated or cooled air. Homeowners should also check for open doors and windows, open fireplace dampers, inadequate weather stripping around windows and doors, and any other openings that can be sealed. Caution must be exercised to provide adequate ventilation, however. Standards vary, depending on the type of occupancy. Ventilation rates specified in the builder guidelines for the American Lung Association's Health House program state that for healthy homes "continuous general ventilation should be at least 1.0 cfm (cubic feet per minute) per 100 ft^2 of floor area plus at least 15 cfm for the first bedroom and 7.5 cfm for each additional bedroom." In addition, intermittent ventilation for the kitchen should be at least 100 cfm. For the bathrooms, rates should be 50 cfm intermittent or 20 cfm continuous. The Health House ventilation rates comply with ASHRAE standard 62.2.

In retrofit or new design projects, the following techniques will save energy:

- Site selection and building orientation
- Building envelope design
- Selection of efficient heating/cooling equipment

Site selection and building orientation are not always under the control of the owner/occupant. Where possible, though, select a site sheltered from temperature extremes and wind. Orient the building (in cold climates) with a maximum southerly exposure to take advantage of direct solar heating in winter. Use earth berms to reduce heat losses on northerly exposed parts of the building. Deciduous trees provide summer shading but permit winter solar heating.

Building envelope design can improve heat absorption and retention in winter and summer coolness. The first requirement is to design a well-insulated, thermally tight structure. Insulation reduces heating and cooling loads by resisting the transfer of heat through ceilings, walls, floors, and ducts. Reductions are usually proportionately higher for heating than for cooling because of generally larger indoor-to-outdoor temperature differences in winter than in summer. Insulation is available in a variety of forms including batts, rolls, boards, blocks, loose-fill, or sprayed foam. The appropriate insulation material is selected on the basis of climate, building type, and recommended R-value. Higher R-values indicate better insulating properties. It is typically cost-effective to use greater than recommended R-values to improve energy efficiency above and beyond standard building practice.

Windows are an important source of heat gain and loss. The heat loss for single pane glazing is around 5–7 W/m^2°C. For double glazing, the comparable value is in the range of 3–4 W/m^2°C. Window technology is constantly improving. Newer windows often have low emissivity (low-E) or spectrally selective coatings to prevent heat gain and/or loss. Low-E windows with an argon gas fill have a heat loss rate of about 2 W/m^2°C. They have a higher visible transmittance, and are available with a low solar heat gain coefficient to reduce cooling loads in the summer. Low-E windows are available with an internal plastic film, which essentially makes them triple glazed. The heat loss rate for these windows is on the order of 1 W/m^2°C.

Windows equipped with vinyl, wood, or fiberglass frames, or aluminum frames with a thermal barrier, provide the best insulation. It is also important to seal windows to prevent infiltration, as well as to use window coverings to minimize heat loss by radiation to the exterior during the evening. The appropriate placement of windows can also save energy by providing daylighting.

In general, the most efficient electric heating and cooling system is the heat pump. Common types are air-to-air heat pumps, either a single package unit (similar to a window air conditioner) or a split system where the air handling equipment is inside the building and the compressor and related equipment are outdoors. Commercially available equipment demonstrates a wide range of efficiency. Heating performance is measured in terms of a heating seasonal performance factor (HSPF), in BTUs of heat added per Watt-hour of electricity input. Typical values are 6.8–10.0 for the most efficient heat pumps. Cooling performance of residential heat pumps, air conditioners, and packaged systems is measured in terms of a seasonal energy efficiency ratio (SEER), which describes the ratio of cooling capacity to electrical power input. Typical values are 10.0–14.5 and even higher for the most efficient systems. The Federal standards set in 2006 for air conditioners, heat pumps, and residential packaged units require a minimum SEER of 13 and a minimum HSPF of 7.7. Many existing older units have SEERs of 6–7, or roughly half the new minimum requirement. Therefore, substantial efficiency improvements are possible by replacing older equipment. In purchasing new equipment, consider selecting systems with the highest HSPF and SEER. The operating savings almost always justify the higher initial cost of these units. In addition, many utilities offer rebates for installing the more efficient units.

Sizing of equipment is important, since the most efficient operation generally occurs at or near full load. Selection of oversized equipment is thus initially more expensive, and will also lead to greater operating costs.

The efficiency of heat pumps declines as the temperature difference between the heat source and heat sink decreases. Since outside air is generally the heat source, heat is most difficult to get when it is most needed. For this reason, heat pumps often have electrical backup heaters for extremely cold weather.

An alternate approach is to design the system using a heat source other than outside air. Examples include heated air (such as is exhausted from a building), a deep well (providing water at a constant year-round temperature), the ground, or a solar heat source. There are a great many variations on solar heating and heat pump combinations.

5.3.2 Nonresidential HVAC

HVAC systems in commercial buildings and other nonresidential installations may involve package rooftop or ground mounted units, or a central plant. Although the basic principles are similar to those discussed earlier in connection with residential systems, the equipment is larger and control more complex.

Efficiency of many existing HVAC systems can be improved. Modifications can reduce energy use by 10%–15%, often with building occupants unaware that changes have been made. Instituting an energy management program that includes evaluating and fine-tuning the operation of HVAC systems to ensure they are operating as originally designed (referred to as retrocommissioning in existing buildings) can uncover many inefficiencies and often yields significant energy savings and improved comfort.

The basic function of HVAC systems is to heat, cool, dehumidify, humidify, and provide air mixing and ventilating. The energy required to carry out these functions depends on the building design, its duty cycle (e.g., 24 h/day use as in a hospital vs. 10 h/day in an office), the type of occupancy, the occupants' use patterns and training in using the HVAC system, the type of HVAC equipment installed, and finally, daily and seasonal temperature and weather conditions to which the building is exposed.

A complete discussion of psychometrics, HVAC system design, and commercially available equipment types is beyond the scope of this chapter.

Energy management strategies will be described in three parts:

- Equipment modifications (control, retrofit, and new designs)
  - Fans
  - Pumps
  - Packaged air conditioning units
  - Chillers
  - Ducts and dampers
  - Systems
- Economizer systems and enthalpy controllers
- Heat recovery techniques

5.3.2.1 Equipment Modifications (Control, Retrofit, and New Designs)

5.3.2.1.1 Fans

All HVAC systems involve some movement of air. The energy needed for this motion can make up a large portion of the total system energy used. This is especially true in moderate weather when the heating or cooling load drops off, but the distribution systems often operate at the same level.

5.3.2.1.1.1 Control Simple control changes can save electrical energy in the operation of fans. Examples include turning off large fan systems when relatively few people are in the building or stopping ventilation a half hour before the building closes. The types of changes that can be made will depend upon the specific facility. Some changes involve more sophisticated controls, which may already be available in the HVAC system.

5.3.2.1.1.2 Retrofit The capacity of the building ventilation system is usually determined by the maximum cooling or heating load in the building. This load may change over time due to reduced outside air requirements, more efficient lighting, and fluctuations in building occupancy. As a result, it may be feasible to decrease air flow in existing commercial buildings as long as adequate indoor air quality is maintained.

The volume rate of air flow through the fan, Q, varies directly with the speed of the impeller's rotation. This is expressed as follows for a fan whose speed is changed from N_1 to N_2.

$$Q_2 = \left(\frac{N_2}{N_1}\right) \times Q_1 \tag{5.1}$$

The pressure developed by the fan, P, (either static or total) varies as the square of the impeller speed.

$$P_2 = \left(\frac{N_2}{N_1}\right)^2 \times P_1 \tag{5.2}$$

The power needed to drive the fan, H, varies as the cube of the impeller speed.

$$H_2 = \left(\frac{N_2}{N_1}\right)^3 \times H_1 \tag{5.3}$$

The result of these laws is that for a given air distribution system (specified ducts, dampers, etc.), if the air flow is to be doubled, eight (2^3) times the power is needed. Conversely, if the air flow is to be cut in half, one-eighth ($\frac{1}{2}^3$) of the power is required. This is useful in HVAC systems, because even a small reduction in air flow (say 10%) can result in significant energy savings (27%).

The manner in which the air flow is reduced is critical in realizing these savings. Sizing the motor exactly to the requirements helps to yield maximum savings. Simply changing pulleys to provide the desired speed will also result in energy reductions according to the cubic law. Note the efficiency of existing fan motors tends to drop off below the half load range.

If variable volume air delivery is required, it may be achieved through inlet vane control, outlet dampers, variable speed drives (VSDs), controlled pitch fans, or cycling. Energy efficiency in a retrofit design is best obtainable with VSDs on motors. This can be seen by calculating the power reduction that would accompany reduced flow using different methods of control, as noted in Table 5.6. Numbers in the table are the percent of full-flow input power:

5.3.2.1.1.3 New Design　The parameters for new design are similar to those for fan retrofit. It is desirable, when possible, to use a varying ventilation rate that will decrease as the load decreases. A system such as variable air volume incorporates this in the interior zones of a building. In some cases, there will be a trade-off between power saved by running the fan slower and the additional power needed to generate colder air. The choices should be determined on a case-by-case basis.

5.3.2.1.2 Pumps

Pumps are found in a variety of HVAC applications such as chilled water, heating hot water, and condenser water loops. They are another piece of peripheral equipment that can use a large portion of HVAC energy, especially at low system loads.

TABLE 5.6

Comparison of Flow Control Alternatives for Fans and Pumps

| | Percent of Full-Flow Input Power | | | | |
| | Fans | | | Pumps | |
Flow (%)	Inlet Vanes	Dampers	VSDs	Throttle Valve	VSDs
100	102	103	102	101	103
90	86	98	76	96	77
80	73	94	58	89	58
70	64	88	43	83	41
60	56	81	31	77	30
50	50	74	22	71	19
40	46	67	15	65	13
30	41	59	9	59	8

5.3.2.1.2.1 Control   The control of pumps is often neglected in medium and large HVAC systems where it could significantly reduce the demand. A typical system would be a three chiller installation where only one chiller is needed much of the year. Two chilled water pumps in parallel are designed to handle the maximum load through all three chillers. Even when only one chiller is on, both pumps are used. When not needed to meet demand, two chillers could be bypassed and one pump turned off. All systems should be reviewed in this manner to ensure that only the necessary pumps operate under normal load conditions.

5.3.2.1.2.2 Retrofit   Pumps follow laws similar to fan laws, the key being the cubic relationship of power to the volume pumped through a given system. Small decreases in flow rate can save significant portions of energy.

In systems in which cooling or heating requirements have been permanently decreased, flow rates may be reduced also. A simple way to do this is by trimming the pump impeller. The pump curve must be checked first, however, because pump efficiency is a function of the impeller diameter, flow rate, and pressure rise. After trimming, one should ensure that the pump will still be operating in an efficient region. This is roughly the equivalent of changing fan pulleys in that the savings follow the cubic law of power reduction.

Another method for decreasing flow rates is to use a *throttle* (pressure reducing) valve. The result is equivalent to that of the discharge damper in the air-side systems. The valve creates an artificial use of energy, which can be responsible for much of the work performed by the pumps. VSDs are more efficient for varying flow, as shown in Table 5.6.

5.3.2.1.2.3 New Design   In a variable load situation, common to most HVAC systems, more efficient systems with new designs are available, rather than the standard constant volume pump. (These may also apply to some retrofit situations.)

One option is the use of several pumps of different capacity so that a smaller pump can be used when it can handle the load and a larger pump used the rest of the time. This can be a retrofit modification as well when a backup pump provides redundancy. Its impeller would be trimmed to provide the lower flow rate.

Another option is to use VSD pumps. While their initial cost is greater, they offer a significant improvement in efficiency over the standard pumps. The economic desirability of this or any similar change can be determined by estimating the number of hours the system will operate under various loads. Many utilities also offer rebates for installing variable speed pumps.

5.3.2.1.3 Package Air Conditioning Units

The most common space conditioning systems for commercial buildings are unitary equipment, either single package systems or split systems. These are used for cooling approximately two-thirds of the air-conditioned commercial buildings in the United States. For very large buildings or building complexes, absorption chillers or central chiller plants are used. (The following section describes Chillers.)

Air conditioner efficiency is typically rated by one or more of three parameters: the energy efficiency ratio (EER), the SEER as described previously, and the integrated energy efficiency ratio (IEER) that replaces a former metric used called the integrated part-load value (IPLV). The EER is easy to understand; it is the ratio of cooling capacity, expressed in BTU/hour (kilojoules/hour) to the power input required in Watts. The SEER is a calculated ratio of the total annual cooling produced per annual electrical energy input in Watt-hours for units rated at less than 65,000 BTU/h. The IEER is a new part-load metric introduced in ASHRAE Standard 90.1-2010 to replace the IPLV for rating commercial unitary loads of more than 65,000 BTU/h.

Great improvements in packaged air conditioner efficiency have been made in the last few decades, and new standards continue to increase efficiency even further. This is illustrated by the 30% increase in minimum SEER requirement that took effect in June 2008 for commercial three-phase central air conditioners and heat pumps under 65,000 BTU/h as the result of the EISA (2007). The current standard is a minimum SEER of 13.0. These commercial units are similar to the residential central air conditioners and heat pumps discussed earlier in the chapter in the residential HVAC section. Standards for larger units (>65,000 BTU/h) also continue to increase. All current standards are published in ASHRAE Standard 90.1-2010. Manufacturers sell systems with a broad range of efficiencies. Units with high EERs are typically more expensive, as the greater efficiency is achieved with larger heat exchange surface, more efficient motors, and so on.

To evaluate the economic benefit of the more efficient units, it is necessary to determine an annual operating profile, which depends in part on the nature of the load and on the weather and temperature conditions at the site where the equipment will be installed. Or, an approximate method can be used. ASHRAE publishes tables that show typical *equivalent full-load operating hours* for different climate zones. These can be used to estimate the savings in electrical energy use over a year and thereby determine if the added cost of a more efficient unit is justified. (It almost always is.)

Since the more efficient unit is almost always more cost-effective (except in light or intermittent load conditions), one might wonder why the less efficient units are sold. The reason is that many commercial buildings are constructed and sold by developers whose principal concern is keeping the initial cost of the building as low as practicable. They do not have to bear the annual operating expense of the building once it is sold, and therefore have less incentive to minimize operating expenses. However, the demand for energy-efficient and LEED-certified buildings is growing, causing many developers to rethink design approaches.*

5.3.2.1.4 Chillers

Chillers are often the largest single energy user in the HVAC system. The chiller cools the water used to extract heat from the building and outside air. Optimizing chiller operation improves the performance of the whole system.

---

* LEED stands for Leadership in Energy & Environmental Design, http://www.usgbc.org/leed.

Two basic types of chillers are found in commercial and industrial applications: absorption and mechanical chillers. Absorption units boil water, the refrigerant, at a low pressure through absorption into a high concentration lithium bromide solution. Mechanical chillers cool through evaporation of a refrigerant at a low pressure after it has been compressed, cooled, and passed through an expansion valve. Hydrochlorofluorocarbon (HCFC) products such as R22 are among the most common refrigerants used; however, they are currently being phased out, because they contribute to ozone depletion. The Environmental Protection Agency (EPA) lists acceptable refrigerant substitutes for various types of chillers on their website. Ammonia is a type of refrigerant that has been increasing in favor due to its efficiency advantages and the fact that it will continue to be available.

There are three common types of mechanical chillers. They have similar thermodynamic properties, but use different types of compressors. Reciprocating and screw-type compressors are both positive displacement units. The centrifugal chiller uses a rapidly rotating impeller to pressurize the refrigerant.

All of these chillers must reject heat to a heat sink outside the building. Some use air-cooled condensers, but most large units operate with evaporative cooling towers. Cooling towers have the advantage of rejecting heat to a lower-temperature heat sink, because the water approaches the ambient wet-bulb temperature, while air-cooled units are limited to the dry-bulb temperature. As a result, air-cooled chillers have a higher condensing temperature, which lowers the efficiency of the chiller. In full-load applications, air-cooled chillers require about 1–1.3 kW or more per ton of cooling, while water-cooled chillers usually require between 0.4 and 0.9 kW/ton. Air-cooled condensers are sometimes used, because they require much less maintenance than cooling towers and have lower installation costs. They can also be desirable in areas of the country where water is scarce and/or water and water treatment costs are high, since they do not depend on water for cooling.

Mechanical cooling can also be performed by direct expansion (DX) units. These are very similar to chillers except that they cool the air directly instead of using the refrigerant as a heat transfer medium. They eliminate the need for chilled water pumps and also reduce efficiency losses associated with the transfer of the heat to and from the water. DX units must be located close (~30 m) to the ducts they are cooling, so they are typically limited in size to the cooling required for a single air handler. A single large chiller can serve a number of distributed air handlers. Where the air handlers are located close together, it can be more efficient to use a DX unit.

5.3.2.1.4.1 Controls Mechanical chillers operate on a principle similar to the heat pump. The objective is to remove heat from a low-temperature building and deposit it in a higher temperature atmosphere. The lower the temperature rise that the chiller has to face, the more efficiently it will operate. It is useful, therefore, to maintain as warm a chilled water loop and as cold a condenser water loop as possible.

Using lower-temperature water from the cooling tower to reject the heat can save energy. However, as the condenser temperature drops, the pressure differential across the expansion valve drops, starving the evaporator of refrigerant. Many units with expansion valves, therefore, operate at a constant condensing temperature, usually 41°C (105°F), even when more cooling is available from the cooling tower. Field experience has shown that in many systems, if the chiller is not fully loaded, it can be operated with a lower cooling tower temperature.

5.3.2.1.4.2 Retrofit Where a heat load exists and the wet-bulb temperature is low, cooling can be done directly with the cooling tower. If proper filtering is available, the cooling tower water can be piped directly into the chilled water loop. Often a direct heat exchanger between the two loops is preferred to protect the coils from fouling. Another technique is to turn off the chiller but use its refrigerant to transfer heat between the two loops. This *thermocycle* uses the same principles as a heat pipe, and only works on chillers with the proper configuration.

A low wet-bulb temperature during the night can also be utilized. It requires a chiller that handles low condensing temperatures and a cold storage tank. This thermal energy storage (TES) technique is particularly desirable for consumers with access to time-based electricity rates that reward peak-shaving or load-shifting.

5.3.2.1.4.3 New Design   In the purchase of a new chiller, an important consideration should be the load control feature. Since the chiller will be operating at partial load most of the time, it is important that it can do so efficiently. Variable speed chillers are an efficient alternative. In addition to control of single units, it is sometimes desirable to use multiple compressor reciprocating chillers. This allows some units to be shut down at partial load. The remaining compressors operate near full load, usually more efficiently. Good opportunities to install a high-efficiency chiller are when an old unit needs to be replaced, or when it is necessary to retire equipment that uses environmentally unacceptable refrigerants.

5.3.2.1.5 Ducting-Dampers

5.3.2.1.5.1 Controls   In HVAC systems using dual ducts, static pressure dampers are often placed near the start of the hot or cold plenum run. They control the pressure throughout the entire distribution system and can be indicators of system operation. In an overdesigned system, the static pressure dampers may never open more than 25%. Reducing the fan speed and opening the dampers fully can eliminate the previous pressure drop. The same volume of air is delivered with a significant drop in fan power.

5.3.2.1.5.2 Retrofit   Other HVAC systems use constant volume mixing boxes for balancing that create their own pressure drops as the static pressure increases. An entire system of these boxes could be overpressurized by several inches of water without affecting the air flow, but the required fan power would increase. (One inch of water pressure is about 250 N/m^2 or 250 Pa.) These systems should be monitored to ensure that static pressure is controlled at the lowest required value. It may also be desirable to replace the constant volume mixing boxes with boxes without volume control to eliminate their minimum pressure drop of approximately 1 in. of water. In this case, static pressure dampers will be necessary in the ducting.

Leakage in any dampers can cause a loss of hot or cold air. Adding or replacing seals on the blades can slow leakage considerably. If a damper leaks more than 10%, it can be less costly to replace the entire damper assembly with effective positive-closing damper blades rather than to tolerate the loss of energy.

5.3.2.1.5.3 New Design   In the past, small ducts were installed because of their low initial cost despite the fact that the additional fan power required offset the initial cost on a life cycle basis. ASHRAE 90.1 guidelines now set a maximum limit on the fan power that can be used for a given cooling capacity. As a result, the air system pressure drop must be low enough to permit the desired air flow. In small buildings, this pressure drop is often largest across filters, coils, and registers. In large buildings, the duct runs may be responsible for a significant fraction of the total static pressure drop, particularly in high velocity systems. New designs should incorporate ducting that optimizes energy efficiency.

5.3.2.1.6 Systems

The use of efficient equipment is only the first step in the optimum operation of a building. Equal emphasis should be placed upon the combination of elements in a system and the control of those elements. This section discusses some opportunities for equipment modifications. Chapter 6 describes HVAC control systems in greater detail.

5.3.2.1.6.1 Control Some systems use a combination of hot and cold to achieve moderate temperatures. Included are dual duct, multizone, and terminal reheat systems, and some induction, variable air volume, and fan coil units. Whenever combined heating and cooling occurs, the temperatures of the hot and cold ducts or water loops should be brought as close together as possible, while still maintaining building comfort.

This can be accomplished in a number of ways. Hot and cold duct temperatures are often reset on the basis of the temperature of the outside air or the return air. Another approach is to monitor the demand for heating and cooling in each zone. For example, in a multizone building, the demand of each zone is communicated back to the supply unit. At the supply end, hot air and cold air are mixed in proportion to this demand. The cold air temperature should be just low enough to cool the zone calling for the most cooling. If the cold air were any colder, it would be mixed with hot air to achieve the right temperature. This creates an overlap in heating and cooling not only for the zone but for all the zones, because they would all be mixing in the colder air.

If no zone calls for total cooling, then the cold air temperature can be increased gradually until the first zone requires cooling. At this point, the minimum cooling necessary for that multizone configuration is performed. The same operation can be performed with the hot air temperature until the first zone is calling for heating only.

Note that simultaneous heating and cooling is still occurring in the rest of the zones. This is not an ideal system, but it is a first step in improving operating efficiency for these types of systems.

The technique for resetting hot and cold duct temperatures can be extended to the systems that have been mentioned. Ideally, it would be performed automatically with a control system, but it could also be done manually. In some buildings, it will require the installation of more monitoring equipment (usually only in the zones of greatest demand), but the expense should be relatively small and the payback period short.

Nighttime temperature set-back is another control option that can save energy without significantly affecting the comfort level. Energy is saved by shutting off or cycling fans. Building heat loss may also be reduced, because the building is cooler and no longer pressurized.

In moderate climates, complete night shutdown can be used with a morning warm-up period. In colder areas where the overall night temperature is below 4°C (40°F), it is usually necessary to provide some heat during the night. Building set-back temperature is partially dictated by the capacity of the heating system to warm the building in the morning. In some cases, it may be the mean radiant temperature of the building rather than air temperature that determines occupant comfort.

Some warm-up designs use *free heating* from people and lights to help attain the last few degrees of heat. This also provides a transition period for the occupants to adjust from the colder outdoor temperatures.

In some locations during the summer, it is desirable to use night air for a cool-down period. This *free cooling* can decrease the temperature of the building mass that has accumulated heat during the day. In certain types of massive buildings (such as libraries or buildings with thick walls), a long period of night cooling may decrease the building mass temperature by a degree or two. This represents a large amount of cooling that the chiller will not have to perform the following day.

5.3.2.1.6.2 Retrofit   Retrofitting HVAC systems may be an easy or difficult task depending upon the possibility of using existing equipment in a more efficient manner. Often retrofitting involves control or ducting changes that appear relatively minor but will greatly increase the efficiency of the system. Some of these common changes, such as decreasing air flow, are discussed elsewhere in this chapter. This section will describe a few changes appropriate to particular systems.

Both dual duct and multizone systems mix hot and cold air to achieve the proper degree of heating or cooling. In most large buildings, the need for heating interior areas is essentially nonexistent, due to internal heat generation. A modification that adjusts for this is simply shutting off air to the hot duct. The mixing box then acts as a variable air volume box, modulating cold air according to room demand as relayed by the existing thermostat. (It should be confirmed that the low volume from a particular box meets minimum air requirements.)

Savings from this modification come mostly from the elimination of simultaneous heating and cooling, depending on fan control strategies. That is, if fans in these systems are controlled by static pressure dampers in the duct after the fan, they do not unload very efficiently and would represent only a small portion of the savings.

5.3.2.2 Economizer Systems and Enthalpy Controllers

The economizer cycle is a technique for introducing varying amounts of outside air to the mixed air duct. Basically, it permits mixing warm return air at 24°C (75°F) with cold outside air to maintain a preset temperature in the mixed air plenum (typically 10°C–15°C, 50°F–60°F). When the outside temperature is slightly above this set point, 100% outside air is used to provide as much of the cooling as possible. During very hot outside weather, minimum outside air will be added to the system.

A major downfall of economizer systems is poor maintenance. The failure of the motor or dampers may not cause a noticeable comfort change in the building, because the system is often capable of handling the additional load. Since the problem is not readily apparent, corrective maintenance may be put off indefinitely. In the meantime, the HVAC system will be working harder than necessary for any economizer installation.

Typically, economizers are controlled by the dry-bulb temperature of the outside air rather than its enthalpy (actual heat content). This is adequate most of the time, but can lead to unnecessary cooling of air. When enthalpy controls are used to measure wet-bulb temperatures, this cooling can be reduced. However, enthalpy controllers are more expensive and less reliable.

The rules that govern the more complex enthalpy controls for cooling-only applications are as follows:

- When outside air enthalpy is greater than that of the return air or when outside air dry-bulb temperature is greater than that of the return air, use minimum outside air.
- When the outside air enthalpy is below the return air enthalpy and the outside dry-bulb temperature is below the return air dry-bulb temperature but above the cooling coil control point, use 100% outside air.

- When outside air enthalpy is below the return air enthalpy and the outside air dry-bulb temperature is below the return air dry-bulb temperature and below the cooling coil controller setting, the return and outside air are mixed by modulating dampers according to the cooling set point.

These points are valid for the majority of cases. When mixed air is to be used for heating and cooling, a more intricate optimization plan will be necessary, based on the value of the fuels used for heating and cooling.

5.3.2.3 Heat Recovery

Heat recovery is often practiced in industrial processes that involve high temperatures. It can also be employed in HVAC systems.

Systems are available that operate with direct heat transfer from the exhaust air to the inlet air. These are most reasonable when there is a large volume of exhaust air, for example, in once-through systems, and when weather conditions are not moderate.

Common heat recovery systems are broken down into two types: regenerative and recuperative. Regenerative units use alternating air flow from the hot and cold stream over the same heat storage/transfer medium. This flow may be reversed by dampers or the whole heat exchanger may rotate between streams. Recuperative units involve continuous flow; the emphasis is upon heat transfer through a medium with little storage.

The rotary regenerative unit, or heat wheel, is one common heat recovery device. It contains a corrugated or woven heat storage material that gains heat in the hot stream. This material is then rotated into the cold stream where the heat is given off again. The wheels can be impregnated with a desiccant to transfer latent as well as sensible heat. Purge sections for HVAC applications can reduce carryover from the exhaust stream to acceptable limits for most installations.

The heat transfer efficiency of heat wheels generally ranges from 60% to 85% depending upon the installation, type of media, and air velocity. For easiest installation, the intake and exhaust ducts should be located near each other.

Another system that can be employed with convenient duct location is a plate type air-to-air heat exchanger. This system is usually lighter though more voluminous than heat wheels. Heat transfer efficiency is typically in the 60%–75% range. Individual units range from 1,000 to 11,000 SCFM and can be grouped together for greater capacity. Almost all designs employ counterflow heat transfer for maximum efficiency.

Another option to consider for nearly contiguous ducts is the heat pipe. This is a unit that uses a boiling refrigerant within a closed pipe to transfer heat. Since the heat of vaporization is utilized, a great deal of heat transfer can take place in a small space.

Heat pipes are often used in double wide coils, which look very much like two steam coils fastened together. The amount of heat transferred can be varied by tilting the tubes to increase or decrease the flow of liquid through capillary action. Heat pipes cannot be *turned off*, so bypass ducting is often desirable. The efficiency of heat transfer ranges from 55% to 75%, depending upon the number of pipes, fins per inch, air face velocity, etc.

Runaround systems are also used for HVAC applications, particularly when the supply and exhaust plenums are not physically close. Runaround systems involve two coils (air-to-water heat exchangers) connected by a piping loop of water or glycol solution and a small pump. The glycol solution is necessary if the air temperatures in the inlet coils are below freezing. Standard air conditioning coils can be used for the runaround system. Precaution should be used when the exhaust air temperature drops below 0°C (32°F), which would cause freezing of the condensed water on its fins. A three-way bypass valve will maintain the temperature of the solution entering the coil at just above 0°C (32°F). The heat transfer efficiency of this system ranges from 60% to 75% depending upon the installation.

Another system similar to the runaround in layout is the desiccant spray system. Instead of using coils in the air plenums, it uses spray towers. The heat transfer fluid is a desiccant (lithium chloride) that transfers both latent and sensible heat—desirable in many applications. Tower capacities range from 7,700 to 92,000 SCFM; multiple units can be used in large installations. The enthalpy recovery efficiency is in the range of 60%–65%.

5.3.2.4 Thermal Energy Storage

TES systems are used to reduce the on-peak electricity demand caused by large cooling loads. TES systems utilize several different storage media, with chilled water, ice, or eutectic salts being most common. Chilled water requires the most space, with the water typically being stored in underground tanks. Ice storage systems can be aboveground insulated tanks with heat exchanger coils that cause the water to freeze, or can be one of several types of ice-making machines.

In a typical system, a chiller operates during off-peak hours to make ice (usually at night). Since the chiller can operate for a longer period of time than during the daily peak, it can have a smaller capacity. Efficiency is greater at night, when the condensing temperature is lower than it is during the day. During daytime operation, chilled water pumps circulate water through the ice storage system and extract heat. Systems can be designed to meet the entire load, or to meet a partial load, with an auxiliary chiller as a backup.

This system reduces peak demand and can also reduce energy use. With ice storage, it is possible to deliver water at a lower temperature than is normally done. This means that the chilled water piping can be smaller and the pumping power reduced. A low-temperature air distribution system will allow smaller ducts and lower capacity fans to deliver a given amount of cooling. Careful attention must be paid to the system design to ensure occupant comfort in conditioned spaces. Some government agencies and electric utilities offer incentives to customers installing TES systems. The utility incentives could be in the form of a set rebate per ton of capacity or per kW of deferred peak demand, or a time-of-use pricing structure that favors TES.

5.3.3 Water Heating

5.3.3.1 Residential Systems

Residential storage water heaters typically range in size from 76 L (20 gal) to 303 L (80 gal). Electric units generally have one or two immersion heaters, each rated at 2–6 kW depending on tank size. Energy input for water heating is a function of the temperature at which water is delivered, the supply water temperature, and standby losses from the water heater, storage tanks, and piping.

Tankless water heaters are an energy-efficient alternative that has achieved greater market penetration over the last few years. These systems can be electric or gas-fired and provide hot water on demand, eliminating energy losses associated with a storage tank.

The efficiency of water heaters is referred to as the energy factor (EF). Higher EF values equate to more efficient water heaters. Typical EF values range from about 0.9–0.95 for electric resistance heaters, 0.6–0.86 for natural gas units, 0.5–0.85 for oil units, and 1.5–2.2

for heat pump water heaters. The higher efficiency values for each fuel type represent the more advanced systems available, while the lower efficiency values are for the more conventional systems.

In single tank residential systems, major savings can be obtained by

- Thermostat temperature set back to 60°C (140°F)
- Automated control
- Supplementary tank insulation
- Hot water piping insulation

The major source of heat loss from electric water theaters is standby losses through the tank walls and from piping, since there are no flame or stack losses in electric units. The heat loss is proportional to the temperature difference between the tank and its surroundings. Thus, lowering the temperature to 60°C will result in two savings: (1) a reduction in the energy needed to heat water and (2) a reduction in the amount of heat lost. Residential hot water uses do not require temperatures in excess of 60°C; for any special use which does, it would be advantageous to provide a booster heater to meet this requirement when needed, rather than maintain 100–200 L of water continuously at this temperature with associated losses. The temperature should be set back even lower when occupants are away during long periods of times.

When the tank is charged with cold water, both heating elements operate until the temperature reaches a set point. After this initial rise, one heating element thermostatically cycles on and off to maintain the temperature, replacing heat that is removed by withdrawing hot water or that is lost by conduction and convection during standby operation.

Experiments indicate that the heating elements may be energized only 10%–20% of the time, depending on the ambient temperature, demand for hot water, water supply temperature, etc. By carefully scheduling hot water usage, this time can be greatly reduced. In one case, a residential water heater was operated for 1 h in the morning and 1 h in the evening. The morning cycle provided sufficient hot water for clothes washing, dishes, and other needs. Throughout the day, the water in the tank, although gradually cooling, still was sufficiently hot for incidental needs. The evening heating cycle provided sufficient water for cooking, washing dishes, and bathing. Standby losses were eliminated during the night and much of the day. Electricity use was cut to a fraction of the normal amount. This method requires the installation of a time clock or other type of control to regulate the water heater. A manual override can be provided to meet special needs.

Supplementary tank insulation can be installed at a low cost to reduce standby losses. The economic benefit depends on the price of electricity and the type of insulation installed. However, paybacks of a few months up to a year are typical on older water heaters. Newer units have better insulation and reduced losses. Hot water piping should also be insulated, particularly when hot water tanks are located outside or when there are long piping runs. If copper pipe is used, it is particularly important to insulate the pipe for the first 3–5 m where it joins the tank, since it can provide an efficient heat conduction path.

Since the energy input depends on the water flow rate and the temperature difference between the supply water temperature and the hot water discharge temperature, reducing either of these two quantities reduces energy use. Hot water demand can be reduced by cold water clothes washing, and by providing hot water at or near the use temperature, to avoid the need for dilution with cold water. Supply water should be provided at the warmest temperature possible. Since reservoirs and underground piping systems are generally warmer than the air temperature on a winter day in a cold climate, supply piping should be buried, insulated, or otherwise kept above the ambient temperature.

Solar systems are available today for heating hot water. Simple inexpensive systems can preheat the water, reducing the amount of electricity needed to reach the final temperature. Alternatively, solar heaters (some with electric backup heaters) are also available, although initial costs may be prohibitively high, depending on the particular installation.

Heat pump water heaters may save as much as 25%–30% of the electricity used by a conventional electric water heater. Some utilities have offered rebates of several thousand dollars to encourage customers to install heat pump water heaters.

The microwave water heater is an interesting technology that is just beginning to emerge in both residential and commercial applications. Microwave water heaters are tankless systems that produce hot water only when needed, thereby avoiding the energy losses incurred by conventional water heaters during the storage of hot water. These heaters consist of a closed stainless steel chamber with a silica-based flexible coil and a magnetron. When there is a demand for hot water, either because a user has opened a tap or because of a heater timing device, water flows into the coil and the magnetron bombards it with microwave energy at a frequency of 2450 MHz. The microwave energy excites the water molecules, heating the water to the required temperature.

Heat recovery is another technique for preheating or heating water, although opportunities in residences are limited. This is discussed in more detail under commercial water heating.

Apartments and larger buildings use a combined water heater/storage tank, a circulation loop, and a circulating pump. Cold water is supplied to the tank, which thermostatically maintains a preset temperature, typically 71°C (160°F). The circulating pump maintains a flow of water through the circulating loop, so hot water is always available instantaneously upon demand to any user. This method is also used in hotels, office buildings, etc.

Adequate piping and tank insulation is even more important here, since the systems are larger and operate at higher temperature. The circulating hot water line should be insulated, since it will dissipate heat continuously otherwise.

5.3.3.2 Heat Recovery in Nonresidential Systems

Commercial/industrial hot water systems offer many opportunities for employing heat recovery. Examples of possible sources of heat include air compressors, chillers, heat pumps, refrigeration systems, and water-cooled equipment. Heat recovery permits a double energy savings in many cases. First, recovery of heat for hot water or space heating reduces the direct energy input needed for heating. The secondary benefit comes from reducing the energy used to dissipate waste heat to a heat sink (usually the atmosphere). This includes pumping energy and energy expended to operate cooling towers and heat exchangers. Solar hot water systems are also finding increasing use. Interestingly enough, the prerequisites for solar hot water systems also permit heat recovery. Once the hot water storage capacity and backup heating capability has been provided for the solar hot water system, it is economical to tie in other sources of waste heat, for example, water jackets on air compressors.

5.3.4 Lighting

There are seven basic techniques for improving the efficiency of lighting systems:

- Delamping
- Relamping
- Improved controls
- More efficient lamps and devices
- Task-oriented lighting
- Increased use of daylight
- Room color changes, lamp maintenance

The first two techniques and possibly the third are low cost and may be considered operational changes. The last four items generally involve retrofit or new designs. (Chapter 7 contains further details on energy-efficient lighting technologies.)

The first step in reviewing lighting electricity use is to perform a lighting survey. An inexpensive handheld light meter can be used as a first approximation; however, distinction must be made between raw intensities (lux or foot-candles) recorded in this way and illumination quality.

Many variables can affect the *correct* lighting values for a particular task: task complexity, age of employee, glare, etc. For reliable results, consult a lighting specialist or refer to the literature and publications of the Illuminating Engineering Society.

The lighting survey indicates those areas of the building where lighting is potentially inadequate or excessive. Deviations from illumination levels that are adequate can occur for several reasons: over design; building changes; change of occupancy; modified layout of equipment or personnel, more efficient lamps, improper use of equipment, dirt buildup, etc.

Once the building manager has identified areas with potentially excessive illumination levels, he or she can apply one or more of the seven techniques listed previously. Each of these will be described briefly.

Delamping refers to the removal of lamps to reduce illumination to acceptable levels. With incandescent lamps, remove unnecessary bulbs. With fluorescent or high intensity discharge (HID) lamps, remove lamps and disconnect ballasts, since ballasts account for 10%–20% of total energy use. Note that delamping is not recommended if it adversely affects the distribution of the lighting; instead, consider relamping.

Relamping refers to the replacement of existing lamps by lamps of lower wattage and lower lumen output in areas with excessive light levels. More efficient lower wattage fluorescent tubes are available that require 15%–20% less wattage while producing 10%–15% less light. In some types of HID systems, it is possible to substitute a more efficient lamp with lower lumen output directly. However, in most cases, ballasts must also be changed.

Improved controls permit lamps to be used only when and where needed. For example, certain office buildings have all lights for one floor or large area on a single contactor. These lamps will be switched on at 6 a.m. before work begins, and are not turned off until 10 p.m. when maintenance personnel finish their cleanup duties. In such cases, energy usage can be cut by as much as 50% by better control and operation strategies: installing individual switches for each office or work area; installing timers, occupancy sensors, daylighting controls, and/or dimmers; and instructing custodial crews to turn lights on as needed and turn them off when work is complete.

Sophisticated building-wide lighting control systems are also available, and today are being implemented at an increasing rate, particularly in commercial buildings.

There is a great variation in the efficacy (a measure of light output per electricity input) of various lamps. Incandescent lamps have the lowest efficacy, typically 5–20 lm/W. Wherever possible, substitute incandescent lamps with fluorescent lamps, or with other efficient alternatives. This not only saves energy, but offers substantial economic savings as well, since fluorescent lamps last 10–50 times longer. Conventional fluorescent lamps have efficacies in the range of 30–70 lm/W; high-efficiency fluorescent systems yield about 85–100 lm/W.

CFLs and LED lamps are also good substitutes for a wide range of incandescent lamps. They are available in a variety of wattages and will replace incandescent lamps with a fraction of the energy consumption. Typical efficacies are 55–70 lm/W for CFLs and 60–100 lm/W for LEDs. In addition to the energy savings, they have longer rated lifetimes and thus do not need to be replaced as often as incandescent lamps. As mentioned previously, EISA (2007) put standards in place to phase out common incandescent lamps (40–100 W) with more efficient CFL and LED options. Replacement lighting must use at least 27% less energy. The first phase began in January 2012 and affects 100 W bulbs; the second phase took effect in January 2013, affecting 75 W bulbs.

Still greater improvements are possible with HID lighting, particularly metal halide lamps. While they are generally not suited to residential use (high light output and high capital cost), they are increasingly being used in commercial and industrial buildings for their high efficiency and long life. It should be noted that HID lamps are not suited for any area that requires lamps to be switched on-and-off, as they still take several minutes to restart after being turned off.

Improving ballasts is another way of saving lighting energy. One example of a significant increase in efficacy in the commercial sector is illustrated in the transition from T12 (1.5 in. diameter) fluorescent lamps with magnetic ballasts to T8 (1 in. diameter) fluorescent lamps with electronic ballasts. This transition began to occur in the late 1970s and early 1980s. Now T8 electronic ballast systems are the standard for new construction and retrofits. The efficacy improvement, depending on the fixture, is roughly 20%–40% or even more. For example, a two-lamp F34T12 fixture (a fixture with two 1.5 in. diameter, 34 W lamps) with energy-saving magnetic ballast requires 76 W, whereas a two-lamp F32T8 fixture (a fixture with two 1 in. diameter, 32 W lamps) with electronic ballast requires only 59 W, which is an electricity savings of 22%. The savings is attributable to the lower wattage lamps as well as the considerably more efficient ballast.

In recent years, T5 (5/8 in. diameter) fluorescent lamps have been gaining a foothold in the fluorescent market. They are smaller and have a higher optimum operating temperature than T8s, which makes them advantageous in certain applications. One common use is for high bay lighting, where they are displacing HID alternatives because of their better coloring rendering, longer life, shorter warm-up time, and greater lumen maintenance properties.

Task-oriented lighting is another important lighting concept. In this approach, lighting is provided for work areas in proportion to the needs of the task. Hallways, storage areas, and other nonwork areas receive less illumination.

Task lighting can replace the so-called uniform illumination method sometimes used in offices and other types of commercial buildings. The rationale for uniform illumination was based on the fact that the designer could never know the exact layout of desks and equipment in advance, so designs provided for uniform illumination and the flexibility it offers. With today's higher electricity costs, a more customized task lighting approach is often a cost-effective alternative.

Daylighting was an important element of building design for centuries before the discovery of electricity. In certain types of buildings and operations today, daylighting can be utilized to at least reduce (if not replace) electric lighting. Techniques include windows, an atrium, skylights, etc. There are obvious limitations such as those imposed by the need for privacy, 24 h operation, and building core locations with no access to natural light.

The final step is to review building and room color schemes and decor. The use of light colors can substantially enhance illumination without modifying existing lamps.

An effective lamp maintenance program also has important benefits. Light output gradually decreases over lamp lifetime. This should be considered in the initial design and when deciding on lamp replacement. Dirt can substantially reduce light output; simply cleaning lamps and luminaries more frequently can gain up to 5%–10% greater illumination, permitting some lamps to be turned off.

In addition to the lighting energy savings, efficient lighting also reduces energy requirements for cooling since efficient systems produce less heat. This is a yearlong benefit in many commercial buildings, since space conditioning equipment often operates year-round. However, it is an energy penalty when buildings operate heating systems, since the heating systems have to work a little harder.

5.3.5 Refrigeration

The refrigerator, at roughly 40–140 kWh/month depending on the size and age of the model, is among the top four residential users of electricity. In the last 60 years, the design of refrigerator/freezers has changed considerably, with sizes increasing from 5–10 ft^3 to 20–25 ft^3 today. At the same time, the energy input per unit increased up until the oil embargo, after which efforts were made that led to a steady decline in energy use per unit between the mid-1970s through today, despite increases in average refrigerator size. As noted earlier, current refrigerators use about one-quarter the energy (average of ~450 kWh/year in 2011) smaller units did 40 years ago (average of ~1800 kWh/year in 1972).

Energy losses in refrigerators arise from a variety of sources. The largest losses are due to heat gains through the walls and frequent door openings. Since much of the energy used by a refrigerator depends on its design, care should be used in selection. Choose a refrigerator that is the correct size for the application and look for ENERGY STAR models to maximize efficiency. Refrigerators with freezers on the top or bottom are more efficient than side-by-side models. In addition, refrigerators without in-door ice and water dispensers use less energy, as do models with automatic moisture control and manual defrost.

Purchase of a new, more efficient unit is not a viable option for many individuals who have a serviceable unit and do not wish to replace it. In this case, the energy management challenge is to obtain the most effective operation of the existing equipment. Techniques include the following:

- Reducing operation of automatic defrost and antisweat heaters
- Providing a cool location for the refrigerator
- Reducing the number of door openings
- Maintaining temperature settings recommended levels for food safety (not lower)
- Precooling foods before refrigerating

Commercial refrigeration systems are found in supermarkets, liquor stores, restaurants, hospitals, hotels, schools, and other institutions—about one-fifth of all commercial facilities. Systems include walk-in dairy cases, open refrigerated cases, and freezer cases. In a typical supermarket, lighting, HVAC, and miscellaneous uses account for half the electricity use, while refrigerated display cases, compressors, and condenser fans account for the other half. Thus, commercial refrigeration can be an important element of electric energy efficiency.

It is a common practice in some types of units to have the compressor and heat exchange equipment located remotely from the refrigerator compartment. In such systems, try to locate the compressor in a cool location rather than placing it next to other equipment that gives off heat. Modern commercial refrigerators often come equipped with heat recovery systems, which recover compressor heat for space conditioning or water heating.

Walk-in freezers and refrigerators lose energy though door openings; refrigerated display cases have direct transfer of heat. Covers, strip curtains, air curtains, glass doors, or other thermal barriers can help mitigate these problems. In addition, it is important to use the most efficient light sources in large refrigerators and freezers; elimination of 1 W of electricity to produce light also eliminates two additional Watts required to extract the heat. Other potential energy saving improvements include high-efficiency motors, VSDs, more efficient compressors, and improved refrigeration cycles and controls, such as the use of floating head pressure controls.

5.3.6 Cooking

Consumer behavior toward cooking has changed dramatically since the first edition of this handbook. Consumers today are cooking less in the home and dining out or picking up prepared food more often. Consumers are also purchasing more foods that are easier to prepare—convenience is key to the modern family. In response, the food processing industry offers a wide variety of pre-prepared, ready-to-eat products. Recent end use electricity data illustrate this change in behavior. Cooking accounted for about 7% of residential electricity use in 2001; in 2011, it accounted for only 2%.

Though habits are changing and more cooking is occurring outside the home (in restaurants and food processing facilities), reductions in energy use are still important. In general, improvements in energy use efficiency for cooking can be divided into three categories:

- More efficient use of existing appliances
- Use of most efficient existing appliances
- More efficient new appliances

The most efficient use of existing appliances can lead to substantial reductions in energy use. While slanted toward electric ranges and appliances, the following observations also apply to cooking devices using other sources of heat.

First, select the right size equipment for the job. Do not heat excessive masses or large surface areas that will needlessly radiate heat. Second, optimize heat transfer by ensuring that pots and pans provide good thermal coupling to the heat sources. Flat-bottomed pans should be used on electric ranges. Third, be sure that pans are covered to prevent heat loss and to shorten cooking times. Fourth, when using the oven, plan meals so that several dishes are cooked at once. Use small appliances (electric fry pans, *slow* cookers, toaster ovens, etc.) whenever they can be substituted efficiently for the larger appliances such as the oven.

Different appliances perform similar cooking tasks with widely varying efficiencies. For example, the electricity used and cooking time required for common foods items can vary as much as ten to one in energy use and five to one in cooking times, depending on the method. As an example, four baked potatoes require 2.3 kWh and 60 min in an oven (5.2 kW) of an electric range, 0.5 kWh and 75 min in a toaster oven (1.0 kW), and 0.3 kWh and 16 min in a microwave oven (1.3 kW). Small appliances are generally more efficient when used as intended. Measurements in a home indicated that a pop-up toaster cooks two slices of bread using only 0.025 kWh. The toaster would be more efficient than using the broiler in the electric range oven, unless a large number of slices of bread (more than 17 in this case) were to be toasted at once.

If new appliances are being purchased, select the most efficient ones available. Heat losses from some ovens approach 1 kW, with insulation accounting for about 50%; losses around the oven door edge and through the window are next in importance. These losses are reduced in certain models. Self-cleaning ovens are normally manufactured with more insulation. Careful design of heating elements can also contribute to better heat transfer. Typically, household electric ranges require around 2300–3200 W for oven use, 3600 W for broiler use, and 4000 W for use of the self-cleaning feature.

Microwave cooking is highly efficient for many types of foods, since the microwave energy is deposited directly in the food. Energy input is minimized, because there is no need to heat the cooking utensil. Although many common foods can be prepared effectively using a microwave oven, different methods must be used as certain foods are not suitable for micro- wave cooking. A typical microwave oven requires 750–1100 W. Convection ovens and induc- tion cook tops are two additional electric alternatives that can reduce cooking energy use.

Commercial cooking operations range from small restaurants and cafes where methods similar to those described previously for residences are practiced, to large institutional kitchens in hotels and hospitals. Many of the same techniques apply. Careful scheduling of equipment use, and provision of several small units rather than a single large one, will save energy. For example, in restaurants, grills, soup kettles, bread warmers, etc., often operate continuously. Generally, it is unnecessary to have full capacity during off-peak hours; one small grill might handle midmorning and midafternoon needs, permitting the second and third units to be shut down. The same strategy can be applied to coffee warm- ing stations, hot plates, etc.

5.3.7 Residential Appliances

A complete discussion of EMOs associated with all the appliances found in homes is beyond the scope of this chapter. However, the following subsections discuss several of the major ones and general suggestions applicable to the others are provided.

5.3.7.1 Clothes Drying

Clothes dryers typically use about 2.5 kWh or more per load, depending on the unit and size of the load. A parameter called the EF, which is a measure of the lbs of clothing dried per kWh of electricity consumed, can be used to quantify the efficiency of clothes drying. The current minimum EF for a standard capacity electric dryer is 3.01. New standards based on a new metric that incorporates standby energy use (combined energy factor [CEF]) are scheduled to take effect in January 2014. ENERGY STAR certified models are not yet available. In the United States, new dryers are not required to display energy use information, so it is difficult to compare models. In fact, most electric dryers in the market are pretty comparable in their construction and the basic heating technology. However, the actual energy consumption of the dryer varies with the types of controls it has, and how the operator uses those controls. Models with moisture-sensing capability can result in the most energy savings—savings on the order of 15% compared to conventional operation are common.

In addition, electric clothes dryers operate most efficiently when fully loaded. Operating with one-third to one-half load costs roughly 10%–15% in energy efficiency.

Locating clothes dryers in heated spaces could save 10%–20% of the energy used by reducing energy needed for heating up. Another approach is to save up loads and do several loads sequentially, so that the dryer does not cool down between loads.

The heavier the clothes, the greater the amount of water they hold. Mechanical water removal (pressing, spinning, and wringing) generally requires less energy than electric heat. Therefore, be certain the washing machine goes through a complete spin cycle (0.1 kWh) before putting clothes in the dryer.

Solar drying, which requires a clothesline (rope) and two poles or trees, has been practiced for millennia and is very sparing of electricity. The chief limitation is, of course, inclement weather. New technologies such as microwave or heat pump clothes dryers may help reduce clothes drying energy consumption in the future.

5.3.7.2 Clothes Washing

The modified energy factor (MEF) can be used to compare different models of clothes washers. It is a measure of the machine energy required during washing, the water heating energy, and the dryer energy needed to remove the remaining moisture. A higher MEF value indicates a more efficient clothes washer. According to Federal standards, all new clothes washers manufactured or imported after January 2007 are required to have an MEF of at least 1.26. In addition, as of February 2013, to be qualified as an ENERGY STAR unit, residential clothes washers must have an MEF of at least 2.0.

Electric clothes washers are designed for typical loads of 3–7 kg. Surprisingly, most of the energy used in clothes washing is for hot water; the washer itself only requires a few percent of the total energy input. Therefore, the major opportunity for energy management in clothes washing is the use of cold or warm water for washing. Under normal household conditions, it is not necessary to use hot water. Clothes are just as clean (in terms of bacteria count) after a 20°C wash as after a 50°C wash. If there is concern for sanitation (e.g., a sick person in the house), authorities recommend use of chlorine bleach. If special cleaning is required, such as removing oil or grease stains, hot water (50°C) and detergent will emulsify oil and fat. There is no benefit in a hot rinse.

A secondary savings can come from using full loads. Surveys indicate that machines are frequently operated with partial loads, even though a full load of hot water is used.

5.3.7.3 Dishwashers

The two major energy uses in electric dishwashers are the hot water and the dry cycle. Depending on the efficiency of the model and operation, dishwashers use between about 2 and 5 kWh/load.

The water heating often accounts for 80% of the total energy requirement of a dishwasher. The volume of hot water used ranges from about 5 gal for the more efficient units to more than double that for less efficient models. The water volume can be varied on some machines depending on the load.

Since 1990, new models have been required to allow for a no-heat drying option. If you are using a very old unit in which that option is not available, stop the cycle prior to the drying step and let the dishes air-dry. Operating the dishwasher with a full load and using a cold water prerinse are additional ways to minimize energy use.

New standards that took effect in mid-2013 require standard-size residential dishwashers to use a maximum of 307 kWh/year and 5.0 gal/cycle and compact dishwashers to use a maximum of 222 kWh/year and 3.5 gal/cycle. ENERGY STAR models are also available.

5.3.7.4 General Suggestions for Residential Appliances and Electrical Equipment

Many electrical appliances (pool pumps, televisions, stereos, DVD and CD players, electronic gaming systems, aquariums, blenders, floor polishers, hand tools, mixers, etc.) perform unique functions that are difficult to duplicate. This is their chief value.

Attention should be focused on those appliances that use more than a few percent of annual electricity use. General techniques for energy management include

- Reduce use of equipment where feasible (e.g., turn off entertainment systems when not in use).
- Perform maintenance to improve efficiency (e.g., clean pool filters to reduce pumping power).
- Schedule use for off-peak hours (evenings).

The last point requires further comment and applies to large electric appliances such as washers, dryers, and dishwashers as well. Some utilities now offer time-based pricing that includes a premium charge for usage occurring on-peak (when the greatest demand for electricity takes place) and lower energy costs for off-peak electricity use. By scheduling energy-intensive activities for off-peak hours (e.g., clothes washing and drying in the evening), the user helps the utility reduce its peaking power requirement, thereby reducing generating costs. The utility then returns the favor by providing lower rates for off-peak use.

5.4 Closing Remarks

This chapter has discussed the management of electrical energy in buildings. Beginning with a discussion of energy use in buildings, we next outlined the major energy using systems and equipment, along with a brief description of their features that influence energy use and efficiency. A systematic methodology for implementing an energy management program was then described. The procedure has been implemented in a wide variety of situations including individual homes, commercial buildings, institutions, multinational conglomerates, and cities, and has worked well in each case. Following the discussion of how to set up an energy management program, a series of techniques for saving electricity in each major end use were presented. The emphasis has been on currently available, cost-effective technology. There are other techniques available, some of which are provided in other chapters of this handbook, but we have concentrated on those of we know will work in today's economy, for the typical energy consumer.

The first edition of this book was published in 1980. Much of the data in the first edition dated from the 1975 to 1980 time frame, when the initial response to the oil embargo of 1973 was gathering momentum and maturing. It is remarkable to return to those data and look at the progress that has been made. In 1975, total U.S. energy use was 72 quads (1 quad = 10^{15} BTU = 1.055 EJ). Most projections at that time predicted U.S. energy use in excess of 100 quads by 1992; instead, we saw that it only reached 85 quads by 1992. In fact, by 2004, total U.S. energy use had just reached the 100 quad mark. Estimates for 2012 show that usage has actually declined to a value of 95 quads (EIA 2013c). In addition, between 1975 and 2012, energy consumption per real dollar of the U.S. gross domestic product (GDP) decreased by half, from 14.76 to 7.00 thousand BTU per chained 2005 dollar (EIA 2013c). Some of this is due to a decrease in domestic energy-intensive industry and the recent economic recession, but much of it represents a remarkable improvement in overall energy efficiency.

As noted earlier in this chapter (Table 5.1), a significant growth in total energy consumption in the residential and commercial sectors has occurred in the intervening decades, but efficiency improvements have helped to moderate this growth rate considerably. The improvement in energy efficiency in lighting, refrigerators, air conditioning, and other devices has been truly remarkable. Today the local hardware or home improvement store has supplies of energy-efficient devices that were beyond imagination in 1973.

This is a remarkable accomplishment, technically and politically, given the diversity of the residential/commercial market. Besides the huge economic savings this has meant to millions of homeowners, apartment dwellers, and businesses, think of the environmental benefits associated with avoiding the massive additional amounts of fuel, mining, and combustion, which otherwise would have been necessary.

References

EIA. 2012. Annual energy review 2011. DOE/EIA-0384(2011). Washington, DC: U.S. Energy Information Administration. http://www.eia.gov/totalenergy/data/annual/pdf/aer.pdf (accessed March 6, 2015).

EIA. 2013a. Annual energy outlook 2013 with projections to 2040. Commercial sector key indicators and consumption. DOE/EIA-0383(2013). Washington, DC: U.S. Energy Information Administration. http://www.eia.gov/oiaf/aeo/tablebrowser/ (accessed March 6, 2015).

EIA. 2013b. Annual energy outlook 2013 with projections to 2040. Residential sector key indicators and consumption. DOE/EIA-0383(2013). Washington, DC: U.S. Energy Information Administration. http://www.eia.gov/oiaf/aeo/tablebrowser/ (accessed March 6, 2015).

EIA. 2013c. Monthly energy review: July 2013. DOE/EIA-0035(2013/07). Washington, DC: U.S. Energy Information Administration. http://www.eia.gov/totalenergy/data/monthly/pdf/mer.pdf (accessed March 6, 2015).

EPA. 2011. CBECS national median source energy use and performance comparisons by building type. http://www.energystar.gov/ia/business/tools_resources/new_bldg_design/2003_CBECSPerformanceTargetsTable.pdf (accessed August 13, 2013).

SMR Research Corporation. 2011. *The Commercial Building Inventory. Age of Commercial Buildings Spreadsheet*. Hackettstown, NJ: SMR Research Corporation.

U.S. Census Bureau. 2012. Housing vacancies and home ownership. Annual statistics: 2012. (Table 22.11. Estimates of the Total Housing Inventory for the United States: 2011 and 2012.) http://www.census.gov/housing/hvs/data/ann12ind.html (accessed March 6, 2015).

6

Heating, Ventilating, and Air-Conditioning Control Systems

Bryan P. Rasmussen, Jan F. Kreider, David E. Claridge, and Charles H. Culp

CONTENTS

6.1 Introduction: The Need for Control

This chapter describes the essentials of control systems for heating, ventilating, and air conditioning (HVAC) of buildings designed for energy conserving operation. Of course, there are other renewable and energy conserving systems that require control. The principles described herein for buildings also apply with appropriate and obvious modification to these other systems. For further reference, the reader is referred to several standard references in the list at the end of this chapter.

HVAC system controls are the information link between varying energy demands on a building's primary and secondary systems and the (usually) approximately uniform demands for indoor environmental conditions. Without a properly functioning control system, the most expensive, most thoroughly designed HVAC system will be a failure. It simply will not control indoor conditions to provide comfort.

The HVAC control system must

- Sustain a comfortable building interior environment

- Maintain acceptable indoor air quality

- Be as simple and inexpensive as possible and yet meet HVAC system operation criteria reliably for the system lifetime

- Result in efficient HVAC system operation under all conditions

- Be commissioned, including the building, equipment, and control systems

- Be fully documented, so that the building staff successfully operates and maintains the HVAC system

A considerable challenge is presented to the HVAC system designer to design a control system that is energy efficient and reliable. Inadequate control system design, inadequate commissioning, and inadequate documentation and training for the building staff often create problems and poor operational control of HVAC systems. This chapter develops the basics of HVAC control and follows through with the operational needs for successfully maintained operation. The reader is encouraged to review the following references on the subject: ASHRAE (2002, 2003, 2004, 2005), Haines (1987), Honeywell (1988), Levine (1996), Sauer et al. (2001), Stein and Reynolds (2000), and Tao and Janis (2005).

To achieve proper control based on the control system design, the HVAC system must be designed correctly and then constructed, calibrated, and commissioned according to the mechanical and electrical systems drawings. These must include properly sized primary and secondary systems. In addition, air stratification must be avoided, proper provision for control sensors is required, freeze protection is necessary in cold climates, and proper attention must be paid to minimizing energy consumption subject to reliable operation and occupant comfort.

The principal and final controlled variable in buildings is zone temperature (and to a lesser extent humidity and/or air quality in some buildings). This chapter will therefore focus on methods to control temperature. Supporting the zone temperature control, numerous other control loops exist in buildings within the primary and secondary HVAC systems, including boiler and chiller control, pump and fan control, liquid and airflow control, humidity control, and auxiliary system control (e.g., thermal energy storage control). This chapter discusses only *automatic* control of these subsystems. Honeywell (1988)

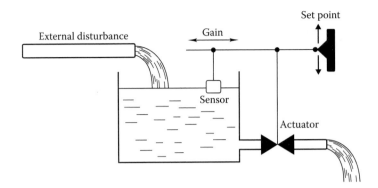

FIGURE 6.1
Simple water level controller. The set point is the full water level; the error is the difference between the full level and the actual level.

defines an automatic control system as "a system that reacts to a change or imbalance in the variable it controls by adjusting other variables to restore the system to the desired balance."

Figure 6.1 defines a familiar control problem with feedback. The water level in the tank must be maintained under varying outflow conditions. The float operates a valve that admits water to the tank as the tank is drained. This simple system includes all the elements of a control system:

Sensor—float; reads the controlled variable, the water level.

Controller—linkage connecting float to valve stem; senses difference between full tank level and operating level and determines needed position of valve stem.

Actuator (controlled device)—internal valve mechanism; sets valve (the final control element) flow in response to level difference sensed by controller.

Controlled system characteristic—water level; this is often termed the *controlled variable*.

This system is called a *closed-loop* or *feedback* system because the sensor (float) is directly affected by the action of the controlled device (valve). In an open-loop system, the sensor operating the controller does not directly sense the action of the controller or actuator. An example would be a method of controlling the valve based on an external parameter such as the time of day, which may have an indirect relation to water consumption from the tank.

There are four common methods of control of which Figure 6.1 shows, but one. In the next section, we will describe each with relation to an HVAC system example.

6.2 Modes of Feedback Control

Feedback control systems adjust an output control signal based on feedback. The feedback is used to generate an error signal, which then drives a control element. Figure 6.1 illustrates a basic control system with feedback. Both off-on (i.e., two-position) control and analog (i.e., variable) control can be used. Numerous methodologies have been developed to implement analog control. These include proportional, proportional–integral (PI), proportional–integral–differential (PID), state feedback control, adaptive control, and predictive control. Proportional and PI controls are currently used for most applications in HVAC control, although more advanced strategies offer the potential for significantly improved performance.

Figure 6.2a shows a steam coil used to heat air in a duct. The simple control system shown includes an air temperature sensor, a controller that compares the sensed temperature to the set point, a steam valve controlled by the controller, and the coil itself. We will use this example system as the point of reference when discussing the various control system types. Figure 6.2b is the control diagram corresponding to the physical system shown in Figure 6.2a.

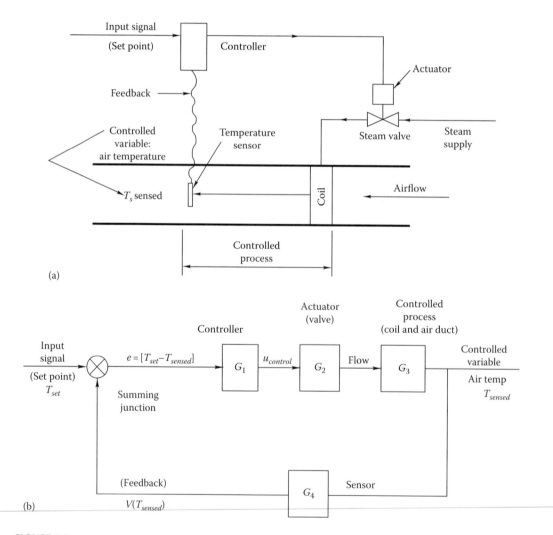

FIGURE 6.2
(a) Simple heating coil control system showing the process (coil and short duct length), controller, controlled device (valve and its actuator), and sensor. The set point entered externally is the desired coil outlet temperature. (b) Equivalent control diagram for heating coil. The Gs represent functions relating the input to the output of each module.

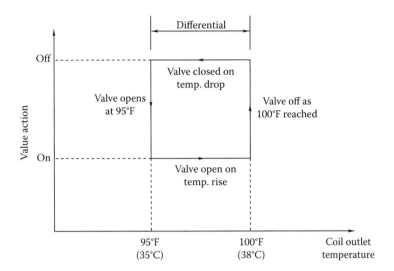

FIGURE 6.3
Two-position (on–off) control characteristic.

Two-position control applies to an actuator that is either fully open or fully closed. This is also known as *on-off, bang-bang,* or *hysteretic* control. In Figure 6.2a, the valve is a two-position valve if two-position control is used. The position of the steam valve is determined by the value of the coil outlet temperature. Figure 6.3 depicts two-position control of the valve. If the air temperature drops below 95°F, the valve opens and remains open until the air temperature reaches 100°F. The differential is usually adjustable, as is the temperature setting itself. Two-position control is the least expensive method of automatic control and is suitable for control of HVAC systems with large time constants and where an oscillatory response is acceptable. Examples include residential space and water heating systems. Systems that are fast reacting should not be controlled using this approach, since overshoot and undershoot may be excessive.

Proportional control adjusts the controlled variable in proportion to the difference between the controlled variable and the set point. For example, a proportional controller would increase the coil heat rate in Figure 6.2 by 10% if the coil outlet air temperature dropped by an amount equal to 10% of the temperature range specified for the heating to go from off to fully on. The following equation defines the behavior of a proportional control loop:

$$u = u_0 + K_p e \tag{6.1}$$

where
  u is the controller output
  u_0 is the constant value of controller output when no error exists
  K_p is the proportional *control gain*; it determines the rate or proportion at which the control signal changes in response to the error
  e is the *error*; in the case of the steam coil, it is the difference between the air temperature set point and the sensed supply air temperature:

$$e = T_{set} - T_{sensed} \tag{6.2}$$

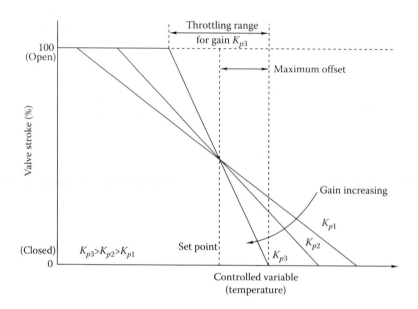

FIGURE 6.4
Proportional control characteristic showing various throttling ranges and the corresponding proportional gains K_p. This characteristic is typical of a heating coil temperature controller.

As coil air outlet temperature drops farther below the set temperature, error e increases leading to increased control action—an increased steam flow rate. Note that the temperatures in Equations 6.1 and 6.2 are often replaced by voltages or other variables, particularly in electronic controllers.

The *throttling range* is the total change in the controlled variable that is required to cause the actuator or controlled device to move between its limits. For example, if the nominal temperature of a zone is 72°F and the heating controller throttling range is 6°F, then the heating control undergoes its full travel between a zone temperature of 69°F and 75°F. This control, whose characteristic is shown in Figure 6.4, is *reverse acting*; that is, as the temperature (controlled variable) increases, the heating valve position decreases.

The throttling range is inversely proportional to the gain as shown in Figure 6.4. Beyond the throttling range, the system is out of control. In actual hardware, one can set the set point and either the gain or the throttling range (most common), but not both of the latter. Proportional control by itself is not capable of reducing the error to zero, since an error is needed to produce the capacity required for meeting a load as we will see in the following example. This unavoidable value of the error in proportional systems is called the offset. From Figure 6.4, it is easy to see that the offset is larger for systems with smaller gains. There is a limit to which one can increase the gain to reduce offset, because high gains can produce control instability.

Example 6.1:   Proportional Gain Calculation

If the steam heating coil in Figure 6.2a has a heat output that varies from 0 to 20 kW as the outlet air temperature varies from 35°C to 45°C in an industrial process, what is the coil gain and what is the throttling range? Find an equation relating the heat rate at any sensed air temperature to the maximum rate in terms of the gain and set point.

Given

$$\dot{Q}_{max} = 20 \text{ kW}$$

$$\dot{Q}_{min} = 0 \text{ kW}$$

$$T_{max} = 45°\text{C}$$

$$T_{min} = 35°\text{C}$$

Figure: See Figure 6.2a.

Assumptions: Steady-state operation

Find: K_p, ΔT_{max}

Solution: The throttling range is the range of the controlled variable (air temperature) over which the controlled system (heating coil) exhibits its full capacity range. The temperature varies from 35°C to 45°C. Assuming that a temperature of 35°C corresponds to heating rate of 0 kW, then the throttling range is

$$\Delta T_{max} = 45°\text{C} - 35°\text{C} = 10°\text{C} \tag{6.3}$$

The proportional gain is the ratio of the controlled system (coil) output to the throttling range. For this example, the Q controller output is $\dot{Q}$ and the gain is

$$K_p = \frac{\dot{Q}_{max} - \dot{Q}_{min}}{\Delta T_{max}} = \frac{(20 - 0) \text{ kW}}{10 \text{ K}} = 2.0 \text{ kW/K} \tag{6.4}$$

The controller characteristic can be found by inspecting Figure 6.4. It is assumed that the average air temperature (40°C) occurs at the average heat rate (10 kW). The equation of the straight line shown is

$$\dot{Q} = K_p(T_{set} - T_{sensed}) + \frac{\dot{Q}_{max} - \dot{Q}_{min}}{2} = K_p e + \frac{\dot{Q}_{max} - \dot{Q}_{min}}{2} \tag{6.5}$$

Note that the quantity $(T_{set} - T_{sensed})$ is the **error** e and a nonzero value indicates that the set temperature is not met. However, the proportional control system used here requires the presence of an error signal to fully open or fully close the valve.

Inserting the numerical values, we have

$$\dot{Q} = 2.0 \text{ kW/K}(40 - T_{sensed}) + 10 \text{ kW} \tag{6.6}$$

Comments: In an actual steam coil control system, it is the steam valve that is controlled directly to indirectly control the heat rate of the coil. This is typical of many HVAC system controls in that the desired control action is achieved indirectly by controlling another variable that in turn accomplishes the desired result. That is why the controller and actuator are often shown separately as in Figure 6.2b.

This example illustrates with a simple system how proportional control uses an error signal to generate an offset and how the offset controls an output quantity. Using a bias value, the error can be set to be zero at one value in the control range. Proportional control generally results in a nonzero error over the remainder of the control range, as shown later.

A common approach for modeling dynamic HVAC systems where only one input–output is considered is a process control model (Equation 6.7). The basic form of this model includes a gain, k_0, a first-order dynamic characterized by the time constant, τ, and a pure time delay, T_d. This basic model is easily augmented with additional time constants as needed:

$$G(s) = \frac{k_0 e^{-T_d s}}{\tau s + 1} \tag{6.7}$$

Assuming the basic feedback control loop (Figure 6.2b) where the controller is selected to be a proportional control and the system model is given as in Equation 6.7, then the relationship between the desired reference signal and the system error is given as follows:

$$\frac{E(s)}{R(s)} = \frac{1}{1 + C(s)G(s)} = \frac{1}{\tau s + 1 + K_p k_0 e^{-T_d s}} \tag{6.8}$$

Assuming a unit step change in the reference signal, then the steady-state value of the system output is given by

$$e_{ss} = \lim_{s \to 0}\left[s\left(\frac{E(s)}{R(s)} \right) R(s) \right] = \frac{1}{1 + C(0)G(0)} = \frac{1}{1 + K_p k_0} \tag{6.9}$$

Thus, using proportional control, the system error (i.e., the difference between the reference signal and system output) becomes smaller as the proportional gain is increased, but the error will never be zero. For more details regarding this type of analysis, see Franklin et al. (2006) for a detailed discussion of transfer functions, final value theorem, and steady-state error analysis.

Real systems also have a dynamic response. This means that proportional control is best suited to slow-response systems, where the throttling range can be set so that the system achieves stability. Typically, slow-responding mechanical systems include pneumatic thermostats for zone control and air handler unit damper control.

Integral control is often added to proportional control to eliminate the offset inherent in proportional-only control. The result, proportional plus integral control, is identified by the acronym PI. Initially, the corrective action produced by a PI controller is the same as for a proportional-only controller. After the initial period, a further adjustment due to the integral term reduces the offset to zero. The rate at which this occurs depends on the timescale of the integration. In equation form, the PI controller is modeled by

$$V = V_0 + K_p e + K_i \int e\, dt \tag{6.10}$$

in which K_i is the integral gain constant. It has units of reciprocal time and is the number of times that the integral term is calculated per unit time. This is also known as the reset rate; reset control is an older term used by some to identify integral control.

Today, most PI control implementations use electronic sensors, analog-to-digital converters (A/Ds), and digital logic to implement the PI control. Integral windup must be taken into account when using PI control, which occurs when actuators reach hardware or software limitations, and the tracking error is nonzero. The integral term will continue to grow, or *wind up,* and creates a large offset. This can prevent the control loop from performing as desired for long periods until the integral term recovers. Various methods exist to minimize or eliminate the windup problem.

The integral term in Equation 6.10 has the effect of adding a correction to the output signal V as long as the error term exists. The continuous offset produced by the proportional-only controller can thereby be reduced to zero because of the integral term. For HVAC systems, the timescale (K_p/K_i) of the integral term is often in the range of 10+ s to 10+ min. Using large integral gains will allow the control system to converge quickly to the desired set point value. However, large gains also tend to increase the oscillations in the response, and thus a balance must be found.

PI control is used for fast-acting systems for which accurate control is needed. Examples include mixed-air controls, duct static pressure controls, and coil controls. Because the offset is eventually eliminated with PI control, the throttling range can be set rather wide to ensure stability under a wider range of conditions than good control would permit with proportional-only control. Hence, PI control is also used on almost all electronic thermostats.

Derivative control is used to speed up the action of PI control. When derivative control is added to PI control, the result is called PID control. The derivative term added to Equation 6.10 generates a correction signal proportional to the time rate of change of error. This term has little effect on a steady proportional system with uniform offset (time derivative is zero) but initially, after a system disturbance, produces a larger correction more rapidly. The derivative control action is anticipatory in nature and can be effective in dampening oscillations caused by an aggressive PI controller. Equation 6.11 includes the derivative term in the mathematical model of the PID controller

$$V = V_0 + K_p e + K_i \int edt + K_d \frac{de}{dt} \qquad (6.11)$$

in which K_d is the derivative gain constant. The timescale (K_d/K_p) of the derivative term is typically in the range of 0.2–15 min. Since HVAC systems do not often require rapid control response, the use of PID control is less common than use of PI control. Since a derivative is involved, any noise in the error (i.e., sensor) signal must be avoided, or this noise will be amplified in the control action, creating undesirable fluctuations. One application in buildings where PID control has been effective is in duct static pressure control, a fast-acting subsystem that has a tendency to be unstable otherwise.

Derivative control has limited application in HVAC systems because it requires correct tuning for each of the three gain constants (Ks) over the performance range that the control loop will need to operate. Another serious limitation centers on the fact that most facility operators lack training and skills in tuning PID control loops.

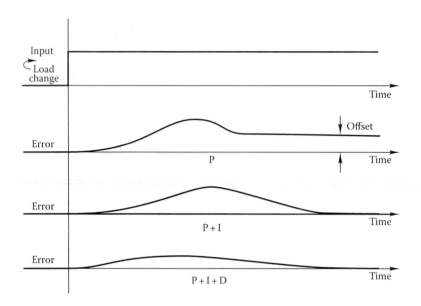

FIGURE 6.5
Performance comparison of P, PI, and PID controllers when subjected to a uniform, input step change.

As a result, many PID controllers are *detuned* and use artificially low control gains. This has the effect of sacrificing good performance, so as to ensure a stable, albeit slow, response.

Figure 6.5 illustrates the loop response for three correctly configured systems when a step function change, or disturbance, occurs. Note that the PI loop achieves the same final control as the PID, and only the PI error signal is larger. An improperly configured PID loop can oscillate from the high value to the low value in a continuous oscillatory manner.

6.3 Basic Control Hardware

In this section, the various physical components needed to achieve the actions required by the control strategies of the previous section are described. Since there are two fundamentally different control approaches—pneumatic and electronic—the following material is so divided. Sensors, controllers, and actuators for principal HVAC applications are described.

6.3.1 Pneumatic Systems

The first widely adopted automatic control systems used compressed air as the signal transmission medium. Compressed air had the advantage that it could be *metered* through various sensors and could power large actuators. The fact that the response of a normal pneumatic loop could take several minutes often worked as an advantage. Pneumatic controls use compressed air (approximately 20 psig in the United States) for the operation of sensors

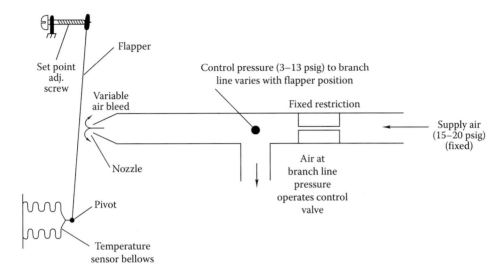

FIGURE 6.6
Drawing of pneumatic thermostat showing adjustment screw used to change temperature setting.

and actuators. Though most new buildings use electronic controls, many existing buildings use pneumatic controls. This section provides an overview of how these devices operate.

Temperature control and damper control comprise the bulk of pneumatic loop controls. Figure 6.6 shows a method of sensing temperature and producing a control signal. Main supply air, supplied by a compressor, enters a branch line through a restriction. The zone thermostat bleeds a variable amount of air out, depending on the position of the flapper, controlled by the temperature sensor bellows. As more air bleeds out, the branch line pressure (control pressure) drops. This reduction in the total pressure to the control element changes the output of the control element. This control can be forward acting or reverse acting. The restrictions typically have hole diameters on the order of a few thousandths of an inch and consume very little air. Typical pressures in the branch lines range between 3 and 13 psig (20–90 kPa). In simple systems, this pressure from a thermostat could operate an actuator such as a control valve for a room heating unit. In this case, the thermostat is both the sensor and the controller—a rather common configuration.

Many other temperature sensor approaches can be used. For example, the bellows shown in Figure 6.6 can be eliminated and the flapper can be made of a bimetallic strip. As temperature changes, the bimetal strip changes curvature, opening or closing the flapper/nozzle gap. Another approach uses a remote bulb filled with either liquid or vapor that pushes a rod (or a bellows) against the flapper to control the pressure signal. This device is useful if the sensing element must be located where direct measurement of temperature by a metal strip or bellows is not possible, such as in a water stream or high-velocity ductwork. The bulb and connecting capillary size may vary considerably by application.

Pressure sensors may use either bellows or diaphragms to control branch line pressure. For example, the motion of a diaphragm may replace that of the flapper in Figure 6.6 to control the bleed rate. A bellows similar to that shown in the same figure may be internally pressurized to produce a displacement that can control air bleed rate. A bellows produces significantly greater displacements than a single diaphragm.

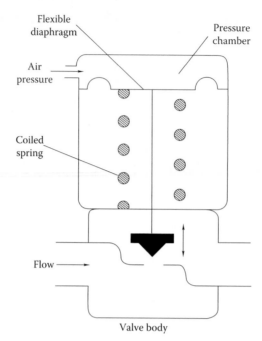

FIGURE 6.7
Pneumatic control valve showing counterforce spring and valve body. Increasing pressure closes the valve.

Humidity sensors in pneumatic systems are made from materials that change size with moisture content. Nylon or other synthetic hygroscopic fibers that change size significantly (i.e., 1%–2%) with humidity are commonly used. Since the dimensional change is relatively small on an absolute basis, mechanical amplification of the displacement is used. The materials that exhibit the desired property include nylon, hair, and cotton fibers. Human hair exhibits a much more linear response with humidity than nylon; however, because the properties of hair vary with age, nylon has much wider use (Letherman, 1981). Humidity sensors for electronic systems are quite different and are discussed in the next section.

An *actuator* converts pneumatic energy to motion—either linear or rotary. It creates a change in the controlled variable by operating control devices such as dampers or valves. Figure 6.7 shows a pneumatically operated control valve. The valve opening is controlled by the pressure in the diaphragm acting against the spring. The spring is essentially a linear device. Therefore, the motion of the valve stem is essentially linear with air pressure. However, this does not necessarily produce a linear effect on flow as discussed later. Figure 6.8 shows a pneumatic damper actuator. Linear actuator motion is converted into rotary damper motion by the simple mechanism shown.

Pneumatic *controllers* produce a branch line (see Figure 6.6) pressure that is appropriate to produce the needed control action for reaching the set point. Such controls are manufactured by a number of control firms for specific purposes. Classifications of controllers include the sign of the output (direct or reverse acting) produced by an error, by the control action (proportional, PI, or two-position), or by number of inputs or outputs. Figure 6.9 shows the essential elements of a dual-input, single-output controller. The two inputs could be the heating system supply temperature and the outdoor temperature sensors, used to control the output water temperature setting of a boiler in a building heating system. This is essentially a boiler *temperature reset* system that reduces heating water temperature with increasing ambient temperature for better system control and reduced energy use.

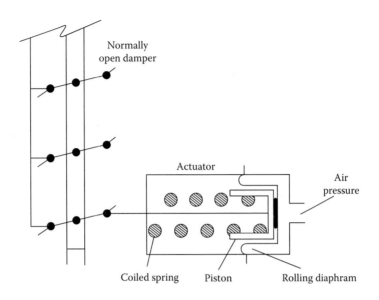

FIGURE 6.8
Pneumatic damper actuator. Increasing pressure closes the parallel-blade damper.

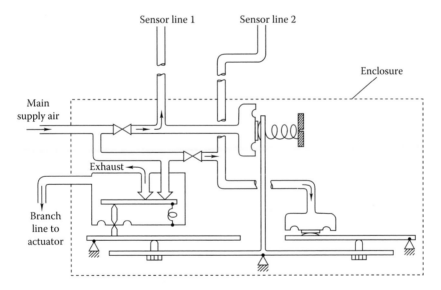

FIGURE 6.9
Example pneumatic controller with two inputs and one control signal output.

The air supply for pneumatic systems must produce very clean, oil-free, dry air. A compressor producing 80–100 psig is typical. Compressed air is stored in a tank for use as needed, avoiding continuous operation of the compressor. The air system should be oversized by 50%–100% of estimated, nominal consumption. The air is then dried to avoid moisture freezing in cold control lines in air-handling units (AHUs) and elsewhere. Dried air should have a dew point of −30°F or less in severe heating climates. In deep cooling climates, the lowest temperature to which the compressed air lines are exposed may be the building cold air supply. Next, the air is filtered to remove water droplets, oil (from the compressor), and any dirt. Finally, the air pressure is reduced in a pressure regulator to the control system operating pressure of approximately 20 psig. Control air piping uses either copper or nylon (in accessible locations).

6.3.2 Electronic Control Systems

Electronic controls comprise the bulk of the controllers for HVAC systems. Direct digital control (DDC) systems began to make inroads in the early 1990s and now make up over 80% of all controller sales. Low-end microprocessors now cost under $0.50 each and are thus very economical to apply. Along with the decreased cost, increased functionality can be obtained with DDC. BACnet has emerged as the standard communication protocol (ASHRAE, 2001) and most control vendors offer a version of the BACnet protocol. In this section, we survey the sensors, actuators, and controllers used in modern electronic control systems for buildings.

DDC enhances the previous analog-only electronic system with digital features. Modern DDC systems use analog sensors (converted to digital signals within a computer) along with digital computer programs to control HVAC systems. The output of this microprocessor-based system can be used to control either electronic, electrical, or pneumatic actuators or a combination. DDC systems have the advantage of reliability and flexibility that others do not. For example, accurately setting control constants in computer software is easier than making adjustments at a controller panel with a screwdriver. DDC systems offer the option of operating energy management systems (EMSs) and HVAC diagnostic, knowledge-based systems since the sensor data used for control are very similar to that used in EMSs. Pneumatic systems do not offer this ability. Figure 6.10 shows a schematic diagram of a DDC controller. The entire control system must include sensors and actuators not shown in this controller-only drawing.

Temperature measurements for DDC applications are made by three principal methods:

1. Thermocouples
2. Resistance temperature detectors (RTDs)
3. Thermistors

Each has its advantages for particular applications. Thermocouples consist of two dissimilar metals chosen to produce a measurable voltage at the temperature of interest

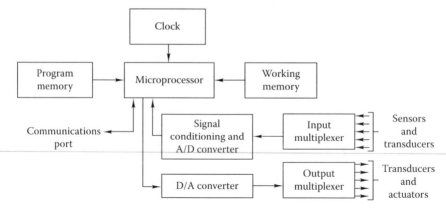

FIGURE 6.10
Block diagram of a DDC controller.

(i.e., Seebeck effect). The voltage output is low (millivolts) but is a well-established function of the junction temperature. By themselves, thermocouples generally produce voltages too small to be useful in most HVAC applications (e.g., a type J thermocouple produces only 5.3 mV at 100°C). However, modern signal conditioning equipment can easily amplify these signals, as well as provide calibration (also known as cold junction compensation).

RTDs use small, responsive sensing sections constructed from metals whose resistance–temperature characteristic is well established and reproducible. To first order,

$$R = R_0(1 + kT) \qquad (6.12)$$

where
  R is the resistance, ohms
  R_0 is the resistance at the reference temperature (0°C), ohms
  k is the temperature coefficient of resistance, $°C^{-1}$
  T is the RTD temperature, °C

This equation is easy to invert to find the temperature as a function of resistance. Although complex higher order expressions exist, their use is not needed for HVAC applications.

Two common materials for RTDs are platinum and Balco (a 40% nickel, 60% iron alloy). The nominal values of k, respectively, are $3.85 \times 10^{-3}°C^{-1}$ and $4.1 \times 10^{-3}°C^{-1}$.

Modern electronics measure current and voltage and then determine the resistance using Ohm's law. The measurement causes power dissipation in the RTD element, raising the temperature and creating an error in the measurement. This Joule self-heating can be minimized by minimizing the power dissipated in the RTD. Raising the resistance of the RTD helps, but the most effective approach requires pulsing the current and making the measurement in a few milliseconds. Since one measurement per second will generally satisfy the most demanding HVAC control loop, the power dissipation can be reduced by a factor of 100 or more. Modern digital controls can easily handle the calculations necessary to implement piecewise linearization and other curve-fitting methods to improve the accuracy of the RTD measurements. In addition, lead wire resistance can cause lack of accuracy for the class of platinum RTDs whose nominal resistance is only 100 ohms because the lead resistance of 1–2 ohms is not negligible by comparison to that of the sensor itself.

Thermistors are semiconductors that exhibit a standard exponential dependence for resistance versus temperature given by

$$R = Ae^{(B/T)} \qquad (6.13)$$

A is related to the nominal value of resistance at the reference temperature (77°F) and is of the order of several thousands of ohms. The exponential coefficient B (a weak function of temperature) is of the order of 5400–7200 R (3000–4000 K). The nonlinearity inherent in thermistors can be reduced by connecting a properly selected fixed resistor in parallel with it. The resulting linearity is desirable from a control system design viewpoint. Thermistors can have a problem with long-term drift and aging; the designer and control manufacturer should consult on the most stable thermistor design for HVAC applications. Some manufacturers provide linearized thermistors that combine both positive and negative resistive dependence on temperature to yield a more linear response function.

Humidity measurements are needed for control of enthalpy economizers or may also be needed to control special environments such as clean rooms, hospitals, and areas housing computers. Relative humidity, dew point, and humidity ratio are all indicators of the moisture content of air. An electrical, capacitance-based approach using a polymer with interdigitated electrodes has become the most common sensor type. The polymer material absorbs moisture and changes the dielectric constant of the material, changing the capacitance of the sensor. The capacitance of the sensor forms part of a resonant circuit so when the capacitance changes, the resonant frequency changes. This frequency then can be correlated to the relative humidity and provide reproducible readings if not saturated by excessive exposure to high humidity levels (Huang, 1991). The response times of tens of seconds easily satisfy most HVAC application requirements. These humidity sensors need frequent calibration, generally yearly. If a sensor becomes saturated or has condensation on the surface, they become uncalibrated and exhibit an offset from their calibration curve. Older technologies used ionic salts on gold grids. These expensive sensors frequently failed.

Pressure measurements are made by electronic devices that depend on a change of resistance or capacitance with imposed pressure. Figure 6.11 shows a cross-sectional drawing

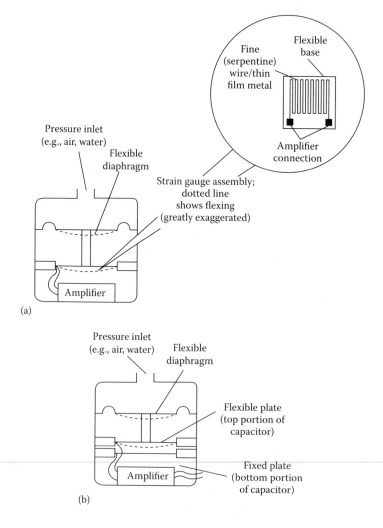

FIGURE 6.11
(a) Resistance- and (b) capacitance-type pressure sensors.

of each. In the resistance type, stretching of the membrane lengthens the resistive element thereby increasing resistance. This resistor is an element in a Wheatstone bridge; the resulting bridge voltage imbalance is linearly related to the imposed pressure. The capacitive-type unit has a capacitance between a fixed and a flexible metal that decreases with pressure. The capacitance change is amplified by a local amplifier that produces an output signal proportional to pressure. Pressure sensors can burst from overpressure or a water hammer effect. Installation needs to carefully follow the manufacturer's requirements.

DDC systems require *flow* measurements to determine the energy flow for air and water delivery systems. Pitot tubes (or arrays of tubes) and other flow measurement devices can be used to measure either air or liquid flow in HVAC systems. Airflow measurements allow for proper flow in variable air volume (VAV) system control, building pressurization control, and outside air control. Water flow measurements enable chiller and boiler control and monitoring and various water loops used in the HVAC system. Some controls only require the knowledge of flow being present. Open–closed sensors fill this need and typically have a paddle that makes a switch connection under the presence of flow. These types of switches can also be used to detect *end of range*, that is, fully open or closed for dampers and other mechanical control elements.

Temperature, humidity, and pressure *transmitters* are often used in HVAC systems. They amplify signals produced by the basic devices described in the preceding paragraphs and produce an electrical signal over a standard range thereby permitting standardization of this aspect of DDC systems. The standard ranges are

Current: 4–20 mA (dc)

Voltage: 0–10 V or 0–5 V (dc)

Although the majority of transmitters produce such signals, the noted values are not universally used.

Figure 6.10 shows the elements of a DDC controller. The heart of the controller is the microprocessor that can be programmed in either a standard or system-specific language. Control algorithms (linear or not), sensor calibrations, output signal shaping, and historical data archiving can all be programmed as the user requires. A number of firms have constructed controllers on standard personal computer platforms. Describing the details of programming HVAC controllers is beyond the scope of this chapter, since each manufacturer uses a different approach. The essence of any DDC system, however, is the same as shown in the figure. Honeywell (1988) discusses DDC systems and their programming in more detail.

Actuators for electronic control systems include

Motors—operate valves, dampers

Variable speed controls—pump, fan, chiller drives

Relays and motor starters—operate other mechanical or electrical equipment (pumps, fans, chillers, compressors), electrical heating equipment

Additional components provide necessary functionality, such as transducers that convert signal types (e.g., electrical to pneumatic) and visual displays that inform system operators of control and HVAC system functions.

Pneumatic and DDC systems have their own advantages and disadvantages. Pneumatics possess increasing disadvantages of cost, hard-to-find replacements, requiring an air compressor with clean oil-free air, sensor drift, imprecise control, and a lack of automated monitoring. The retained advantages include explosion-proof operation and a fail-soft degradation of performance. DDC systems have emerged and have taken the lead for HVAC systems over pneumatics because of the ability to integrate the control system into a large energy management and control system (EMCS), the accuracy of the control, and the ability to diagnose problems remotely. Systems based on either technology require maintenance and skilled operators.

6.4 Basic Control System Design Considerations

This section discusses selected topics in control system design including control system zoning, valve and damper selection, and control logic diagrams. The following section shows several HVAC system control design concepts. Bauman (1998) may be consulted for additional information.

The ultimate purpose of an HVAC control system is to control zone temperature (and secondarily air motion and humidity) to conditions that assure maximum comfort and productivity of the occupants. From a controls viewpoint, the HVAC system is assumed to be able to provide comfort conditions if controlled properly. Basically, a zone is a portion of a building that has loads that differ in magnitude and timing sufficiently from other areas so that separate portions of the secondary HVAC system and control system are needed to maintain comfort.

Having specified the zones, the designer must select the thermostat (and other sensors, if used) location. Thermostat signals are either passed to the central controller or used locally to control the amount and temperature of conditioned air or coil water introduced into a zone. The air is conditioned either locally (e.g., by a unit ventilator or baseboard heater) or centrally (e.g., by the heating and cooling coils in the central air handler). In either case, a flow control actuator is controlled by the thermostat signal. In addition, airflow itself may be controlled in response to zone information in VAV systems. Except for variable speed drives used in variable-volume air or liquid systems, flow is controlled by valves or dampers. The design selection of valves and dampers is discussed next.

6.4.1 Steam and Liquid Flow Control

The flow through valves such as that shown in Figure 6.7 is controlled by valve stem position, which determines the flow area. The variable flow resistance offered by valves depends on their design. The flow characteristic may be linear with position or not. Figure 6.12 shows flow characteristics of the three most common types. Note that the plotted characteristics apply only for *constant valve pressure* drop. The characteristics shown are idealizations of actual valves. Commercially available valves will resemble but not necessarily exactly match the curves shown.

The *linear* valve has a proportional relation between volumetric flow V and valve stem position z:

$$\dot{V} = kz \tag{6.14}$$

The flow in *equal percentage* valves increases by the same fractional amount for each increment of opening. In other words, if the valve is opened from 20% to 30% of full travel, the

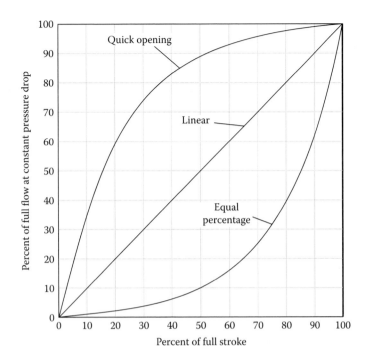

FIGURE 6.12
Quick-opening, linear, and equal percentage valve characteristics.

flow will increase by the same percentage as if the travel had increased from 80% to 90% of its full travel. However, the absolute volumetric flow increase for the latter case is much greater than for the former. The equal percentage valve flow characteristic is given by

$$\dot{V} = Ke^{(kz)} \tag{6.15}$$

where k and K are proportionality constants for a specific valve. Quick-opening valves do not provide good flow control but are used when rapid action is required with little stem movement for on/off control.

Example 6.2: Equal Percentage Valve

A valve at 30% of travel has a flow of 4 gal/min. If the valve opens another 10% and the flow increases by 50% to 6 gal/min, what are the constants in Equation 6.15? What will be the flow at 50% of full travel?

Figure: See Figure 6.15.
Assumptions: Pressure drop across the valve remains constant.

Find: k, K, $\dot{V}_{50}$

Solution: Equation 6.15 can be evaluated at the two flow conditions. If the results are divided by each other, we have

$$\frac{\dot{V}_2}{\dot{V}_1} = \frac{6}{4} = e^{k(z_2 - z_1)} = e^{k(0.4 - 0.3)} \tag{6.16}$$

In this expression, the travel z is expressed as a fraction of the total travel and is dimensionless. Solving this equation for k gives the result

$$k = 4.05 \text{ (no units)}$$

From the known flow at 30% travel, we can find the second constant K:

$$K = \frac{4 \text{ gal/min}}{e^{4.05 \times 0.3}} = 1.19 \text{ gal/min} \qquad (6.17)$$

Finally, the flow is given by

$$\dot{V} = 1.19 e^{4.05z} \qquad (6.18)$$

At 50% travel, the flow can be found from the following expression:

$$\dot{V}_{50} = 1.19 e^{4.05 \times 0.5} = 9.0 \text{ gal/min} \qquad (6.19)$$

Comments: This result can be checked, since the valve is an equal percentage valve. At 50% travel, the valve has moved 10% beyond its 40% setting at which the flow was 6 gal/min. Another 10% stem movement will result in another 50% flow increase from 6 to 9 gal/min, confirming the solution.

The plotted characteristics of all three valve types assume constant pressure drop across the valve. In an actual system, the pressure drop across a valve will not remain constant, but if the valve is to maintain its control characteristics, the pressure drop across it must be the majority of the entire loop pressure drop. If the valve is designed to have a full-open pressure drop equal to that of the balance of the loop, good flow control will exist. This introduces the concept of valve *authority* defined as the valve pressure drop as a fraction of total system pressure drop:

$$A \equiv \frac{\Delta p_{v,open}}{(\Delta p_{v,open} + \Delta p_{system})} \qquad (6.20)$$

For proper control, the full-open valve authority should be at least 0.50. If the authority is 0.5 or more, control valves will have installed characteristics not much different from those shown in Figure 6.12. If not, the valve characteristic will be distorted upward, since the majority of the system pressure drop will be dissipated across the system at high flows.

Valves are further classified by the number of connections or ports. Figure 6.13 shows sections of typical *two-way* and *three-way* valves. Two-port valves control flow through coils or other HVAC equipment by varying valve flow resistance as a result of flow area changes. As shown, the flow must oppose the closing of the valve. If not, near closure, the valve would slam shut or oscillate, both of which cause excessive wear and noise. The three-way valve shown in the figure is configured in the *diverting* mode. That is, one stream is split into two depending on the valve opening.

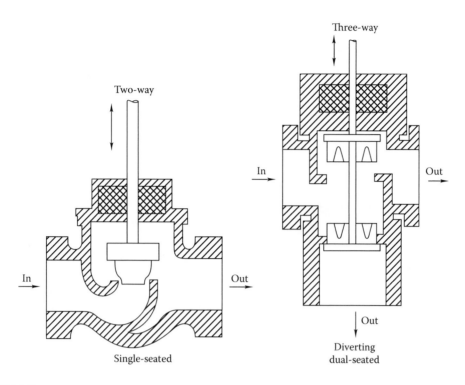

FIGURE 6.13
Cross-sectional drawings of direct-acting, single-seated two-way valve and dual-seated, three-way, diverting valve.

The three-way valve shown is double seated (single-seated three-way valves are also available); it is therefore easier to close than a single-seated valve, but tight shutoff is not possible.

Three-way valves can also be used as *mixing* valves. In this application, two streams enter the valve and one leaves. Mixing and diverting valves *cannot be used interchangeably*, since their internal design is different to ensure that they can each seat properly. Particular attention is needed by the installer to be sure that connections are made properly; arrows cast in the valve body show the proper flow direction. Figure 6.14 shows an example of three-way valves for both mixing and diverting applications.

Valve flow capacity is denoted in the industry by the dimensional flow coefficient C_v defined by

$$\dot{V} \,(\text{gal/min}) = C_v \left[\Delta p \,(\text{psi}) \right]^{0.5} \tag{6.21}$$

Δp is the pressure drop across the fully open valve, so C_v is specified as the flow rate of 60°F water that will pass through the fully open valve if a pressure difference of 1.0 psi is imposed across the valve. If SI units (m³/s and Pa) are used, the numerical value of C_v is 17% larger than in USCS units. Once the designer has determined a value of C_v, manufacturer's tables can be consulted to select a valve for the known pipe size. If a fluid other than water is to be controlled, the C_v, found from Equation 6.21, should be multiplied by the square root of the fluid's specific gravity.

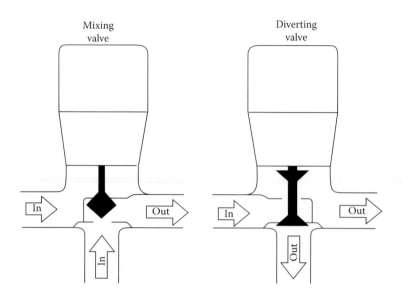

FIGURE 6.14
Three-way mixing and diverting valves. Note the significant difference in internal construction. Mixing valves are more commonly used.

Steam valves are sized using a similar dimensional expression:

$$\dot{m} \, (\text{lb/h}) = 63.5 C_v \left[\frac{\Delta p \, (\text{psi})}{v \, (\text{ft}^3/\text{lb})} \right]^{0.5} \tag{6.22}$$

in which v is the steam-specific volume. If the steam is highly superheated, multiply C_v found from Equation 6.22 by 1.07 for every 100°F of superheat. For wet steam, multiply C_v by the square root of the steam quality. Honeywell (1988) recommends that the pressure drop across the valve to be used in the equation be 80% of the difference between steam supply and return pressures (subject to the sonic flow limitation discussed later). Table 6.1 can be used for preliminary selection of control valves for either steam or water.

The type of valve (linear or not) for a specific application must be selected so the controlled system is as nearly linear as possible. Control valves are very commonly used to control the heat transfer rate in coils. For a linear system, the combined characteristic of the actuator, valve, and coil should be linear. This will require quite different valves for hot water and steam control, for example, as we shall see.

Figure 6.15 shows the part load performance of a hot water coil used for air heating; at 10% of full flow, the heat rate is 50% of its peak value. The heat rate in a cross-flow heat exchanger increases roughly in exponential fashion with flow rate, a highly nonlinear characteristic. This heating coil nonlinearity follows from the longer water residence time in a coil at reduced flow and the relatively large temperature difference between air being heated and the water heating it.

However, if one were to control the flow through this heating coil by an equal percentage valve (positive exponential increase in flow with valve position), the combined valve plus the coil characteristic would be roughly linear. Referring to Figure 6.15, we see that 50% of stem travel corresponds to 10% flow. The third graph in the figure is the combined characteristic. This approximately linear subsystem is much easier to control than if a linear valve were used with the highly nonlinear coil. Hence the general rule: use equal percentage valves for heating coil control.

TABLE 6.1

Quick Sizing Chart for Control Valves

	Steam Capacity (lb/h)						Water Capacity (gal/min)						
	Vacuum Return Systems			Atmospheric Return Systems			Differential Pressure (psig)						
	2 psi Supply Press.	5 psi Supply Press.	10 psi Supply Press.	2 psi Supply Press.	5 psi Supply Press.	10 psi Supply Press.							
C_v	3.2 psi Press. Drop[b]	5.6 psi Press. Drop[b]	9.6 psi Press. Drop[b]	1.6 psi Press. Drop[b]	4.0 psi Press. Drop[b]	8.0 psi Press. Drop[b]	2	4	6	8	10	15	20
0.33	7.7	11.0	16.0	5.4	9.3	14.6	0.41	0.66	0.81	0.93	1.04	1.27	1.47
0.63	14.6	20.9	30.5	10.4	17.7	27.8	0.89	1.26	1.54	1.78	1.99	2.4	2.81
0.73	17.0	24.3	35.4	12	20.5	32.2	1.0	1.46	1.78	2.06	2.3	2.8	3.25
1.0	23.0	33.2	48.5	16.4	28	44	1.4	2.0	2.44	2.82	3.16	3.9	4.46
1.6	37.09	53.1	77.6	26.8	45	70.6	2.25	3.2	3.9	4.51	5.06	6.2	7.13
2.5	58.25	82.9	121.2	41.9	70.25	110.25	3.53	5.0	6.1	7.05	7.9	9.68	11.15
3.0	69.9	99.5	145.5	50.2	84.3	132.3	4.23	6.0	7.32	8.46	9.48	11.61	13.38
4.0	93.2	132.2	194.0	67	112.4	177.4	5.6	8.0	9.76	11.28	12.6	15.5	17.87
5.0	116.2	165.2	242.5	82.7	140.5	220.5	7.1	10.0	12.2	14.1	15.8	19.4	22.3
6.0	139	200	291.0	99	168	265	8.5	12.0	14.6	16.92	18.9	23.2	27.0
6.3	146	209	311.5	104	177	278	8.9	12.6	15.4	17.78	19.9	24.4	28.1
7.0	162	233	339.5	115	196	309	9.9	14.0	17.1	19.74	22.1	27.1	31
8.0	186.5	264.4	388.0	131.2	224.8	352.8	11.3	16.0	19.5	22.56	25.3	31.6	35.7
10.0	232	332	485.0	164	281	441	14.1	20	24.4	28.2	31.6	38.7	44.6
11.0	256	366	533.5	181	309	486	15.5	22	27	31.02	34.4	42.5	49
13.0	303	434	630.5	213.7	365.3	373.3	18.3	27	31.7	36.7	41.1	50.3	58
14.0	326	465	679.0	232	393	617	19.7	28	34	39	44	54	62
15.0	349.3	497.6	727.5	246	421.5	661.5	21.1	30	36.6	42.3	47.4	58	66.9
16.0	370.9	531	776.0	268	450	706	22.5	32	39	45.1	50.6	62	71.3
18.0	419	597	873.0	301	505	794	25	36	44	51	57	70	80
20.0	466	664	970.0	335	562	882	28	40	49	56	63	77	89
23.0	541	763	1,115	385	646	1,014	32	46	56	65	73	89	103
25.0	582.5	829	1,212	419	702.5	1,102.5	35.3	50	61	70.5	79	96.8	111.5

(Continued)

TABLE 6.1 (Continued)

Quick Sizing Chart for Control Valves

| | Steam Capacity (lb/h) | | | | | | Water Capacity (gal/min) | | | | | | |
| | Vacuum Return Systems[a] | | | Atmospheric Return Systems | | | Differential Pressure (psig) | | | | | | |
C_v	2 psi Supply Press. 3.2 psi Press. Drop[b]	5 psi Supply Press. 5.6 psi Press. Drop[b]	10 psi Supply Press. 9.6 psi Press. Drop[b]	2 psi Supply Press. 1.6 psi Press. Drop[b]	5 psi Supply Press. 4.0 psi Press. Drop[b]	10 psi Supply Press. 8.0 psi Press. Drop[b]	2	4	6	8	10	15	20
38.0	885	1,257	1,833	636	1,069	1,676	53	76	93	107	120	147	169
40.0	932	1,322	1,940	670	1,124	1,764	56	80	97.6	112.8	126	155	178.7
50.0	1,162	1,652	2,425	827	1,405	2,205	71	100	122	141	158	194	223
56.0	1,305	1,851	2,716	938	1,574	2,469	79	112	137	158	177	217	250
63.0	1,460	2,090	3,056	1,043	1,770	2,778	89	126	154	178	199	244	281
75.0	1,748	2,481	3,637	1,230	2,107	3,307	106	150	183	212	237	290	335
80.0	1,865	2,644	3,880	1,312	2,248	3,528	113	160	195	225.6	253	316	357
90.0	2,096	2,980	4,365	1,476	2,529	3,969	127	180	220	254	284	348	401
97.0	2,229	3,204	4,703	1,590	2,725	4,277	137	196	231	274	307	375	432
100.0	2,330	3,319	4,850	1,640	2,816	4,410	141	200	244	282	316	387	446
105.0	2,442	3,481	5,092	1,722	2,950	4,630	148	210	256	296	332	406	468
130.0	3,030	4,340	6,305	2,137	3,653	5,733	183	270	317	367	411	503	580
150.0	3,493	4,976	7,275	2,460	4,215	6,615	211	300	366	423	474	280	699
160.0	3,709	5,310	7,760	2,680	4,500	7,060	225	320	390	451	560	620	713
170.0	3,960	5,642	8,245	2,788	4,777	7,497	240	340	415	479	537	658	758
190.0	4,450	6,310	9,215	3,116	5,339	8,379	268	360	464	536	600	735	847
244.0	5,670	7,930	11,834	4,001	6,856	10,760	344	488	595	688	771	944	1,088
250.0	5,825	8,290	12,125	4,190	7,025	11,025	353	500	610	705	790	968	1,115
270.0	6,282	8,960	13,095	4,525	7,587	11,907	381	540	659	761	853	1,045	1,204
300.0	6,990	9,950	14,550	5,025	8,430	13,230	423	600	732	846	948	1,161	1,338
350.0	8,160	11,590	16,975	5,860	9,835	15,435	494	700	854	987	1,106	1,355	1,561

(Continued)

TABLE 6.1 (*Continued*)

Quick Sizing Chart for Control Valves

| | Steam Capacity (lb/h) | | | | | | Water Capacity (gal/min) | | | | | | |
| | Vacuum Return Systems[a] | | | Atmospheric Return Systems | | | Differential Pressure (psig) | | | | | | |
C_v	2 psi Supply Press. / 3.2 psi Press. Drop[b]	5 psi Supply Press. / 5.6 psi Press. Drop[b]	10 psi Supply Press. / 9.6 psi Press. Drop[b]	2 psi Supply Press. / 1.6 psi Press. Drop[b]	5 psi Supply Press. / 4.0 psi Press. Drop[b]	10 psi Supply Press. / 8.0 psi Press. Drop[b]	2	4	6	8	10	15	20
480.0	11,180	15,860	23,280	8,045	13,408	21,168	677	960	1,171	1,353	1,517	1,858	2,141
640.0	14,910	21,180	31,040	10,496	17,984	28,224	902	1,280	1,561	1,805	2,022	2,477	2,854
760.0	17,700	25,120	36,860	12,464	21,356	33,516	1,071	1,520	1,854	2,143	2,401	2,941	3,390
1,000.0	23,300	33,190	48,500	16,400	28,160	44,100	1,410	2,000	2,440	2,820	3,160	3,870	4,460
1,200.0	27,150	39,790	58,200	19,680	33,720	52,920	1,692	2,400	2,928	3,384	2,792	4,644	5,352
1,440.0	33,290	47,160	69,840	23,616	40,464	63,504	2,030	2,880	3,514	4,061	4,550	5,573	6,422

Source: From Honeywell, Inc., *Engineering Manual of Automatic Control*, Honeywell, Inc., Minneapolis, MN, 1988.

[a] Assuming a 4–8 in. vacuum.

[b] Pressure drop across fully open valve taking 80% of the pressure difference between supply and return main pressures.

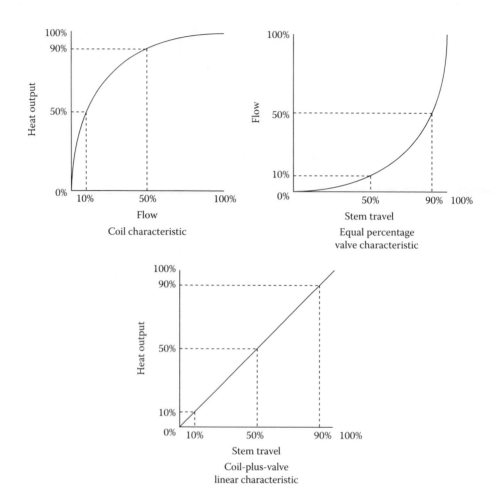

FIGURE 6.15
Heating coil, equal percentage valve, and combined coil-plus-valve linear characteristics.

Linear, two-port valves are to be used for steam flow control to coils, since the transfer of heat by steam condensation is a linear, constant temperature process—the more steam supplied, the greater the heat rate in exact proportion. Note that this is a completely different coil flow characteristic than for hot water coils. However, steam is a compressible fluid, and the sonic velocity sets the flow limit for a given valve opening when the pressure drop across the valve is more than 60% of the steam supply line absolute pressure. As a result, the pressure drop to be used in Equation 6.22 is the *smaller* of (1) 50% of the absolute stream pressure upstream of the valve and (2) 80% of the difference between the steam supply and return line pressures. The 80% rule gives good valve modulation in the subsonic flow regime (Honeywell, 1988).

Chilled water control valves should also be linear, since the performance of chilled water coils (smaller air–water temperature difference than in hot water coils) is more similar to steam coils than to hot water coils.

Either two- or three-way valves can be used to control flow at part load through heating and cooling coils as shown in Figure 6.16. The control valve can either be controlled from coil outlet water or air temperature. Two- or three-way valves achieve the same local result at the coil when used for part load control. However, the designer must consider the effects on the balance of the secondary system when selecting the valve type.

In essence, the two-way valve flow control method results in variable flow (tracking variable loads) with constant coil water temperature change, whereas the three-way valve approach results in roughly constant secondary loop flow rate but smaller coil water temperature change (beyond the local coil loop itself). In large systems, a primary/secondary design with two-way valves is preferred, unless the primary equipment can handle the range of flow variation that will result without a secondary loop. Since chillers and boilers require that flow remain within a restricted range, the energy and cost savings that could accrue due to the two-way valve, variable-volume system are difficult to achieve in small

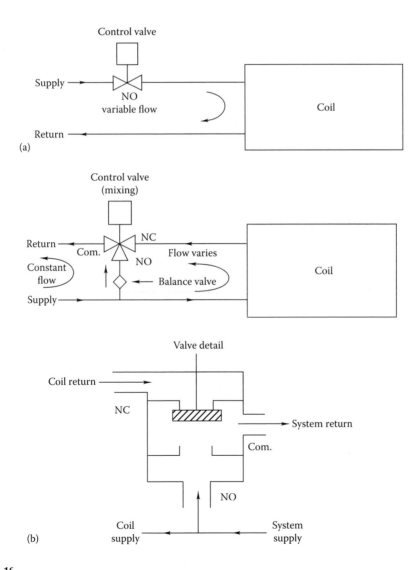

FIGURE 6.16
Various control valve piping arrangements: (a) two-way valve, (b) three-way mixing valve.   (*Continued*)

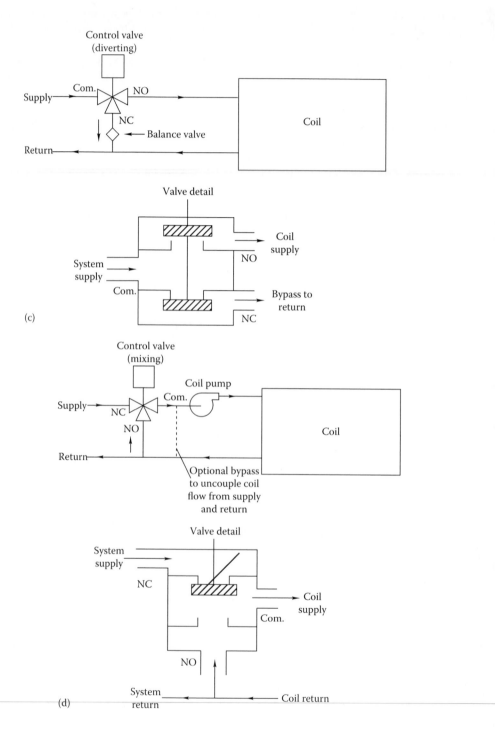

FIGURE 6.16 (Continued)
Various control valve piping arrangements: (c) three-way diverting valve, and (d) pumped coil with three-way mixing valve.

systems unless a two-pump, primary/secondary loop approach is employed. If this dual-loop approach is not used, the three-way valve method is required to maintain required boiler or chiller flow.

The location of the three-way valve at a coil must also be considered by the designer. Figure 6.16b shows the valve used downstream of the coil in a mixing, bypass mode. If a balancing valve is installed in the bypass line and set to have the same pressure drop as the coil, the local coil loop will have the same pressure drop for both full and zero coil flows. However, at the valve mid-flow position, overall flow resistance is less, since two parallel paths are involved, and the total loop flow increases to 25% more than that at either extreme.

Alternatively, the three-way valve can also be used in a diverting mode as shown in Figure 6.16c. In this arrangement, essentially the same considerations apply as for the mixing arrangement discussed earlier.* However, if a circulator (small pump) is inserted as shown in Figure 6.16d, the direction of flow in the branch line changes and a mixing valve is used. The reason that pumped coils are used is that control is improved. With constant coil flow, the highly nonlinear coil characteristic shown in Figure 6.15 is reduced, since the residence time of hot water in the coil is constant independent of load. However, this arrangement appears to the external secondary loop the same as a two-way valve. As load is decreased, flow into the local coil loop also decreases. Therefore, the uniform secondary loop flow normally associated with three-way valves is not present unless the optional bypass is used.

For HVAC systems requiring precise control, high-quality control valves are required. The best controllers and valves are of *industrial quality*; the additional cost for these valves compared to conventional building hardware results in more accurate control and longer lifetime.

6.4.2 Airflow Control

Dampers are used to control airflow in secondary HVAC air systems in buildings. In this section, we discuss the characteristics of dampers used for flow control in systems where constant-speed fans are involved. Figure 6.17 shows cross sections of the two common types of dampers used in commercial buildings. Parallel-blade dampers use blades that rotate in the same direction. They are most often applied to two position locations—open or closed. Use for flow control is not recommended. The blade rotation changes airflow direction, a characteristic that can be useful when airstreams at different temperatures are to be effectively blended.

Opposed-blade dampers have adjacent counterrotating blades. Airflow direction is not changed with this design, but pressure drops are higher than for parallel blading. Opposed-blade dampers are preferred for flow control. Figure 6.18 shows the flow characteristics of these dampers to be closer to the desired linear behavior. The parameter α on the curves is the ratio of system pressure drop to fully open damper pressure drop.

---

* A little-known disadvantage of three-way valve control has to do with the *conduction* of heat from a closed valve to a coil. For example, the constant flow of hot water through two ports of a *closed* three-way heating coil control valve keeps the valve body hot. Conduction from the closed, hot valve mounted close to a coil can cause sufficient air heating to actually decrease the expected cooling rate of a downstream cooling coil during the cooling season. Three-way valves have a second practical problem; installers often connect three-way valves incorrectly given the choice of three pipe connections and three pipes to be connected. Both of these problems are avoided by using two-way valves.

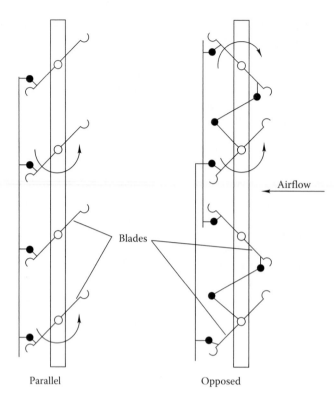

FIGURE 6.17
Diagram of parallel and opposed-blade dampers.

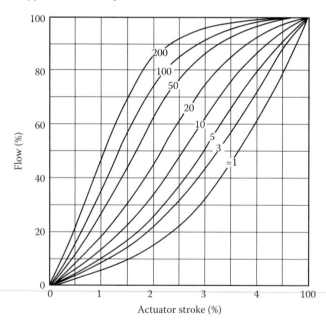

FIGURE 6.18
Flow characteristics of opposed-blade dampers. The parameter α is the ratio of system resistance (not including the damper) to damper resistance. An approximately linear damper characteristic is achieved if this ratio is about 10 for opposed-blade dampers.

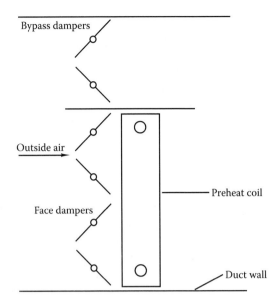

FIGURE 6.19
Face and bypass dampers used for preheating coil control.

A common application of dampers controlling the flow of outside air uses two sets in a *face and bypass* configuration as shown in Figure 6.19. For full heating, all air is passed through the coil and the bypass dampers are closed. If no heating is needed in mild weather, the coil is bypassed (for minimum flow resistance and fan power cost, flow through fully open face and bypass dampers can be used if the preheat coil water flow is shut off). Between these extremes, flow is split between the two paths. The face and bypass dampers are sized so that the pressure drop in full bypass mode (damper pressure drop only) and full heating mode (coil plus damper pressure drop) is the same.

6.5 Example HVAC System Control Systems

Several widely used control configurations for specific tasks are described in this section. These have been selected from the hundreds of control system configurations that have been used for buildings. The goal of this section is to illustrate how control components described previously are assembled into systems and what design considerations are involved. For a complete overview of HVAC control system configurations, see ASHRAE (2002, 2003, 2004), Grimm and Rosaler (1990), Tao and Janis (2005), Sauer et al. (2001), and Honeywell (1988). The illustrative systems in this section are drawn in part from the latter reference.

In this section, we will discuss seven control systems in common use. Each system will be described using a schematic diagram, and its operation and key features will be discussed in the accompanying text.

6.5.1 Outside Air Control

Figure 6.20 shows a system for controlling outside and exhaust air from a central AHU equipped for economizer cooling when available. In this and the following diagrams, the following symbols are used:

C—cooling coil

DA—discharge air (supply air from fan)

DX—direct-expansion coil

E—damper controller

EA—exhaust air

H—heating coil

LT—low-temperature limit sensor or switch; must sense the lowest temperature in the air volume being controlled

M—motor or actuator (for damper or valve), variable speed drive

MA—mixed air

NC—normally closed

NO—normally open

OA—outside air

PI—proportional plus integral controller

R—relay

RA—return air

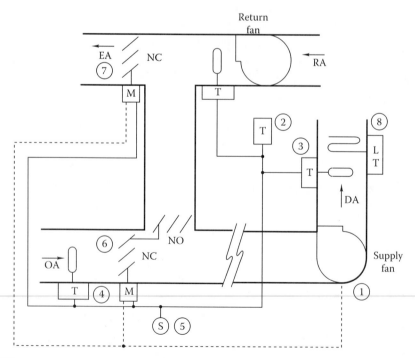

FIGURE 6.20
Outside air control system with economizer capability.

S—switch

SP—static pressure sensor used in VAV systems

T—temperature sensor; must be located to read the average temperature representative of the air volume being controlled

This system is able to provide the minimum outside air during occupied periods, to use outdoor air for cooling when appropriate by means of a temperature-based economizer cycle, and to operate fans and dampers under all conditions. The numbering system used in the figure indicates the sequence of events as the air-handling system begins operation after an off period:

1. The fan control system turns on when the fan is turned on. This may be by a clock signal or a low- or high-temperature space condition.

2. The space temperature signal determines if the space is above or below the set point. If above, the economizer feature will be activated if the OA temperature is below the upper limit for economizer operation and control the outdoor and mixed-air dampers. If below, the outside air damper is set to its minimum position.

3. The discharge air PI controller controls both sets of dampers (OA/RA and EA) to provide the desired mixed-air temperature.

4. When the outdoor temperature rises above the upper limit for economizer operation, the outdoor air damper is returned to its minimum setting.

5. Switch S is used to set the minimum setting on outside and exhaust air dampers manually. This is ordinarily done only once during building commissioning and flow testing.

6. When the supply fan is off, the outdoor air damper returns to its NC position and the return air damper returns to its NO position.

7. When the supply fan is off, the exhaust damper also returns to its NC position.

8. Low temperature sensed in the duct will initiate a freeze-protect cycle. This may be as simple as turning on the supply fan to circulate warmer room air. Of course, the OA and EA dampers remain tightly closed during this operation.

6.5.2 Heating Control

If the minimum air setting is large in the preceding system, the amount of outdoor air admitted in cold climates may require preheating. Figure 6.21 shows a preheating system using face and bypass dampers. (A similar arrangement is used for DX cooling coils.) The equipment shown is installed upstream of the fan in Figure 6.20. This system operates as follows:

1. The preheat subsystem control is activated when the supply fan is turned on.

2. The preheat PI controller senses temperature leaving the preheat section. It operates the face and bypass dampers to control the exit air temperature between 45°F and 50°F.

3. The outdoor air sensor and associated controller controls the water valve at the preheat coil. The valve may be either a modulating valve (better control) or an on-off valve (less costly).

4. The low-temperature sensors (LTs) activate coil freeze protection measures including closing dampers and turning off the supply fan.

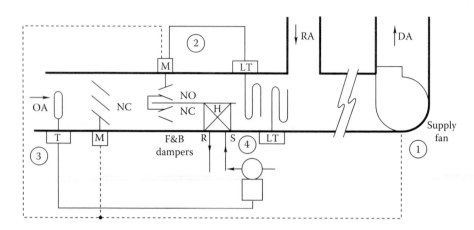

FIGURE 6.21
Preheat control system. Counterflow of air and hot water in the preheat coil results in the highest heat transfer rate.

Note that the preheat coil (as well as all coils in this section) is connected so that the hot water (or steam) flows counter to the direction of airflow. Counterflow provides a higher heating rate for a given coil than does parallel flow. Mixing of heated and cold bypass air must occur upstream of the control sensors. Stratification can be reduced by using sheet metal *air blenders* or by propeller fans in the ducting. The preheat coil should be located in the bottom of the duct. Steam preheat coils must have adequately sized traps and vacuum breakers to avoid condensate buildup that could lead to coil freezing at light loads.

The face and bypass damper approach enables air to be heated to the required system supply temperature without endangering the heating coil. (If a coil were to be as large as the duct—no bypass area—it could freeze when the hot water control valve cycles are opened and closed to maintain the discharge temperature.) The designer should consider pumping the preheat coil as shown in Figure 6.19d to maintain water velocity above the 3 ft/s needed to avoid freezing. If glycol is used in the system, the pump is not necessary but heat transfer will be reduced.

During winter, in heating climates, heat must be added to the mixed airstream to heat the outside air portion of mixed air to an acceptable discharge temperature. Figure 6.22 shows a common heating subsystem controller used with central air handlers. (It is assumed that the mixed-air temperature is kept above freezing by the action of the preheat coil, if needed.) This system has the added feature that coil discharge temperature is adjusted for ambient temperature, since the amount of heat needed decreases with increasing outside temperature. This feature, called coil discharge reset, provides better control and can reduce energy consumption. The system operates as follows:

1. During operation, the discharge air sensor and PI controller control the hot water valve.

2. The outside air sensor and controller reset the *set point* of the discharge air PI controller up as ambient temperature drops.

3. Under sensed low-temperature conditions, freeze protection measures are initiated as discussed earlier.

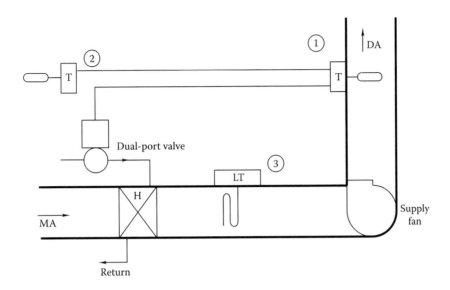

FIGURE 6.22
Heating coil control subsystem using two-way valve and optional reset sensor.

Reheating at zones in VAV or other systems uses a system similar to that just discussed. However, boiler water temperature is reset and no freeze protection is normally included. The air temperature sensor is the zone thermostat for VAV reheat, not a duct temperature sensor.

6.5.3 Cooling Control

Figure 6.23 shows the components in a cooling coil control system for a single-zone system. Control is similar to that for the heating coil discussed earlier except that the zone thermostat (not a duct temperature sensor) controls the coil. If the system were a central system serving several zones, a duct sensor would be used. Chilled water supplied to the coil

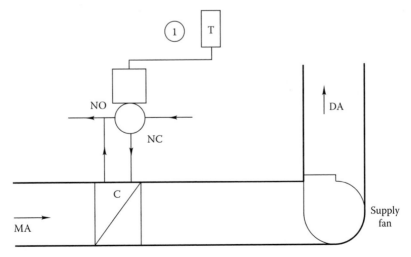

FIGURE 6.23
Cooling coil control subsystem using three-way diverting valve.

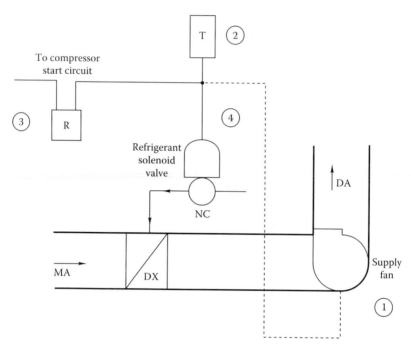

FIGURE 6.24
DX cooling coil control subsystem (on-off control).

partially bypasses and partially flows through the coil, depending on the coil load. The use of three- and two-way valves for coil control has been discussed in detail earlier. The valve NC connection is used as shown so that valve failure will not block secondary loop flow.

Figure 6.24 shows another common cooling coil control system. In this case, the coil is a DX refrigerant coil and the controlled medium is refrigerant flow. DX coils are used when precise temperature control is not required, since the coil outlet temperature drop is large whenever refrigerant is released into the coil because refrigerant flow is not modulated; it is most commonly either on or off. The control system sequences as follows:

1. The coil control system is energized when the supply fan is turned on.
2. The zone thermostat opens the two-position refrigerant valve for temperatures above the set point and closes it in the opposite condition.
3. At the same time, the compressor is energized or deenergized. The compressor has its own internal controls for oil control and pumpdown.
4. When the supply fan is off, the refrigerant solenoid valve returns to its NC position and the compressor relay to its NO position.

At light loads, bypass rates are high and ice may build up on coils. Therefore, control is poor at light loads with this system.

6.5.4 Complete Systems

The preceding five example systems are actually control subsystems that must be integrated into a single control system for the HVAC system's primary and secondary systems. In the

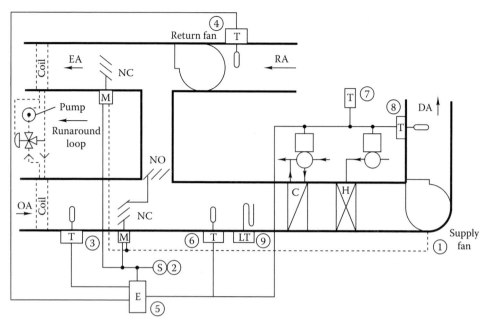

FIGURE 6.25

Control for a complete, constant-volume HVAC system. Optional runaround heat recovery system is shown to left by dashed lines.

remainder of this section, we will describe briefly two complete HVAC control systems widely used in commercial buildings. The first is a constant-volume system, whereas the second is a VAV system.

Figure 6.25 shows a constant-volume central system air-handling system equipped with supply and return fans, heating and cooling coils, and economizer for a single-zone application. If the system were to be used for multiple zones, the zone thermostat shown would be replaced by a discharge air temperature sensor. This constant-volume system operates as follows:

1. When the fan is energized, the control system is activated.
2. The minimum outside air setting is set (usually only once during commissioning as described earlier).
3. The OA temperature sensor supplies a signal to the damper controller.
4. The RA temperature sensor supplies a signal to the damper controller.
5. The damper controller positions the dampers to use outdoor or return air depending on which is cooler.
6. The mixed-air low-temperature controller controls the outside air dampers to avoid excessively low-temperature air from entering the coils. If a preheating system were included, this sensor would control it.
7. The space temperature sensor resets the coil discharge air PI controller.
8. The discharge air controller controls the
   a. Heating coil valve
   b. Outdoor air damper
   c. Exhaust air damper

    d. Return air damper

    e. Cooling coil valve after the economizer cycle upper limit is reached

  9. The low-temperature sensor initiates freeze protection measures as described previously.

A method for reclaiming either heating or cooling energy is shown by dashed lines on the left side of Figure 6.25. This so-called *runaround* system extracts energy from exhaust air and uses it to precondition outside air. For example, the heating season exhaust air may be at 75°F while the outdoor air is at 10°F. The upper coil in the figure extracts heat from the 75°F exhaust and transfers it through the lower coil to the 10°F intake air. To avoid icing of the air intake coil, the three-way valve controls this coil's liquid inlet temperature to a temperature above freezing. In heating climates, the liquid loop should also be freeze protected with a glycol solution. Heat reclaiming systems of this type can also be effective in the cooling season, when the outdoor temperatures are well above the indoor temperature.

A VAV system has additional control features including a motor speed (or inlet vanes in some older systems) control and a duct static pressure control. Figure 6.26 shows a VAV system serving both perimeter and interior zones. It is assumed that the core zones always require cooling during the occupied period. The system shown has a number of options and does not include every feature present in all VAV systems. However, it is representative of VAV design practice. The sequence of operation during the heating season is as follows:

  1. When the fan is energized, the control system is activated. Prior to activation during unoccupied periods, the perimeter zone baseboard heating is under control of room thermostats.

  2. Return and supply fan interlocks are used to prevent pressure imbalances in the supply air ductwork.

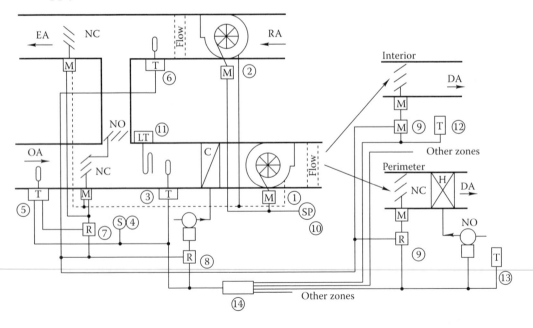

FIGURE 6.26
Control for complete VAV system. Optional supply and return flow stations shown by dashed lines.

3. The mixed-air sensor controls the outdoor air dampers (and/or preheat coil not shown) to provide proper coil air inlet temperature. The dampers will typically be at their minimum position at about 40°F.

4. The damper minimum position controls the minimum outdoor airflow.

5. As the upper limit for economizer operation is reached, the OA dampers are returned to their minimum position.

6. The return air temperature is used to control the morning warm-up cycle after night setback (option present only if night setback is used).

7. The outdoor air damper is not permitted to open during morning warm-up by the action of the relay shown.

8. Likewise, the cooling coil valve is deenergized (NC) during morning warm-up.

9. All VAV box dampers are moved full open during morning warm-up by the action of the relay override. This minimizes the warm-up time. Perimeter zone coils and baseboard units are under control of the local thermostat.

10. During operating periods, the PI static pressure controller controls both the supply and return fan speeds (or inlet vane positions) to maintain approximately 1.0 in. WG of static pressure at the pressure sensor location (or optionally to maintain building pressure). An additional pressure sensor (not shown) at the supply fan outlet will shut down the fan if fire dampers or other dampers should close completely and block airflow. This sensor overrides the duct static pressure sensor shown.

11. The low-temperature sensor initiates freeze protection measures.

12. At each zone, room thermostats control VAV boxes (and fans if present); as zone temperature rises, the boxes open more.

13. At each perimeter zone room, thermostats close VAV dampers to their minimum settings and activate zone heat (coil and/or perimeter baseboard) as zone temperature falls.

14. The controller, using temperature information for all zones (or at least for enough zones to represent the characteristics of all zones), modulates outdoor air dampers (during economizer operation) and the cooling control valve (above the economizer cycle cutoff) to provide air sufficiently cooled to maintain acceptable zone humidity and meet the load of the warmest zone.

The duct static pressure controller is critical to the proper operation of VAV systems. The static pressure controller must be of PI design, since a proportional-only controller would permit duct pressure to drift upward as cooling loads drop due to the unavoidable offset in P-type controllers. In addition, the control system should position inlet vanes (if present) closed during fan shutdown to avoid overloading on restart.

Return fan control is best achieved in VAV systems by an actual flow measurement in supply and return ducts as shown by dashed lines in the figure. The return airflow rate is the supply rate less local exhausts (fume hoods, toilets, etc.) and exfiltration needed to pressurize the building.

VAV boxes are controlled locally, assuming that adequate duct static pressure exists in the supply duct and that supply air is at an adequate temperature to meet the load (this is the function of the controller described in item 14). Figure 6.27 shows a *local* control system used with a series-type, fan-powered VAV box. This particular system delivers a

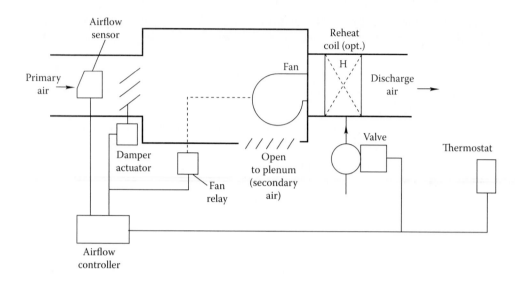

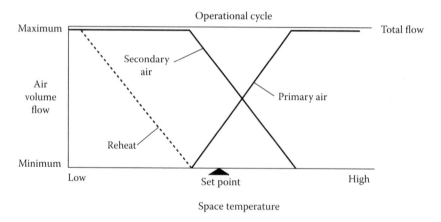

FIGURE 6.27
Series-type, fan-powered VAV box control subsystem and primary flow characteristic. The total box flow is constant at the level identified as *maximum* in the figure. The difference between primary and total airflow is secondary air recirculated through the return air grille. Optional reheat coil requires airflow shown by dashed line.

constant flow rate to the zone, to assure proper zone air distribution, by the action of the airflow controller. Primary air varies with cooling load as shown in the lower part of the figure. Optional reheating is provided by the coil shown.

6.5.5 Other Systems

This section has not covered the control of central plant equipment such as chillers and boilers. Most primary system equipment controls are furnished with the equipment and as such do not offer much flexibility to the designer. However, Braun et al. (1989) have shown that considerable energy savings can be made by properly sequencing cooling tower stages on chiller plants and by properly sequencing chillers themselves in multiple chiller plants.

Fire and smoke control are important for life safety in large buildings. The design of smoke control systems is controlled by national codes. The principal concept is to eliminate smoke from the zones where it is present while keeping adjacent zones pressurized to avoid smoke infiltration. Some components of space conditioning systems (e.g., fans) can be used for smoke control, but HVAC systems are generally not smoke control systems by design.

Electrical systems are primarily the responsibility of the electrical engineer on a design team. However, HVAC engineers must make sure that the electrical design accommodates the HVAC control system. Interfaces between the two occur where the HVAC controls activate motors on fans or chiller compressors, pumps, electrical boilers, or other electrical equipment.

In addition to electrical specifications, the HVAC engineer often conveys electrical control logic using a *ladder diagram*. An example is shown in Figure 6.28 for the control of the supply and return fans in a central system. The electrical control system is shown at the bottom and operates on low voltage (24 or 48 VAC) from the control transformer shown. The supply fan is manually started by closing the *start* switch. This activates the motor starter coil labeled 1M, thereby closing the three contacts labeled 1M in the supply fan circuit. The fourth 1M contact (in parallel with the start switch) holds the starter closed after the start button is released.

The hand–off–auto switch is typical and allows both automatic and manual operations of the return fan. When switched to the *hand* position, the fan starts. In the *auto* position, the fans will operate only when the adjacent contacts 3M are closed. Either of these actions activates the relay coil 2M, which in turn closes the three 2M contacts in the return fan motor starter. When either fan produces actual airflow, a flow switch is closed in the ducting, thereby completing the circuit to the pilot lamps L. The fan motors are protected by fuses and thermal overload heaters. If motor current draw is excessive, the heaters shown in the figure produce sufficient heat to open the normally closed thermal overload contacts.

This example ladder diagram is primarily illustrative and is not typical of an actual design. In a fully automatic system, both fans would be controlled by 3M contacts actuated by the HVAC control system. In a fully manual system, the return fan would be activated by a fifth 1M contact, not by the 3M automatic control system.

6.6 Commissioning and Operation of Control Systems

This chapter emphasizes the importance of making sound decisions in the design of HVAC control systems. Ensuring that the control system be commissioned and used properly is extremely important. The design process requires many assumptions about the building and its use. The designer must be sure that the systems will provide comfort under extreme conditions, and the sequence of design decisions and construction decisions often leads to systems that are substantially oversized. Operation at loads far below design conditions is generally much less efficient than at larger loads. Normal control practice can be a major contributor to this inefficiency. For example, it is quite common to see variable-volume air handler systems that operate at minimum flow as constant-volume systems almost all the time due to design flows that sometimes are twice as large as the maximum flow used in the building.

Hence it is very important that following construction, the control system and the rest of the HVAC system be *commissioned*. This process (ASHRAE, 2005) normally seeks to ensure that the control system operates according to design intent. This is really a minimum requirement to be sure that the system functions as designed. However, after construction, the control system setup can be modified to meet the loads actually present in

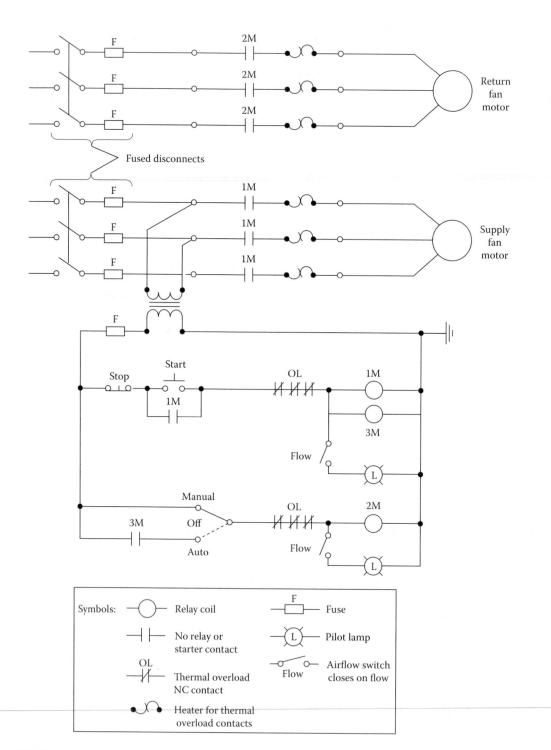

FIGURE 6.28
Ladder diagram for supply and return fan control. Hand–off–auto switch permits manual or automatic control of the return fan.

the building and fit the way the building is actually being used rather than basing these decisions on the design assumptions. If the VAV system is designed for more flow than is required, minimum flow settings of the terminal boxes can be reduced below the design value so that the system will operate in the VAV mode most of the time. Numerous other adjustments may be made as well. Such adjustments, as commonly made during the version of commissioning known as *Continuous Commissioning*®* (CC®), can frequently reduce the overall building energy use by 10% or more (Liu et al., 2002). If the process is applied to an older building where control practices have drifted away from design intent and undetected component failures have further eroded system efficiency, energy savings often exceed 20% (Claridge et al., 2004).

6.6.1 Control Commissioning Case Study

A case study in which this process was applied to a major army hospital facility located in San Antonio, TX (Zhu et al., 2000a–c), is provided. The Brooke Army Medical Center (BAMC) was a relatively new facility when the CC process was begun. The facility was operated for the army by a third-party company, and it was operated in accordance with the original design intent (Figure 6.29).

BAMC is a large, multifunctional medical facility with a total floor area of 1,349,707 ft². The complex includes all the usual in-patient facilities as well as out-patient and research areas. The complex is equipped with a central energy plant, which has four 1200-ton water-cooled electric chillers. Four primary pumps (75 hp each) are used to pump water through the chillers. Two secondary pumps (200 hp each), equipped with VFDs, supply chilled water from the plant to the building entrance. Fourteen chilled water risers equipped with 28 pumps totaling 557 hp are used to pump chilled water to all of the

FIGURE 6.29
The BAMC in San Antonio, TX.

---

* Continuous Commissioning and CC are registered trademarks of the Texas A&M Engineering Experiment Station.

AHUs and small fan coil units. All of the chilled water riser pumps are equipped with VFDs. There are four natural gas-fired steam boilers in this plant. The maximum output of each boiler is 20 MMBtu/h. Steam is supplied to each building at 125 psi (prior to CC) where heating water is generated.

There are 90 major AHUs serving the whole complex with a total fan power of 2570 hp. VFDs are installed on 65 AHUs, while the others are constant-volume systems. There are 2700 terminal boxes in the complex of which 27% are dual-duct variable-volume boxes, 71% are dual-duct constant-volume boxes, and 2% are single-duct variable-volume boxes.

The HVAC systems (chillers, boilers, AHUs, pumps, terminal boxes, and room conditions) are controlled by a DDC system. Individual controller-field panels are used for the AHUs and water loops located in the mechanical rooms. The control program and parameters can be changed either by the central computers or by the field panels.

6.6.1.1 Design Conditions

The design control program was being fully utilized by the EMCS. It included the following features:

1. Hot deck reset control for AHUs
2. Cold deck reset during unoccupied periods for some units
3. Static pressure reset between high and low limits for VAV units
4. Hot water supply temperature control with reset schedule
5. VFD control of chilled water pumps with ΔP set point (no reset schedule)
6. Terminal box level control and monitoring

It was also determined that the facility was being well maintained by the facility operator in accordance with the original design intent. The building is considered energy efficient for a large hospital complex.

The commissioning activities were performed at the terminal box level, AHU level, loop level, and central plant level. Several different types of improved operation measures and energy solutions were implemented in different HVAC systems due to the actual function and usage of the areas and rooms. Each measure will be discussed briefly, starting with the AHUs.

6.6.1.2 Optimization of AHU Operation

EMCS trending complemented by site measurements and use of short-term data loggers found that many supply fans operated above 90% of full speed most of the time. Static pressures were much higher than needed. Wide room temperature swings due to AHU shutoff lead to hot and cold complaints in some areas. Through field measurements and analysis, the following opportunities to improve the operation of the two AHUs were identified:

1. Zone air balancing and determination of new static pressure set points for VFDs
2. Optimize the cold deck temperature set points with reset schedules
3. Optimize the hot deck temperature reset schedules
4. Control of outside air intake and relief dampers during unoccupied periods to reduce ventilation during these periods

5. Optimized time schedule for fans to improve room conditions

6. Improve the preheat temperature set point to avoid unnecessary preheating

Implementation of these measures improved comfort and reduced heating, cooling, and electric use.

6.6.1.3 Optimization at the Terminal Box Level

Field measurements showed that many VAV boxes had minimum flow settings that were higher than necessary, and some boxes were unable to supply adequate hot air due to specific control sequences. A new control logic was developed, which increased hot air capacity by 30% on average, in the full heating mode, and reduced simultaneous heating and cooling. During unoccupied periods, minimum flow settings on VAV boxes were reduced to zero and flow settings were reduced in constant-volume boxes.

During commissioning, it was found that some terminal boxes could not provide the required airflow either before or after the control program modification. Specific problems were identified in about 200 boxes, with most being high flow resistance due to kinked flex ducts.

6.6.1.4 Water Loop Optimization

There are 14 chilled water risers equipped with 28 pumps that provide chilled water to the entire complex. During the commissioning assessment phase, the following were observed:

1. All the riser pumps were equipped with VFDs and they were running from 70% to 100% of full speed.
2. All the manual balancing valves on the risers were only 30%–60% open.
3. The ΔP sensor for each riser was located 10–20 ft from the far-end coil of the AHU on the top floor.
4. Differential pressure set points for each riser ranged from 13 to 26 psi.
5. There is no control valve on the return loop.
6. Although most of the cold deck temperatures were holding well, there were 13 AHUs whose cooling coils were 100% open but could not maintain cold deck temperature set points.

Since the risers are equipped with VFDs, traditional manual balancing techniques are not appropriate. All the risers were rebalanced by initially opening all of the manual balancing valves. The actual pressure requirements were measured for each riser, and it was determined that the ΔP for each riser could be reduced significantly. Pumping power requirements were reduced by more than 40%.

6.6.1.5 Central Plant Measures

Boiler system: Steam pressure was reduced from 125 to 110 psi and one boiler was operated instead of two during summer and swing seasons.

Chilled water loop: Before the commissioning, the blending valve separating the primary and secondary loops at the plant was 100% open. The primary and secondary pumps were both running. The manual valves were partially open for the secondary loop although the secondary loop pumps are equipped with VFDs. After the commissioning assessment and investigations, the following were implemented:

1. Open the manual valves for the secondary loop.
2. Close the blending stations.
3. Shut down the secondary loop pumps.

As a result, the primary loop pumps provide required chilled water flow and pressure to the building entrance for most of the year, and the secondary pumps stay offline for most of the time. The operator drops the online chiller numbers according to the load conditions and the minimum chilled water flow can be maintained to the chillers. At the same time, the chiller efficiency is also increased.

6.6.1.6 Results

For the 14-month period following initial CC implementation, measured savings were nearly $410,000, or approximately $30,000/month, for a reduction in both electricity and gas use of about 10%. The contracted cost to meter, monitor, commission, and provide a year's follow-up services was less than $350,000. This cost does not include any time for the facilities operating staff who repaired kinked flex ducts, replaced failed sensors, implemented some of the controls and subroutines, and participated in the commissioning process.

6.6.2 Commissioning Existing Buildings

The savings achieved from commissioning HVAC systems in older buildings are even larger. In addition to the opportunities for improving efficiency similar to those in new buildings, opportunities come from the following:

1. Control changes that have been made to *solve* problems, often resulting in lower operating efficiency
2. Component failures that compromise efficiency without compromising comfort
3. Deferred maintenance that lowers efficiency

Mills et al. (2004, 2005) surveyed 150 existing buildings that had been commissioned and found median energy cost savings of 15%, with savings in one-fourth of the buildings exceeding 29%. Over 60% of the problems corrected were control changes, and another 20% were related to faulty components that prevented proper control.

This suggests that relatively few control systems are operated to achieve the efficiency they are capable of providing.

6.7 Advanced Control System Design Topics

Modern control techniques can offer significant benefits compared to basic control algorithms such as PID. This section provides a short introduction to two approaches that are being used with increasing frequency to improve the stability and performance of HVAC control systems. These include nonlinear compensation and model predictive control (MPC).

6.7.1 Nonlinear Compensation

Most HVAC and refrigeration systems exhibit nonlinear dynamics. The physics of these systems are described by nonlinear thermo-fluid relationships that result in dynamic behavior that changes based on operating condition, external conditions, and controllable inputs. Despite the many factors that contribute to the system nonlinearity, the dominant source tends to be the system actuators that manipulate the flow of mass; compressors, valves, pumps, and fans all are intended to vary the flow of primary/secondary fluids. In contrast, the control objectives are generally posed in terms of regulating the flow or state of energy, such as regulating chiller outlet water temperature or specifying a desired amount of cooling. As mass flow devices, HVAC actuators are typically designed to provide a linear relationship between control input and flow rate. However, the remaining portion of the system generally exhibits a nonlinear relationship between flow rate and system outputs (temperatures, pressures, etc.).

There are many established methods or ways to account for these dynamic nonlinearities. Although these can be approximated with linear models, they are typically only valid for a small operating range. Common approaches to extend model validity include Hammerstein/Wiener models (i.e., identified linear models with a nonlinear input or output function, Figure 6.30) and linear parameter-varying models where parameter values are determined at multiple conditions and then scheduled based on a measured signal:

$$G(s,\theta) = \frac{k(\theta)}{\tau(\theta)s + 1} \tag{6.23}$$

Fixed controllers (such as PID) are typically tuned for a single operating condition and thus perform poorly at off-design conditions because the fundamental behavior of the system is different. Oscillations, hunting behavior, and sluggish responses are all symptomatic of this problem. Common nonlinear control techniques include the following:

- *Gain-scheduled control*: An approach where multiple linear controllers are designed to cover the range in operating conditions and then interpolated appropriately (Khalil, 1996). Examples of this approach applied to HVAC systems include Outtagarts et al. (1997) and Finn and Doyle (2000).

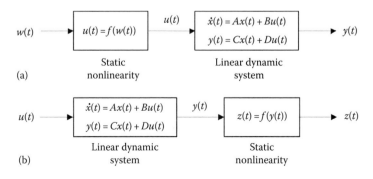

FIGURE 6.30
Nonlinear process models: (a) Hammerstein model and (b) Wiener model.

- *Adaptive control*: A technique for updating the dynamic model using real-time measurements and then using a model-based control strategy that is likewise updated (e.g., Astrom and Wittenmark, 1995).

- *Static nonlinearity compensation*: For systems whose dominant nonlinearity is a change in system gain, for example, $k(\theta)$. Various methods have been proposed, including inverse modeling (Franklin et al., 2006) and nonlinear mapping (Singhal and Salsbury, 2007). The interested reader can find a survey of linearization through feedback in Guardabassi and Savaresi (2001).

- *Cascaded feedback loops*: Using multiple feedback loops has immediate practical benefits in terms of inherent robustness (Skogestad and Postlethwaite, 1996) and can partially compensate for nonlinear system gains (Elliott and Rasmussen, 2010).

6.7.1.1 Case Study: Nonlinear Compensation for Air Conditioning Expansion Valves

In this section, we present an example of a particularly simple and effective technique for compensating for the static nonlinearities present in many HVAC systems. The system being considered is a vapor compression cycle, which is used extensively for air conditioning and refrigeration systems. During operation, an expansion valve modulates the refrigerant flow through the evaporator and regulates the temperature at the evaporator outlet (i.e., superheat). In these systems, superheat must not only be kept low to ensure efficient operation, but also remain high enough to prevent liquid refrigerant from entering the compressor.

As with many HVAC actuators, the relationship between valve input and system outputs is highly nonlinear, with the system gain, $k(\theta)$, varying several hundred percent between low-flow and high-flow conditions. This can lead to poor performance in practice. If fixed-gain controllers (e.g., PID) are designed for the high-flow conditions, then the system will oscillate at low- and medium-flow conditions (i.e., valve hunting). However, if the controller gains are selected based on the low-flow conditions, the performance will be extremely sluggish at medium- and high-flow conditions.

One simple method for compensating for the nonlinear system gain is to utilize cascaded control loops (Figure 6.31). An inner loop controller utilizes a high proportional

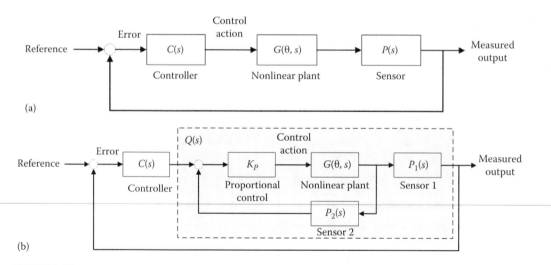

FIGURE 6.31
Feedback control architectures: (a) standard feedback control and (b) cascaded feedback control.

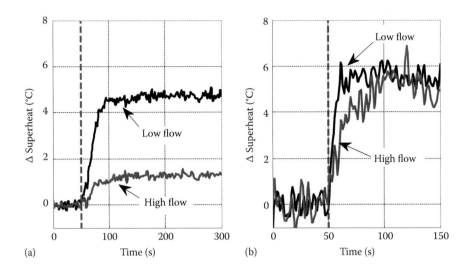

FIGURE 6.32
Superheat response to step change in valve position for high and low flows: (a) standard feedback control and (b) cascaded feedback control. Note the difference in speed of response, as well as the nonlinear compensation.

gain, which significantly lessens the sensitivity of the resulting closed-loop system, $Q(s)$, to changes in the system gain. A standard fixed controller, such as PID, is then designed and implemented in cascade. The resulting system generally exhibits improved performance over a wide range of operating conditions. This approach was advocated by Elliott et al. (2009) and Elliott and Rasmussen (2010) who analyzed the effects for nonlinear electronic expansion valve (EEV) controlled evaporator using temperature and pressure sensors. The resulting step responses under different operating conditions are shown in Figure 6.32 and clearly show that the use of cascaded control loops improves the response time, as well as virtually eliminates the dynamic differences due to system nonlinearities.

6.7.2 Model Predictive Control

MPC is an overarching term for a suite of control strategies first developed for the process industries during the 1970s and 1980s. With the increased computational capabilities of embedded controllers, MPC is increasingly being used in a wide range of control applications. MPCs use an explicit dynamic model to optimize a user-defined cost function. System constraints, such as actuator limitations, can be explicitly accounted for in the optimization, which permits better operation than mere saturation of actuator signals. Additionally, output constraints can be imposed upon the controller, which will keep the system operating in a safe range. Since for many industrial systems, the most efficient operating point is at or near a set of operating constraints, MPC has been successfully implemented as a cost and energy minimization technology (Garcia et al., 1989; Qin and Badgwell, 2003).

MPC is a discrete-time control approach. At each time step, the controller uses a system model to predict outputs over a finite-time prediction horizon (Figure 6.33a) and then determine the optimal choice of the control inputs for a defined control horizon (Figure 6.33b). The controller can also account for the effect of external disturbances, input constraints, and, to some extent, output constraints. The controller then implements the first

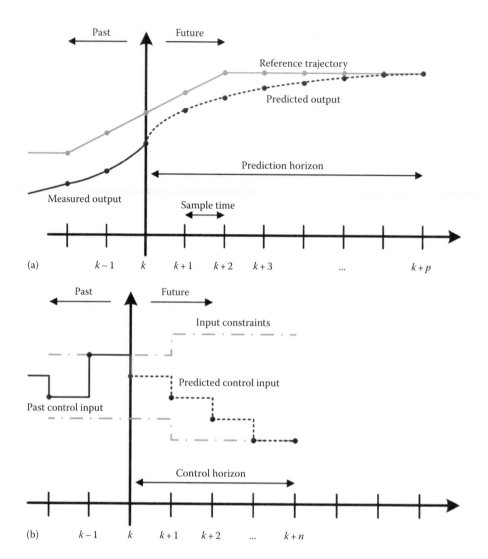

FIGURE 6.33
MPC: (a) prediction horizon and (b) control horizon.

control action in this optimal sequence. At the next time step, the process is repeated, accounting for any new information. The ability to anticipate future outcomes and optimize the response while conforming to constraints is unique to MPC and is not found in basic PID controllers, or even fixed optimal controllers such as the linear quadratic regulator.

The majority of MPC techniques assume a linear model of the system dynamics, usually a discrete-time model based on empirical data. A representative model is defined in Equation 6.24, where x are the dynamic states of the system, u are the control inputs, w and v are the external disturbances, y is the measured output, r is the desired reference, e is the tracking error, and k is the sampling index. The controller uses this model to determine the sequence of control actions that minimizes the cost function $J(x,u)$ (Equation 6.25), without violating the constraints (Equation 6.26). In this formulation, Q and R are matrices that define the relative penalty of tracking errors and actuator effort, p and n are the prediction and control horizons, and T and P are the matrices that define the hard actuator constraints and soft state constraints, respectively:

$$x(k+1) = Ax(k) + B_u u(k) + B_w w(k)$$

$$y(k) = Cx(k) + Dv(k)$$

$$e(k) = r(k) - y(k) \tag{6.24}$$

$$J(x,u) = \sum_0^p e(k)^T Q e(k) + \sum_0^n u(k)^T R u(k) \tag{6.25}$$

$$Tu \le b$$

$$Px \le c \tag{6.26}$$

MPC is attractive for many applications because of its capability to optimize performance while explicitly handling constraints. MPC is a mature technology with extensive research literature addressing issues of stability, constrained feasibility, robustness, and nonlinearities. Since optimal operating conditions often lie at the intersection of constraints, MPC offers a safe way to drive the system to the optimum while not violating performance or actuation limitations. The interested reader is referred to Garcia et al. (1989), Clarke (1994), Kothare et al. (1996), Scokaert and Rawlings (1999), Mayne et al. (2000), Qin and Badgwell (2003), Rossiter (2003), Camacho and Bordons (2004), Allgower (2005), and Rawlings and Mayne (2009), and the references therein for additional details on each of these topics.

Given these advantages, it is not surprising that MPC quickly found success beyond its origins in the chemical industries and is applied with increasing frequency to HVAC systems, including the following:

- Direct application of MPC to basic HVAC components, such as AHUs, VAVs, or heating systems (e.g., MacArthur and Woessner, 1993; He et al., 2005; Yuan and Perez, 2006; Xi et al., 2007; Freire et al., 2008; Huang et al., 2010; Xu et al., 2010; May-Ostendorp et al., 2011; Privara et al., 2011)
- Indirect application as a method of tuning standard HVAC control algorithms (e.g., Dexter and Haves, 1989; MacArthur and Woessner, 1993; Sousa et al., 1997; Xu et al., 2005)
- In the form of the generalized predictive control algorithm, which is perhaps most widely used in HVAC applications (Clarke, 1987)
- Vapor compression systems (e.g., Leducq et al., 2006; Elliott and Rasmussen, 2012, 2013)
- As a supervisory control for HVAC applications (Wang and Ma, 2008)

6.7.2.1 Case Study: MPC for Air Conditioning Expansion Valves

The ability to explicitly account for constraints when determining appropriate control actions is a unique capability of MPC. This capability has numerous potential applications for HVAC equipment. This section presents an experimental example of MPC's constraint handling.

Again we examine the case of expansion valve control for vapor compression systems. Although most valves are designed to regulate superheat to a fixed level, the exact superheat set point is not of particular importance for system efficiency; as long as superheat is kept in a reasonable band, the coefficient of performance (COP) does not vary significantly. By using an MPC-based controller, the superheat can be kept in a band around an optimal point without exerting a large amount of actuator effort, achieving efficient operation while minimizing actuator wear. The following experimental results demonstrate that an MPC controller can be used with an EEV to keep superheat in a band, exerting little effort until superheat begins to leave the defined range.

Figures 6.33 and 6.34 present the results of experiments conducted on a residential air conditioning system. For these tests, an MPC controller is used to keep the evaporator superheat between 5°C and 13°C, with a set point of 9°C. As long as the superheat does not leave the designated band, the controller will not react. While similar tracking results can be had with a very nonaggressive PID controller, the construction of the MPC controller allows the actuator to react very strongly to large disturbances while responding slowly to small disturbances, so long as the user-defined constraints are not violated. This capability is not

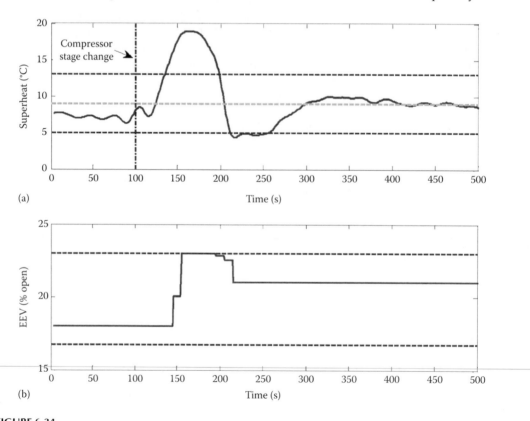

(a)

(b)

FIGURE 6.34
Superheat response using MPC of the EEV with compressor stage change: (a) superheat response and soft output constraints, (b) EEV actuation with hard input constraints.

available under a PID control paradigm. An illustration of this is provided in the test displayed in Figure 6.33a. At 100 s, the compressor changes from stage 1 (low cooling) to stage 2 (high cooling). As the superheat leaves the prescribed band, the EEV reacts strongly, opening up to the EEV maximum constraint, which is set to 23% open. As the superheat drops down toward its minimum constraint, the valve reacts again by closing, bringing superheat back within the desired band. As Figure 6.33b shows, very little actuator effort is expended, even for such a large disturbance as a compressor step change.

Figure 6.34 displays the system efficiency under two different control paradigms. The left-hand side shows the superheat, actuator effort, evaporator fan speed, and system COP under the same MPC controller as that used in Figure 6.33. The right-hand side shows the same data for a standard PID controller. The evaporator fan speed oscillated as shown, with a period of 20 min. In the MPC-controlled case, the superheat oscillates within the prescribed limits, with the valve only moving as necessary to respect the constraints. In the PID-controlled case, the superheat is kept within a much tighter band, as is expected. The bottom figures show that the mean COP that each control paradigm achieves is virtually the same (4.9 for MPC, 4.8 for PID). Additionally, the MPC controller requires much less motion from the EEV—a total of 14 steps as opposed to 254 steps over 2500 s of test time. Despite the relatively poor set point tracking of the MPC controller, neither system efficiency nor compressor safety is compromised, and the actuator is used much less heavily, implying a longer service life for the valve (Figure 6.35).

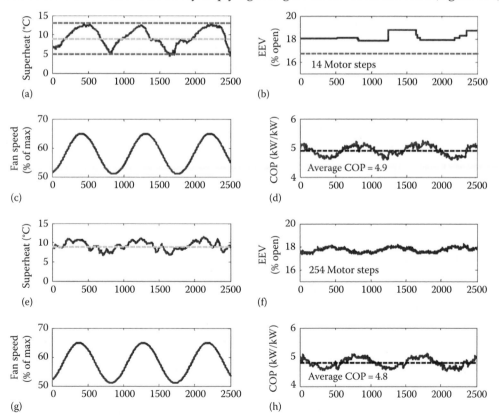

FIGURE 6.35
Comparison of superheat response for MPC- and PID-controlled EEV : (a–d) evaporator superheat, EEV actuation, fan speed, and COP using MPC control; (e–h) evaporator superheat, EEV actuation, fan speed, and COP using standard PID control

6.8 Summary

This chapter has introduced the important features of properly designed control systems for HVAC applications. Sensors, actuators, and control methods have been described. The method for determining control system characteristics either analytically or empirically has been discussed.

The following rules (ASHRAE, 2003) should be followed to ensure that the control system is as energy efficient as possible:

1. Operate HVAC equipment only when the building is occupied or when heat is needed to prevent freezing.
2. Consider the efficacy of night setback vis-à-vis building mass. Massive buildings may not benefit from night setback due to overcapacity needed for the morning pickup load.
3. Do not supply heating and cooling simultaneously. Do not supply humidification and dehumidification at the same time.
4. Reset heating and cooling air or water temperature to provide only the heating or cooling needed.
5. Use the most economical source of energy first, the most costly last.
6. Minimize the use of outdoor air during the deep heating and cooling seasons subject to ventilation requirements.
7. Consider the use of *dead-band* or *zero-energy* thermostats.
8. Establish control settings for stable operation to avoid system wear and to achieve proper comfort.

Finally, consider the use of nonlinear compensation and MPC techniques where appropriate.

References

Allgower, F. (2005). Nonlinear model predictive control. *IEE Proceedings—Control Theory and Applications* **152**(3): 257–258.

ASHRAE (2001). ANSI/ASHRAE standard 135-2001, BACnet: A data communication protocol for building automation and control networks. ASHRAE, Atlanta, GA.

ASHRAE (2002). *HVAC Applications*. ASHRAE, Atlanta, GA.

ASHRAE (2003). *HVAC Systems and Equipment*. ASHRAE, Atlanta, GA.

ASHRAE (2004). *Handbook of Fundamentals*. Chapter 15, Fundamentals of control. ASHRAE, Atlanta, GA.

ASHRAE (2005). Guideline 0, commissioning. ASHRAE, Atlanta, GA.

Astrom, K. J. and B. Wittenmark (1995). *Adaptive Control*. Addison-Wesley, Reading, MA.

Bauman, H. D. (2004). *Control Valve Primer*, 4th edn. Instrument Society of America, North Carolina.

Braun, J. E., S. A. Klein et al. (1989). Applications of optimal control to chilled water systems without storage. *ASHRAE Transactions* **95**(Pt. 1): 663–675.

Camacho, E. F. and C. Bordons (2004). *Model Predictive Control*. Springer, New York.

Claridge, D. E., W. D. Turner et al. (2004). Is commissioning once enough? *Energy Engineering* **101**(4): 7–19.

Clarke, D. (1987). Generalized predictive control—Part I. The basic algorithm. *Automatica* **23**(2): 137–148.

Clarke, D. (1994). *Advances in Model-Based Predictive Control*. Oxford University Press, New York.

Dexter, A. and P. Haves (1989). A robust self-tuning predictive controller for HVAC applications. *ASHRAE Transactions* **95**(2): 431–437.

Elliott, M., B. Bolding et al. (2009). Superheat control: A hybrid approach. *HVAC&R Research* **15**(6): 1021–1044.

Elliott, M. S. and B. P. Rasmussen (2010). On reducing evaporator superheat nonlinearity with control architecture. *International Journal of Refrigeration* **33**(3): 607–614.

Elliott, M. S. and B. P. Rasmussen (2012). Neighbor-communication model predictive control and HVAC systems. *American Control Conference (ACC)*, Montreal, Quebec, Canada, June 27–29, 2012.

Elliott, M. S. and B. P. Rasmussen (2013). Decentralized model predictive control of a multi-evaporator air conditioning system. *Control Engineering Practice* **21**(12): 1665–1677.

Finn, D. P. and C. J. Doyle (2000). Control and optimization issues associated with algorithm-controlled refrigerant throttling devices. *ASHRAE Transactions* **106**(1): 524–533.

Franklin, G. F., J. D. Powell et al. (2006). *Feedback Control of Dynamic Systems*. Pearson Prentice Hall, Upper Saddle River, NJ.

Freire, R., G. Oliveira et al. (2008). Predictive controllers for thermal comfort optimization and energy savings. *Energy and Buildings* **40**: 1353–1365.

Garcia, C. E., D. M. Prett et al. (1989). Model predictive control: Theory and practice—A survey. *Automatica* **25**(3): 335–348.

Grimm, N. R. and R. C. Rosaler (1990). *Handbook of HVAC Design*. McGraw-Hill, New York.

Guardabassi, G. O. and S. M. Savaresi (2001). Approximate linearization via feedback—An overview. *Automatica* **37**: 1–15.

Haines, R. W. (1987). *Control Systems for Heating, Ventilating and Air Conditioning*, 4th edn. Van Nostrand Reinhold, New York.

He, M., W. Cai et al. (2005). Multiple fuzzy model-based temperature predictive control for HVAC systems. *Information Sciences* **169**: 155–174.

Honeywell, Inc. (1988). *Engineering Manual of Automatic Control*. Honeywell, Inc., Minneapolis, MN.

Huang, G., S. Wang et al. (2010). Robust model predictive control of VAV air-handling units concerning uncertainties and constraints. *HVAC&R Research* **16**(1): 15–33.

Huang, P. H. (1991). Humidity measurements and calibration standards. *ASHRAE Transactions* **97**(Pt. 2): 298–304.

Khalil, H. K. (1996). *Nonlinear Systems*. Prentice-Hall Inc., Upper Saddle River, NJ.

Kothare, M., V. Balakrishnan et al. (1996). Robust constrained model predictive control using linear matrix inequalities. *Automatica* **32**(10): 1361–1379.

Leducq, D., J. Guilpart et al. (2006). Non-linear predictive control of a vapour compression cycle. *International Journal of Refrigeration* **29**: 761–772.

Letherman, K. M. (1981). *Automatic Controls for Heating and Air Conditioning*. Pergamon Press, New York.

Levine, W. S. (1996). *The Control Handbook*. CRC Press Inc., Boca Raton, FL.

Liu, M., D. E. Claridge et al. (2002). *Continuous Commissioning SM Guidebook: Maximizing Building Energy Efficiency and Comfort*. Federal Energy Management Program, U.S. Department of Energy, 144pp.

MacArthur, J. and M. Woessner (1993). Receding horizon control: A model-based policy for HVAC applications. *ASHRAE Transactions* **99**(1): 139–148.

May-Ostendorp, P., G. Henze et al. (2011). Model-predictive control of mixed-mode buildings with rule extraction. *Building and Environment* **46**(2): 428–437.

Mayne, D., J. Rawlings et al. (2000). Constrained model predictive control: Stability and optimality. *Automatica* **36**(6): 789–814.

Mills, E., N. Bourassa et al. (2005). The cost-effectiveness of commissioning new and existing commercial buildings: Lessons from 224 buildings. *Proceedings of the 13th National Conference on Building Commissioning*, New York, May 4–6, CD.

Mills, E., H. Friedman et al. (2004). The cost-effectiveness of commercial-buildings commissioning: A meta-analysis of energy and non-energy impacts in existing buildings and new construction in the United States. Lawrence Berkeley National Laboratory Report No. 56637. Available at http://eetd.lbl.gov/emills/PUBS/Cx-Costs-Benefits.html (accessed April 6, 2015).

Outtagarts, A., P. Haberschill et al. (1997). Transient response of an evaporator fed through an electronic expansion valve. *International Journal of Energy Research* **21**(9): 793–807.

Privara, S., J. Siroky et al. (2011). Model predictive control of a building heating system: The first experience. *Energy and Buildings* **43**: 564–572.

Qin, S. and T. Badgwell (2003). A survey of model predictive control technology. *Control Engineering Practice* **11**(7): 733–764.

Rawlings, J. and D. Mayne (2009). *Model Predictive Control: Theory and Design*. Nob Hill, Madison, WI.

Rossiter, J. A. (2003). *Model-Based Predictive Control: A Practical Approach*. CRC Press, Boca Raton, FL.

Sauer, H. J., R. H. Howell et al. (2001). *Principles of Heating, Ventilating and Air Conditioning*. American Society of Heating, Refrigerating and Air-Conditioning Engineers, Inc., Atlanta, GA.

Scokaert, P. and J. Rawlings (1999). Feasibility issues in linear model predictive control. *AIChE Journal* **45**(8): 1649–1659.

Singhal, A. and T. Salsbury (2007). Characterization and cancellation of static nonlinearity in HVAC systems. *ASHRAE Transactions* **113**(1): 391–399.

Skogestad, S. and I. Postlethwaite (1996). *Multivariable Feedback Control: Analysis and Design*. John Wiley & Sons, New York.

Sousa, J., R. Babuska et al. (1997). Fuzzy predictive control applied to an air-conditioning system. *Control Engineering Practice* **5**(10): 1395–1406.

Stein, B. and J. S. Reynolds (2000). *Mechanical and Electrical Equipment for Buildings*. John Wiley & Sons, Inc., New York.

Tao, W. K. Y. and R. R. Janis (2005). *Mechanical and Electrical Systems in Buildings*. Pearson Prentice Hall, Upper Saddle River, NJ.

Wang, S. and Z. Ma (2008). Supervisory and optimal control of building HVAC systems: A review. *HVAC&R Research* **14**(1): 3–32.

Xi, X.-C., A.-N. Poo et al. (2007). Support vector regression model predictive control on a HVAC plant. *Control Engineering Practice* **15**: 897–908.

Xu, M., S. Li et al. (2005). Practical receding-horizon optimization control of the air handling unit in HVAC systems. *Industrial & Chemical Engineering Research* **44**(8): 2848–2855.

Xu, X., S. Wang et al. (2010). Robust MPC for temperature control of air-conditioning systems concerning on constraints and multitype uncertainties. *Building Services Engineering Research & Technology* **31**(1): 39–55.

Yuan, S. and R. Perez (2006). Model predictive control of supply air temperature and outside air intake rate of a VAV air-handling unit. *ASHRAE Transactions* **112**(2): 145–161.

Zhu, Y., M. Liu et al. (2000a). Integrated commissioning for a large medical facility. *Proceedings of the 12th Symposium on Improving Building Systems in Hot and Humid Climates*, San Antonio, TX, May 15–16, 2000.

Zhu, Y., M. Liu et al. (2000b). A simple and quick chilled water loop balancing for variable flow systems. *Proceedings of the 12th Symposium on Improving Building Systems in Hot and Humid Climates*, San Antonio, TX, May 15–16, 2000.

Zhu, Y., M. Liu et al. (2000c). Optimization control strategies for HVAC terminal boxes. *Proceedings of the 12th Symposium on Improving Building Systems in Hot and Humid Climates*, San Antonio, TX, May 15–16, 2000.

7

Energy-Efficient Lighting Technologies and Their Applications in the Residential and Commercial Sectors

Karina Garbesi, Brian F. Gerke, Andrea L. Alstone,
Barbara Atkinson, Alex J. Valenti, and Vagelis Vossos

CONTENTS

7.1 Introduction

Electricity for lighting accounts for approximately 15% of global power consumption and 5% of worldwide greenhouse gas (GHG) emissions [1]. According to the United Nations Environment Programme's en.lighten initiative,* replacing all inefficient grid-connected lighting globally would reduce global electricity consumption by approximately 5% and CO_2 emissions by 490 million tons per year [2]. Grid-connected *space* lighting accounts for 99% of total lighting energy use, with vehicle lighting (0.9%) and off-grid fuel-based lighting (0.1%) constituting the remainder [3]. Together, residential and commercial lighting constitute about three quarters of global grid-connected lighting (Figure 7.1).

Per-capita consumption of artificial light averages about 20 megalumen-hours per year globally but varies widely. North Americans consume about 100 megalumen-hours per year on average. By comparison, Chinese consume about 10, and Indians consume only 3, megalumen-hours per year [3]. Given global demographic trends, lighting efficiency improvements not only offer significant energy savings in developed countries, but more importantly, the opportunity to contain large growth in lighting energy use in less developed countries.

Figure 7.2 shows the historic efficacies of different light source technologies (solid lines). Incandescent lamps, including halogen incandescent lamps, are the lowest efficiency light source type. (Lighting efficiency, referred to as efficacy, is quantified in units of lm/W.)

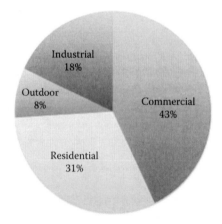

FIGURE 7.1

Global grid-connected lighting consumption by sector in 2005. (From IEA, *Light's Labour's Lost*, International Energy Agency, Paris, France, 2006, http://www.iea.org/publications/freepublications/publication/light2006.pdf.)

---

* The en.lighten initiative was created in 2009 as a partnership between the United Nations Environment Programme, OSRAM AG, Philips Lighting, and the National Lighting Test Centre of China, with the support of the Global Environment Facility.

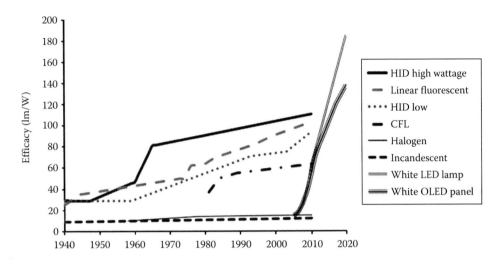

FIGURE 7.2
Historic best-on-market lamp efficacy trends by light source and lamp type and projected trends for solid-state lighting. (From DOE, Solid-state lighting research and development: Multi-year Program Plan, U.S. Department of Energy, Energy Efficiency & Renewable Energy Building Technologies Program, Washington, DC, April 2012.)

Compact fluorescent lamps (CFLs) are the next highest with efficacies approximately four times those of incandescent lamps. Linear fluorescents have trended with an efficacy intermediate that of low and high wattage high-intensity discharge (HID) lamps. Light-emitting diodes (LEDs) are now starting to exceed the efficacy of fluorescent lamps, and expected to reach about 200 lm/W by 2020—approximately twice that of today's most efficacious conventional lighting technologies.

Up to the present, different lighting technologies have dominated different sectors. For example, in 2010 in the United States, an estimated 86% of outdoor lighting was provided by HID lamps, 72% of commercial lighting was provided by linear fluorescent lamps, and 79% of residential lighting was provided by incandescent lamps [5]. Given the large differences in efficiency of the different light sources (Figure 7.2), this has resulted in large sectoral differences in efficiency. Solid-state lighting is poised to change that equation with LED rapidly entering all sectors of the lighting market.

LEDs are expected to replace essentially all other lighting technologies in the coming decades. Navigant Research anticipates LED's global market share of replacement lamps to grow from 5% in 2013 to 63% by 2021 [6]. The U.S. Department of Energy expects LEDs to represent 70% of the U.S. residential and commercial lighting markets by 2030 (Figure 7.3). In combination with an expected doubling in LED lamp efficacy anticipated by 2020 (also shown in Figure 7.3), the total energy savings are expected to be very large. This is particularly true in the residential sector, where the dominant incumbent technology (incandescent lamps) has an efficacy of only about 15 lm/W. In addition to the trend toward LED lighting, increased use of lighting controls is expected to further reduce energy use.

Because LED lights are significantly longer lived than conventional residential and commercial lamps, global lamp sales are expected to peak and decline as a result of the LED takeover of global lighting markets. Table 7.1 compares the typical lifetimes of various lamp technologies currently on the market. Figure 7.4 shows projected revenues from technology-specific lamps sales through 2021. This implies that, not only will consumers

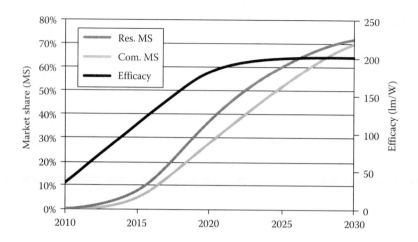

FIGURE 7.3

The projected market share of LED lamps in the U.S. residential and commercial markets and projected LED lamp efficacy. (From Navigant Consulting, Inc., Energy savings potential of solid-state lighting in general illumination applications, U.S. Department of Energy, Solid State Lighting Program, Navigant Consulting, Inc., January 2012, http://apps1.eere.energy.gov/buildings/publications/pdfs/ssl/ssl_energy-savings-report_10-30.pdf.)

TABLE 7.1

Typical Lifetimes and Approximate Best-on-Market Efficacies in 2012, by Lamp Type

Lamp Type	Lifetime (h)	Efficacy (lm/W)
Incandescent	1,000	15
Halogen incandescent	5,000	22
Compact fluorescent	12,000	65
Linear fluorescent	25,000	120
HID (high wattage)	30,000	120
LED (commercial)	50,000	130
LED (residential)	25,000	100

Sources: DOE, Solid-state lighting research and development: Multi-year program plan, U.S. Department of Energy, Energy Efficiency & Renewable Energy Building Technologies Program, Washington, DC, April 2012; The U.S. Department of Energy's LED Lighting Facts Database, http://www.lightingfacts.com/Products.

save money as a result of efficiency-induced electricity savings resulting from increasing use of LEDs, they will also save as a result of reduced lamps purchases.

This paper examines these lighting efficiency trends, focusing on opportunities in the residential and commercial sectors, which includes indoor and outdoor lighting associated with residential and commercial buildings. Because there is some sectoral overlap in lighting technologies, much of the technology discussion also applies to stationary source applications for industrial and roadway lighting. Section 7.2 introduces lighting concepts used throughout the remainder of the chapter. Section 7.3 discusses the different major lamp technologies by light source type, focusing on factors that affect their energy efficiency and related market trends. Section 7.4 discusses the cost-effectiveness of the different light source technologies from both the societal and consumer perspectives. Section 7.5 describes policies to stimulate energy lighting efficiency improvements. Section 7.6 presents conclusions.

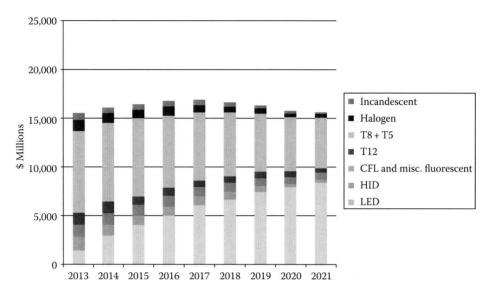

FIGURE 7.4
Projected revenues in world lamps market by technology type. The bar segments in each bar are ordered from top to bottom at shown in the legend. (From Navigant Research, Inc., Energy efficient lighting for commercial markets—LED lighting adoption and global outlook for LED, fluorescent, halogen and HID lamps and luminaires in commercial buildings: Market analysis and forecasts (Executive Summary), Navigant Consulting, 2013, http://www.navigantresearch.com/research/energy-efficient-lighting-for-commercial-markets. With permission.)

7.2 Lighting Concepts

7.2.1 Principles of Light Production

Visible light is made up of electromagnetic waves whose wavelengths fall within a narrow band of the electromagnetic spectrum, between 390 and 700 nm. The perceived color of visible light corresponds to its wavelength, with violet and blue light having shorter wavelengths, while orange and red light have longer wavelengths. Light with wavelengths longer than visible light, from 700 nm up to 1 mm, is called *infrared radiation* and is generally perceived as heat. Light with wavelengths shorter than visible light, from 390 nm down to 10 nm is called *ultraviolet radiation*. On an atomic level, light emission occurs in discrete units called *photons*, which are wave packets that carry energy that is inversely proportional to their wavelength.

Common electric light sources use one of two basic physical mechanisms to produce visible light: thermal and atomic emission. The mechanism that has been used longest is *thermal emission*, which is electromagnetic radiation emitted by all objects with temperatures above absolute zero. The spectrum of thermal emission for most objects is very well approximated by the spectrum from a *blackbody*, which is a theoretical object that perfectly absorbs all incident light and emits only thermal radiation. *Blackbody radiation* has a broad spectrum, with an intensity that scales as the fourth power of the object's temperature and a peak wavelength that varies inversely as the temperature. Objects at room temperature emit most of their thermal radiation in the infrared part of the spectrum, and the visible light emitted is too faint to be readily detected. As temperature is increased, the total power radiated increases rapidly, and the peak of the spectrum shifts toward shorter wavelengths. An object with a temperature of 5000 K will glow brightly, with the peak of its thermal spectrum at visible wavelengths, but since the thermal emission spectrum is very broad, objects with temperatures as low as 750 K emit detectable visible light even though the bulk of their emission occurs in the infrared. The visible light emitted via thermal emission is called *incandescence*.

Atomic emission occurs when electrons in atoms are excited to a high-energy configuration via some mechanism and then transition back to a lower-energy configuration via the emission of a photon. Because atomic electrons can only occupy a discrete set of energy levels, the light that is emitted in a particular atomic transition is monochromatic, with a well-defined wavelength that depends on the particular element and the specific energy levels involved in the transition. By mixing various elements and exciting a variety of energy-level transitions, it is possible to create light at a range of wavelengths via atomic emission. The spectral emission peaks can also be broadened by the Doppler effect induced by random motion of the atoms, for example, in a gas. Nevertheless, atomic emission generally produces light with a spectrum made up of one or a number of relatively narrow wavelength peaks (called *emission lines*) in contrast to the relatively broad, smooth spectrum of thermal emission.

Figure 7.5 shows example spectra for blackbody radiation, atomic emission, and, for reference, daylight. Solid lines show the theoretical spectrum of a blackbody at various different temperatures (from bottom to top, 3750 K, 4500 K, 5250 K, and 6000 K). As the temperature rises, the source intensity increases, and the spectral peak shifts to shorter wavelengths. The filled light gray region shows the spectrum of sunlight, whose broad, smooth shape is similar to a blackbody spectrum. The filled dark gray region shows atomic emission from a mercury vapor (MV) lamp, characterized by narrow peaks (emission lines) at characteristic wavelengths. The curves plotted are not on the same scale, except for the relative intensities of the different blackbody temperatures. (Spectral data for sunlight and MV lamp taken from McNamara [8].)

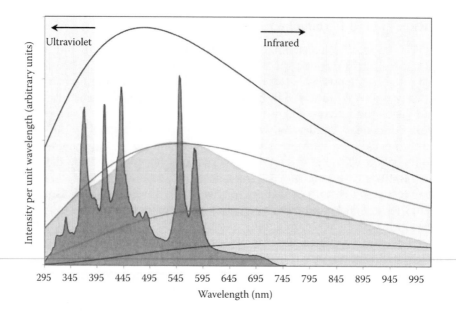

FIGURE 7.5
The spectra of different types of light sources at visible and neighboring infrared and ultraviolet wavelengths.

Atomic emission can take many forms, depending on the mechanism by which the electrons are excited. *Fluorescence* occurs when an atom absorbs light at a particular wavelength, which excites an electron, and then emits light at the same or a different wavelength as that electron relaxes to some lower energy level. If the electron relaxes to a higher energy than it had initially, the emitted photon will have a longer wavelength than the absorbed photon; this mechanism is commonly used to convert ultraviolet to visible light, for example, in fluorescent lamps. *Plasma discharge emission* occurs when an electric current passes through plasma (i.e., an ionized gas) in a phenomenon known as an *electric arc*. The free electrons collide with atoms in the gas and excite their electrons, which then transition to lower energy levels and emit photons. This is the primary light-generation method for fluorescent and HID lamps. *Diode emission* from a LED occurs when an electrical current is applied to a semiconductor made of two adjoining materials with different electrical properties. Excited electrons, which are free to move through the semiconducting material, are added to one material and removed from the other by the current. Those electrons that are conducted across the junction between the two materials then relax to the empty energy levels in the electron-depleted atoms, emitting photons in the process.

7.2.2 Design of Energy-Efficient Lighting Systems

Residential and commercial lighting systems typically are designed to provide both functional lighting and aesthetic value at the same time. The aim of energy-efficient lighting design is to provide the desired quality of light and other aesthetic properties of the lighting system with the least energy use. Design choices made for aesthetic reasons may in some cases reduce efficiency. For example, adding a diffuser to a luminaire to reduce glare will reduce luminaire efficacy. In other cases, these choices may improve efficiency. For examples, lighting controls used to avoid the delivery of undesirable excess light are aesthetically preferable and save energy.

The goal of energy-efficient lighting design is to provide only as much artificial light as is desirable at the locations where it is needed, when it is needed. The optimal system incorporates daylighting, provides proper illumination for the design application, and addresses changes in occupancy. In the future, systems may even modulate the spectral distribution throughout the day to improve human performance and well-being (see Section 7.2.3).

The key elements of efficient lighting design are (1) making maximum use of natural light, (2) using the most efficient artificial lighting technologies, (3) incorporating automated lighting controls, and (4) controlling high-illumination task lighting separately from lower-illumination ambient lighting. While energy-efficient lighting technologies and controls are addressed at length in the following, this section provides additional background on the benefits and pitfalls of daylighting. Proper daylighting design requires climate-sensitive building design to avoid potential adverse consequences. Specifically, excessive direct penetration of sunlight can result in discomfort from glare and heat loading and increase cooling energy demand, which could offset or outweigh lighting energy savings. On the other hand, because energy-efficient lights produce less heat, they can reduce cooling energy demand, which is an advantage in regions where peak electric loads are dominated by cooling. Achieving an optimal balance requires building energy modeling.

The potential to offset the need for artificial light with daylight is greatly affected by the overall architectural design of a building and the design of the interior space. Passive solar building design, which optimizes building orientation and window placement and minimizes extreme thermal excursions using appropriately located shading devices, thermal mass, and phase change materials, can facilitate daylighting while avoiding excessive heat loading. Skylights, reflective surfaces inside of and outside of the building, and other measures can be used to bounce natural light farther into a building. Light pipes can be used to provide natural light deeper to the interior. Light pipes may even incorporate a hybrid design that includes artificial lights when natural light is insufficient. Advanced window technologies with dynamic solar control can limit heat losses in winter and avoid excessive heat loading in summer while allowing natural light to enter buildings. The newest technologies incorporate zonal controls that can block glare to building occupants while allowing unaltered light to pass where it will not cause discomfort to occupants. Smart blinds are being developed that automatically adjust to address light level, glare, and thermal loading. Interior design and furnishings that place high light-need activities near windows and that allows the maximum penetration of light from windows into the building interior, can also reduce the need for artificial light during the day.

Another environmentally promising design option that has recently emerged in the market is solar-assisted lighting—products that have integrated photovoltaic (PV) power supply into lighting systems. NexTek Power Systems pioneered this approach with efficient commercial lighting that runs directly off of the direct current (DC) generated by the PV system when PV power is available, and off of rectified alternating current (AC) power when it is not. The approach avoids DC to AC to DC power conversion losses as well as allowing for flexible lighting design. The lights run off of a 24 V bus integrated into the metal grid that supports ceiling tiles in standard commercial drop ceiling (Armstrong Ceiling's DC Flexzone system). The design allows lights to be removed and installed without an electrician and to be placed anywhere on the grid, reducing the cost and greatly enhancing the flexibility of commercial lighting design. At the opposite end of the spectrum, small solar-assisted lighting systems that include a small PV module, a battery, and LED lights (e.g., Fenix International's ReadySet Solar Kit) are now in the market. The systems were designed for developing country markets. While particularly appropriate for off-grid areas, they may also become popular in the many regions with highly unreliable grid power.

7.2.3 Lighting, Human Health, and the Environment

Exposure to light, both natural and artificial, has both beneficial and adverse effects on human health that depend strongly on the wavelength of the light, the timing of exposure, and the organs exposed [9]. Exposure to the natural diurnal cycle of darkness and light promotes mental and physical health by maintaining the body's circadian rhythms [9]. Exposure to blue light in the morning triggers the production of an array of hormones that control stress and impulsive behavior, hunger, and metabolism, and promote reproductive health and alertness. Lack of circadian blue light in the morning can cause sadness, food cravings, decreased energy and libido, anxiety, and depression. The risks of cancer and infectious disease are also increased due to insufficient blue light exposure in the morning and the absence of darkness at night because of disruption of immune response pathways that destroy viruses and cancer.

In addition to being damaging to vision [10], exposure to blue light at night is particularly disruptive to well-being [11–13]. Exposure to darkness stimulates the body's production of hormones that promote sleep, reduce blood pressure, and cellular repair. Eliminating the exposure of geriatric and Alzheimer patients to blue light at night was demonstrated to improve sleep, reduce illness and blood pressure, reduce aggression, and improve lucidity and social participation in both populations [14].

While to date, lighting design has tended to ignore the health impacts of light color, avoidance of light pollution has become a major issue. The concept is that light should hit its intended target, and it should not shine elsewhere. The BUG System, which quantifies *backlight*, *uplight*, and *glare*, is a system that quantifies undesirable stray light from outdoor lighting [15]. The system was adopted into the Illuminating Engineering Society's Model Lighting Ordinance [16].

While high-end lighting designers have begun in recent years to incorporate spectral dynamics in their lighting systems to alter the mood of spaces throughout the day, off-the-shelf technologies are now becoming available, enabling residential customers to control the hue of lighting. For example, the new Philips Hue lighting system is designed to be controlled by the Apple iPhone. The technologies are likely to become more popular with increased awareness of the impact of lighting on human health and productivity.

7.3 Lighting Technologies

The following sections discuss the technologies and changing markets of the four major lamp types: incandescent lamps, including halogen (Section 7.3.1); fluorescent lamps (Section 7.3.2), plasma lamps (Section 7.3.3), and LEDs and organic light-emitting diodes (OLEDs) (Section 7.3.4). Following that, two additional technologies are described in the context of their impact on energy efficiency: lighting controls (Section 7.3.5) and luminaires (Section 7.3.6).

7.3.1 Incandescent

An *incandescent lamp* produces light by passing an electric current through a conductive filament, which is heated by electrical resistance to the point of incandescence. The filament is generally housed in a glass vessel (bulb) that has been evacuated and typically filled with an inert gas. The bulb typically has a metal base that can be attached to an electrical fixture, most often via a screw-in connection.

7.3.1.1 Technology

Incandescent lamps are the oldest electric lighting technology, having been invented independently by Thomas Edison in the United States and Joseph Swann in England in the late 1800s. In modern lamps, the filament is made of tungsten, although filaments of carbon, tantalum, and osmium were used in early incandescent lamps produced around the turn of the twentieth century. (Upon their introduction, tungsten filaments represented a substantial improvement in efficacy over these earlier filament materials.) Incandescent lamp filaments produce light whose spectrum is well approximated by a blackbody spectrum. Typical incandescent lamps have operating temperatures at which the overwhelming majority of the radiated power (typically 90% or more) falls in the infrared part of the spectrum, where it is sensed as radiant heat rather than visible light. This makes incandescent lamps significantly less efficacious than other lamp types. Nevertheless, incandescent lamps have remained very popular, especially in residential applications.

The continuing popularity of incandescent lamps owes partly to their familiarity but also to a wide range of desirable features. Incandescent lamps have warm color temperatures and excellent color rendering index. They are easily dimmed with no adverse effect on the lamp. Compared with other lighting technologies, incandescent lamps are inexpensive, small, lightweight, and can be used with inexpensive fixtures with no need for ballasts or other controllers. They work equally well on AC or DC power, and they have no need for high-quality power input. In a properly designed fixture, they permit excellent optical control. In addition, incandescent lamps are easy to install and maintain, produce no audible noise, create no electromagnetic interference, and contain few toxic chemicals, allowing their disposal in the general waste stream.

The two primary types of incandescent lamps are general-service and reflector lamps. General-service lamps (also known as A-lamps) are pear-shaped, common household lamps. Reflector lamps are typically conical in shape with a reflective coating applied to part of the bulb surface that has been specially contoured to control the light direction and distribution. Common types of reflector lamps include flood and spot lights, which are often used to illuminate outdoor areas, highlight indoor retail displays and artwork, and improve the optical efficiency of track lights or downlights.

Halogen lamps operate on the same principle as standard incandescent lamps, by heating a filament to the point of incandescence, but they are made more efficacious by the utilization of the tungsten-halogen cycle. During operation of a standard incandescent lamp, tungsten is evaporated from the filament and deposited on the interior surface of the glass envelope. This process blackens the bulb over time and causes the filament to grow thinner until it eventually fails. Operating the lamp at a higher temperature would accelerate this process. In a halogen lamp, the filament is surrounded by quartz capsule filled with a small amount of halogen gas, such as iodine or bromine. At moderate temperatures, the halogen gas binds to the tungsten that has evaporated from the filament, preventing deposition onto the quartz capsule. At the higher temperatures found in the vicinity of the filament, the tungsten-halogen bond is broken, and the tungsten is redeposited onto the filament. This tungsten-halogen cycle thus allows halogen lamps to be operated at higher temperatures without adversely affecting their lifetimes (indeed, many halogen lamps have longer lifetimes than standard incandescent lamps). The higher-temperature blackbody spectrum thus emitted by the filament has a larger overlap with the response curve of the human eye, so that more lumens are produced at a fixed power draw, resulting in a higher luminous efficacy.

Even more efficacious than the standard tungsten-halogen lamp is the *halogen infrared reflecting (HIR) lamp*. In such lamps, the halogen capsule, or the lamp reflector, is coated with an optically transparent but infrared-reflective coating, which reflects some portion of the infrared radiation back onto the filament. This increases the operating temperature of the filament without the need for additional watts of electrical power. Because the increased temperature provides more lumens at a fixed wattage, HIR lamps have a higher luminous efficacy than non-HIR halogens. HIR lamps have historically had small market share due to their high initial cost, even though they generally have lifetimes that are two to three times longer than standard halogens.

Because they are operating at higher temperature, halogen lamps produce bright white light with color temperatures slightly higher than those of standard incandescent lamps, with similarly high CRI values. In addition, they tend to have longer rated lifetimes, can be much more compact, are slightly more efficacious, and have better lumen maintenance than standard incandescent lamps. Halogen technology has historically been used heavily in reflector lamps; minimum efficiency standards have led them to capture an increasing share of this market in the United States and elsewhere. General-service halogen lamps are also available in the market. Historically, these have seen limited use, but recent improvements in their efficacy coupled with new regulatory standards are expected to increase their adoption in the near term.

7.3.1.2 Changing Market

General-service incandescent lamps have long been the workhorse of residential lighting applications, so much so that they are commonly referred to as household light bulbs. Incandescent and halogen reflector lamps have also been widely used in the residential sector, especially for outdoor lighting and directional indoor lighting. They have also seen significant use in the commercial sector in applications that require directional lighting, such as retail display.

There are now a variety of significantly more efficacious substitutes for incandescent and halogen lamp technology. The past two decades have seen a substantial portion of the market for general-service incandescent lamps shift to CFL technology. However, CFLs' unfamiliar appearance, higher price, inferior CRI, and limited capacity for dimming have limited the further growth of their market share. In recent years, LED-based lamps are entering the market intended to be direct replacements for many incandescent lamps. The directional nature of LED sources lends itself to reflector-lamp applications, and such lamps have gained a small foothold in the commercial sector. Recent advances in omnidirectional LED lamps have produced products with similar efficacy and color rendering to CFLs; these can be used as replacements for general-service incandescent lamps. Their very high prices have significantly limited their adoption to date, but as discussed in Section 7.4, LED prices have been falling rapidly. As discussed in Section 7.3.4, the color rendering and efficacy of LED lamps have simultaneously been improving, and LED lamps are expected to capture much of the current market for incandescent lamps over the coming decades [7]. As mentioned earlier, recent efficiency regulations in many markets are phasing out traditional general-service incandescent lamps; so a faster shift toward halogen, CFL, and LED technologies is expected in the general-service market.

7.3.2 Fluorescent

A typical fluorescent lamp system consists of a lamp and an electrical regulating ballast. The lamp is comprised of a glass tube, two electrodes, and a lamp base. The glass tube contains low pressure MV and an inert gas, which is usually argon, and a phosphor coating on the inside of the tube, which determines the light spectrum of the fluorescent lamp and thus its CCT and CRI. The electrodes are made of tungsten wire and are coated with a mix of alkaline oxides that, when heated, emit the electrons that excite and ionize the mercury atoms inside the tube. The base holds the lamp in place and provides the electrical connection to the lamp.

7.3.2.1 Technology

7.3.2.1.1 Ballasts

A fluorescent ballast provides the necessary voltage to start and maintain the arc in the lamp tube, because the electrical resistance of a plasma goes down as its temperature increases and a ballast is required to regulate the electric current through the lamp. Some ballast types are also designed to provide a specified amount of energy in the form of heat to the lamp electrodes. This minimizes stress on the electrode coating materials and thus extends lamp lifetime. The current delivered by the ballast to the lamp determines its light output. For a particular lamp-ballast system, light output is characterized by the ballast factor, which is the lumen output of the lamp (or lamps) operated with the ballast, relative to the lumen output of the same lamp (or lamps) operated with a reference ballast. Ballasts can be integral to the lamp or external. For example, medium screw base CFLs have integral ballasts, whereas in most linear fluorescent lamps, the ballast is external to the lamp.

There are two main types of fluorescent ballasts: magnetic and electronic. Magnetic ballasts operate at the electric power system frequency, while electronic ballasts operate at higher frequency and have many advantages over magnetic ballasts. These include higher efficiency, the ability to drive multiple lamps in series or parallel, lower weight, reduced lamp flicker, and quieter operation. These characteristics, as well as regulatory measures, have led to the phase-out of most types of magnetic fluorescent ballasts.

Ballasts are usually categorized by their starting method. Typical starting methods for electronic ballasts are instant start, rapid start, and programmed start. Instant start ballasts use a high voltage to strike the arc of the lamp, without preheating of the electrodes, having a negative impact on lamp life. As a result, instant start ballasts are very energy efficient but are more appropriate for applications with less frequent switching cycles (on–off). Rapid start ballasts heat the electrodes and apply starting voltage at the same time. They use more energy compared to instant start ballasts (~2 W/lamp), but they allow more lamp switching cycles (on–off) before lamp failure. Programmed start ballasts delay the starting voltage until the electrodes are heated to an optimum temperature to minimize the impact of starting on lamp lifetime. They are most suitable for frequent starting applications, such as areas with occupancy sensors. Table 7.2 summarizes characteristics of electronic ballast starting types.

Fluorescent lamps can be dimmed with the use of dimmable ballasts. Dimmable ballasts operate by reducing the electric current through the lamp. While doing so, the ballast must maintain the electrode and starting voltages and regulate electrode heating to maintain rated lamp lifetime. The power required to maintain these functions under dimming conditions can lead to higher ballast energy use at dimmed levels compared to full output levels. However, dimming still typically reduces the overall energy consumption of the lamp-ballast system because of the reduced power consumption of the lamp.

Because fluorescent lamps cannot operate without ballasts, their energy use must account for energy losses in the ballast. Ballast energy use is characterized by two metrics: ballast efficacy factor (BEF) and ballast luminous efficiency (BLE). The BEF, also sometimes referred to as the ballast efficiency factor, is the ratio of a lamp's ballast factor to the ballast input power. BEF is a lamp-ballast performance metric, which accounts for the efficiency of the lamp-ballast system compared to other systems with the same type and number of lamps.

TABLE 7.2

Electronic Ballast Starting Type Characteristics

Ballast Starting Type	Start Time (s)	Lamp Switch Cycles
Instant start	<0.1	10,000–15,000
Rapid start	0.5–1.0	15,000–20,000
Programmed start	1.0–1.5	50,000

Source: Philips, The ABCs of electronic fluorescent ballasts, A guide to fluorescent ballasts, Philips Advance, 2011, http://www.siongboon.com/projects/2010-08-22 electronic ballast/ABC of electronic fluorescent ballast.pdf.

BEF requires the measurement of several (lamp-related) parameters, which can lead to inaccuracies in its measurement. To reduce measurement variation and testing burden, BLE was introduced in 2012. BLE is immaterial to the lamp and only considers ballast characteristics. It is the total lamp arc power divided by the ballast input power and multiplied by a frequency adjustment factor, which depends on the ballast operating frequency.

7.3.2.1.2 Lamps

The two main categories of fluorescent lamps are linear fluorescent lamps and CFLs. Linear fluorescent lamps, or fluorescent tubes, are typically categorized based on their tube diameter. The most common types of these lamps are T12, T8, and T5 lamps (with tube diameters of 12/8, 8/8, and 5/8 of 1 in., respectively). They are usually available in lengths of 4 ft (1.2 m) lamps and 8 ft (2.4 m) lamps, although 2, 3, 5, and 6 ft lengths can also be found. CFLs have small diameter tubes (typically ¼ of 1 in.) that are bent into two to six sections or into a spiral shape. The tube(s) are sometimes covered with a diffuser that makes the assembly look more like a general-service incandescent lamp. They have a compact size and various base types, including common household-type screw bases. This allows them to substitute for incandescent lamps in many fixtures.

The physical characteristics (length, diameter, and shape) of fluorescent lamps can influence their efficacy. Several other lamp characteristics, such as the emissive quality of the electrode coatings and the ability of the inert gas fill to improve the mobility of the mercury atoms, can also impact fluorescent lamp efficacy. Another important factor that can significantly affect efficacy, as well as color quality and lumen maintenance, is the phosphor coating. High efficacy fluorescent lamps include *triband* phosphors, which contain oxides of certain rare earth elements—lanthanum, cerium, europium, terbium, and yttrium. Rare earth oxides account for a significant portion of the manufacturing cost of fluorescent lamps. This percentage is higher in high efficacy fluorescent lamps (e.g., 800-series T8 and T5 linear fluorescent), which use 100% triband phosphor coatings, whereas lower efficacy linear fluorescents (700-series) contain only about 30% triband phosphor. Between 2010 and 2011, the price of rare earth oxides increased sharply, causing concerns about the resulting price impacts on high efficacy fluorescent lamps, but has since been reduced considerably. The future trajectory of rare earth oxide prices is highly uncertain and will depend on global supply and demand of rare earth elements.

7.3.2.2 Changing Market

Linear fluorescent lighting is used predominantly in the commercial (and industrial) building sectors. In the residential sector, linear fluorescent lighting is primarily found in specific areas, such as kitchens, bathrooms, garages, and workshops. T12 lamp/ballast systems are less efficacious and more costly to operate compared to T8 and T5 lamp/ballast systems, and thus are currently being phased out of the market. CFLs have seen increased adoption in recent years for residential applications as replacements of incandescent and halogen technologies because of their decreasing first costs, higher efficacies, longer life times, and improved color quality.

Continuous performance improvements and reducing costs of LED-based technologies, in particular in LED troffer luminaires as replacements for linear fluorescents [18], and in reflector and omnidirectional LED lamps as replacements for CFLs, are causing a market shift in the fluorescent lamp market to LEDs. LEDs are projected to replace more than 60% of the linear fluorescent market and to reach 70% market penetration in residential applications by 2030 [7].

7.3.3 Plasma Lighting

Plasma lamps emit light via plasma discharge emission from electrically excited gases contained within an arc tube. A typical lamp design consists of a ballast, igniter, electrodes, an alumina glass arc tube filled with pressurized gases, and a lamp base. The operation of a plasma lamp is unique, as it involves a warm-up period of 2–10 min and, when power is interrupted, a restrike time of up to 10 min depending on the technology.

7.3.3.1 Technology

7.3.3.1.1 Ballasts

Plasma lighting requires a ballast to initiate the lamp and to regulate the current during operation, similar to fluorescent ballasts. In addition to a ballast, some plasma lamps require an igniter to initiate the necessary high voltage during start-up. As with fluorescent lighting, there are both magnetic and electronic ballast technologies and similarly, electronic ballasts offer improved lamp lifetime, lumen maintenance, ballast efficiency, higher lamp efficacy, and lower ballast losses, offering better light quality and energy savings. For more description on basic ballast function, see fluorescent Section 7.3.2.

Electronic ballast technology offers higher efficiencies for HID technologies over magnetic ballasts, nearly a 15% higher ballast efficiency for some lamp and ballast types [19]. Industry experts agree that there are limited efficacy gains remaining in HID ballast technology, that only a few percent gains are technologically feasible but may be practically infeasible to attain.

In lieu of further efficiency gains, the use of electronic ballasts allows for the implementation of lighting controls, which reduce hours of use and in turn energy consumption. Due to the nature of HID lamp technology, these control systems utilize a stepwise dimming function where discreet dimming levels are designated, instead of continuous systems, as in incandescent and fluorescent lamp systems. Some systems offering dimming capabilities near 40% of total light output, which in certain applications translates to nearly a 30% reduction in energy consumption.

7.3.3.1.2 Lamps

The most common plasma lamps are mercury vapor (MV), high-pressure sodium (HPS), and metal halide (MH); these are classified as HID. Low-pressure sodium (LPS) lamps are also plasma lamps and are often grouped together with HID, since they are high lumen output technologies used in similar applications (exterior and large interior commercial and industrial spaces). LPS lamp technology is not discussed in depth, as it is an older technology that constitutes a very small fraction of commercial and residential markets.*

MV lamps, first developed in 1901, use ionized mercury as the primary discharge element, producing a bluish light with better CRI than sodium lamps, but with relatively low efficacy. Phosphors can be used to improve the color and CRI of MV lamps' relatively blue light, but they yield relatively poor color rendering compared to other HID technologies. These lamps are now used primarily in legacy lighting designs or special applications, and are being replaced by MH lamps that have twice the efficacy and far better CRI.

HPS lamps and LPS lamps both utilize mixtures of ionized sodium and mercury gases, and this difference in pressure leads to unique light quality characteristics. LPS lamps

---

* For example, in the United States, LPS constitutes less than half of 1% of the total lumen output of HID lamps [5].

TABLE 7.3

Typical Characteristics of Plasma Lamps

Lighting Type	Typical Efficacies (lm/W)	Lifetime (h)	CRI	CCT (K)	Application
LPS	60–150	12,000–18,000	−44 (very poor)	1,800 (warm)	Exterior
HPS	50–140	16,000–24,000	25 (poor)	2,100 (warm)	Exterior
MV	25–60	16,000–24,000	50 (poor to fair)	3,200–7,000 (warm to cold)	Exterior
MH	70–115	5,000–20,000	70 (fair)	3,700 (cool)	Interior/exterior
Light-emitting plasma	60–130	25,000–100,000	70–95 (good)	2,000–10,000 (warm to cold)	Interior/exterior

Sources: luxim.com; Topanga.com; http://www.eere.energy.gov/basics/buildings/low_pressure_sodium.html; http://www.eere.energy.gov/basics/buildings/high_intensity_discharge.html.

offer the highest efficacy, but the monochromatic condition of the light spectra reduces the visual acuity in mesopic (relatively low-light) conditions, thus reducing the effectiveness of the light in outdoor applications. HPS produces a broader light spectrum than LPS by subjecting the gas to higher pressures. This broadening of spectral emission creates a better quality light than LPS but with reduced efficacy. The characteristics of available lamps are highlighted in Table 7.3.

MH lamps are similar to MV lamps, but MH gases are added to the mercury gas in the discharge tube, yielding higher lumen output and efficacy with higher CRI. Improvements in the starting technology of MH lamps have had considerable effect on lamp quality. Innovation from probe start, which involves a third electrode to initiate the lamp, to pulse-start, which utilizes an exterior igniter, allows for higher pressures in the arc tube, which leads to a shorter start-up and restrike time, longer lamp life, lumen maintenance, better cold weather operation, and up to a 20% increase in efficacy compared to traditional MH lamps. Ceramic metal halide (CMH) lamps, utilizing a ceramic glass arc tube and the pulse starting method, have some of the highest efficacies for MH lamps.

A new HID technology, known as light-emitting plasma, uses induction starting to excite the gas vapors instead of an electrical discharge. This electrodeless induction lamp utilizes fill gases that are similar to those in MH lamps but uses a solid-state radio frequency driver to excite the gases inside the lamp. Manufacturers report long rated lifetimes of up to 100,000 h, which far exceed those of other HID technologies. Another advantage of this technology is the relatively small size of the lamp capsule, which allows for greater directionality of the light from the luminaire, leading to significantly higher luminaire efficacies than traditional HID luminaires. The induction starting technique allows for a quicker start-up time (45 s) and restrike time (2 min) than other HID lamps, which allows for easier integration of occupancy controls. Light-emitting plasma (LEP) also has dimming functionality that could allow for significant energy savings compared to other HID technologies.

7.3.3.2 Changing Market

While HID lamps are primarily used for exterior lighting, they are also widely used for large indoor spaces in the commercial and industrial sectors and in rare cases in the residential sector. In the United States in 2010, 24% of the installed stock of HID lamps served the commercial sector, while only 1% served the residential sector [5]. Commercial lighting applications include exterior security lighting, warehouse high bay and low bay lighting, walkway lighting, and, more recently, retail lighting due to an influx of smaller sized CMH lamps with high CRI. Residential HID lighting consists mainly of exterior security lighting with rare specialty interior lighting utilizing CMH lamps.

Advancements in HID efficacy have been leveling off in recent years, and with the advancement of LED replacement technology (see Figure 7.2), HID market share is projected to decrease. LED replacement efficacies are projected to double from 100 lm/W to over 200 lm/W in the next 20 years (Figure 7.2), offering significant energy savings [4]. LED replacements also have the potential for spectral control, are amenable to occupancy-based controls, and are more conducive to dimming. But in applications where an intense point-source light is needed, LEDs may never replace plasma lamps. With the persistence of these high-output applications and the limitations of current LED technology, continued improvements are expected for MH lamp technology, both traditional and induction, whereas there is no expectation of the same for MV, LPS, and HPS lamps [20].

7.3.4 Solid-State Lighting (LED and OLED)

LEDs and OLEDs turn electricity into narrow-band light via diode emission and can produce varying wavelengths ranging from infrared through the visible spectrum to ultraviolet. LEDs consist of a semiconducting die, typically a form of silicon or germanium, a circuit board, and a lens or diffuser. In some cases, phosphors and a heat sink are also utilized depending on the application and desired light output. OLEDs consist of a wide variety of organic light-emitting compounds, electrodes (one usually being transparent), and typically a plastic or glass envelope that holds all of the components in place and acts as a lens or diffuser for the unit.

7.3.4.1 Technology

There are currently three general methods for creating the white light necessary for general-service lighting from LEDs: Phosphor conversion (PC) where phosphors are used in conjunction with blue or violet LEDs to produce white light via fluorescence, color mixing (CM) where LEDs with discrete colors (red, green, blue, and sometimes amber) are mixed in an array to create white light, and hybrid where PC and CM are combined to create white light.

LED lamps utilizing PC are the most common on the market today. Typically, phosphor (or a mixture of phosphors) is deposited directly onto a blue or near-ultraviolet LED die and encapsulated into the LED package. Locating phosphors away from the LED package, known as remote phosphor conversion, allows for better light dispersion and can create larger uniform light-emitting surfaces that better accommodate current form factors of existing luminaires. Phosphor-converted LEDs can have a range of color temperatures ranging from cool to warm white light. Color rendering of PC-LEDs ranges from moderate to excellent, with CRIs ranging from 70 to greater than 90, depending on the phosphors used. Higher CRIs are achieved at the cost of efficacy. Efficacies have been demonstrated as high as 140 lm/W for cool white emitters and 110 lm/W for warm white LEDs [4]. In 2013, lamps were available in the market with efficacies ranging from 75 to 92 lm/W.

Currently, cool-colored LEDs are about 20% more efficacious than warm-colored LEDs, for both CM and PC LEDs, but the gap is decreasing and it is expected to become negligible over the next decade [18]. White light achieved by CM has the potential for higher efficacies owing to fewer conversion losses compared to PC-LED. However, the green and sometimes amber colored LEDs used for CM currently have a lower efficiency than blue LEDs, which means less light output at a given input power compared to most other LED colors. Improvements made to the efficiency of green and amber LEDs could lead to large improvements in the efficacy of CM white LED lamps. Current mixed-color lamps on the market have efficacies around 90 lm/W. Color rendering of mixed-color lamps is good to excellent: mixed-color lamps currently exist in the market, with CRI greater than 90.

Another unique feature of CM lamps is the ability to tune the CCT of the output light. This ability can allow for color correcting as lamps age, since LEDs of different colors have different rates of lumen depreciation. It can also allow users to change the color of lamps via appropriately integrated controls.

The third approach, the hybrid method, uses PC with the addition of amber LED to fill in gaps in the PC-LED spectrum. This enables high-CRI lamps with greater efficacies. The light mixing can take place at the LED die level or in the luminaire, depending on the application.

The advantages of LEDs compared to incumbent lighting technologies include the following:

- High efficacies, resulting in lower operating costs; long lifetimes, with reported lifetime values of 30,000 h and projections of up to 50,000 h, which lowers annualized life-cycle costs
- Improved operation in low-temperature applications, since LEDs are more efficient at lower temperatures
- Dimmability, with compatible dimmers designed to work with LED drivers
- Ability to tune colors with CM technologies

Disadvantages of LEDs compared to incumbent technologies include the following:

- High first cost, although LED prices have been dropping very quickly
- Changes in color temperature over time
- Heat management issues: LED lifetimes and efficacies are greatly reduced when operating at high temperature, limiting their ability to be used in existing enclosed fixtures
- Incompatibility with incumbent dimmers: Dimmers that operate by varying the incoming line current to an LED lamp or fixture often do not work well with LED lamps
- Flicker problems with some LED lamps, especially when dimming
- High glare, due to the compact nature of LED packages, though this can be remedied by use of a diffuser

Despite the limitations of current LED technology, none of the current disadvantages appear to be technologically insurmountable. Industry is working, often with the support of governments, to address them.

OLEDs use a thin film of organic materials to emit diffuse light over a planar surface. They have been used mainly for displays in handheld electronic devices such as smart phones, but hold potential for general and niche lighting applications, in particular for diffuse lighting. The main difference between OLEDs and LEDs is form factor, the way light emission is distributed. OLEDs produce relatively diffuse light over a large area compared to the high-intensity compact source of light given off by an LED. OLEDs can be deposited on a number of different substrates including glass, plastic, and metal. This flexibility allows for the creation of luminaires with a wide variety of shapes and sizes. However, there are few OLED luminaires currently on the market and those that are available are not cost-competitive with other lighting technologies.

Currently, OLEDs lag behind LEDs in many performance metrics; in 2012, the best reported efficacies for OLEDs were approximately 60 lm/W. Color rendering by OLEDs is good to excellent, with CRIs up to 95. However, the lifetime of an OLED is only 10,000 h and they are very sensitive to moisture and heat. Care must be taken in the manufacturing process to properly seal OLEDs or their lifetime is shortened dramatically [4,21].

7.3.4.2 Changing Market

Single colored LEDs, developed in the 1960s, were the original technology used in nongeneral lighting applications such as traffic signals, exit signs, and outdoor displays. In 1993, innovations in blue LEDs made it possible to create products that could serve general purpose lighting applications that require white light as it became possible to create a white light by CM with the newly invented blue LED. As indicated in the introduction, LEDs are poised to become the dominant lighting technology for all general lighting applications as technology improves and prices continue to drop. Manufacturers' commitments to improving incumbent technologies are waning with most of their research and development efforts shifting toward LED lighting and lighting controls. It is currently unclear the role that OLEDs will play in that market.

7.3.5 Lighting Controls

Lighting controls are used to reduce energy consumption, to enhance security, and for aesthetic design. Lighting controls systems can include motion sensors, infrared sensors, photo sensors, timers, transmitters and receivers, and computer control systems. They can range in complexity from simple on/off switches that control a single luminaire to sophisticated integrated lighting control systems that control the lighting of entire buildings or campuses of buildings.

7.3.5.1 Technology

There are three main strategies for lighting control: manual controls, sensor-based controls, and scheduled controls. Manual controls are the most simple; they allow users to control light levels either with on/off switches or dimmers to desired levels. Sensor-based controls rely on sensors to initiate changes in light levels. Scheduled controls adjust light levels according to a predetermined schedule.

Although manual controls, in the form of simple on/off switches, have been used since the inception of electric lighting, they are sometimes incorporated in more advanced lighting strategies (e.g., providing the capacity for individuals in large commercial work places to control the light in their own work space). Multiple strategies may also be included in a single luminaire: for example, stair-well lighting with multi-light-level switching controlled by occupancy sensors.

Sensor technologies fall into two main categories: occupancy sensors and photosensors. Occupancy sensors use several different strategies to determine whether a space is occupied.

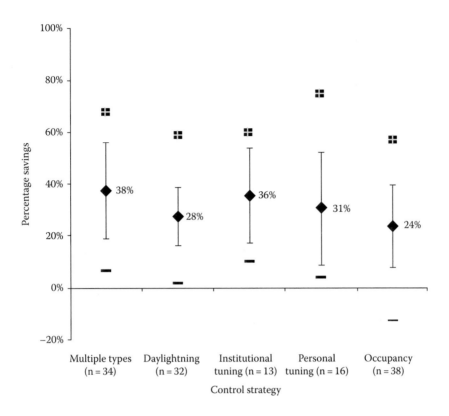

FIGURE 7.6
Energy savings for lighting controls and lighting energy in actual installations. The symbols indicate the maximum, average, and minimum values found in the study. The vertical bars indicated one standard deviation. (From Williams, A. et al., *LEUKOS* 8(3), 161, 2012.)

Passive infrared (PIR) sensors react to moving emitters of infrared radiation at wavelengths near the peak thermal emission given off by people. Ultrasonic and microwave sensors use the Doppler effect to detect movement, by emitting ultrasonic waves and microwaves, respectively, and measuring frequency shifts in the reflected signal to detect movement. Dual technology sensors use both PIR and ultrasonic technology to detect occupancy, which reduces the false triggering events that can occur with either technology used singly.

A recent metastudy of energy savings from lighting controls [22] found that providing the capacity for manual dimming can save as much energy as more advanced lighting strategies. As shown in Figure 7.6, based on field studies of actual energy savings, personal tuning (using dimmers) saved about the same amount of lighting energy (31%, on average) compared to automated controls, including occupancy sensors (saved 24%, on average) and daylight sensors (28%, on average). While applying multiple strategies simultaneously saved somewhat more energy (38%, on average), the biggest savings are obtained by the first control installed.

7.3.5.2 Changing Market

The market for lighting controls is expected to grow rapidly in the coming decades. Building codes have recently begun to incorporate requirements for lighting controls as a means to reduce lighting energy use beyond the more traditional lighting power density limits.

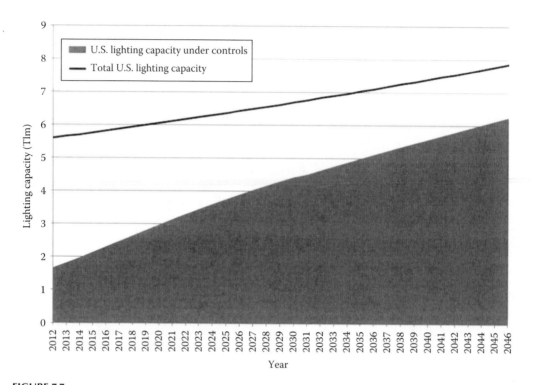

FIGURE 7.7

U.S. Department of Energy projections of the total annual commercial lighting capacity and the capacity projected to be under lighting controls. (From DOE, GSFL-IRL preliminary analysis technical support document, Appendix 9A, Lighting controls, U.S. Department of Energy, 2013, http://www.regulations.gov/#!documentDetail;D=EERE-2011-BT-STD-0006-0022, accessed February 28, 2013.)

For example, based primarily on building code requirements now in place, the United States Department of Energy estimates that lighting controls will cover approximately 75% of U.S. commercial floorspace by 2050, up from approximately 30% in 2012, see Figure 7.7 [23].

7.3.6 Luminaires

A number of factors can affect lighting efficiency at the luminaire level: the choice of light source type (as described earlier), the design of luminaire optical elements (reflectors, lenses, shades, and louvers), and the use of luminaire-based controls. In addition to energy efficiency, light quality must be considered in luminaire design (e.g., the need for glare control in certain applications). While there can be trade-offs between efficiency and light quality—for example, diffusers, which control glare, also reduce efficiency—advanced lighting technologies may eliminate such trade-offs (e.g., by using an inherently diffuse and efficiency light sources, like OLEDs). Similarly, the incorporation of increasingly advanced controls may allow the customization of light levels that reduce energy use.

Various metrics have been developed to quantify luminaire efficiency, and they have become increasingly sophisticated. The simplest metric is luminaire efficiency, the ratio of light leaving the fixture to the light emitted by its light source(s), which accounts for light absorbed in the luminaire. The luminaire efficacy rating (LER) system, developed by the National Electrical Manufacturers Association (NEMA), also accounts for light absorbed in the luminaire, but the metric is quantified as the light output from the luminaire (lumens) divided by the power input to the luminaire.* In 2008, NEMA introduced its target effi-cacy rating (TER) system, which was designed to supersede LER [24]. TER, also quantified in lumens per watt, credits only light that hits its intended target surface. TER covers 22 widely used types of commercial, residential, and industrial luminaires.

Because luminaire efficacy depends on lamp efficacy, luminaire efficacy must be mea-sured with lamps in place. If those lamps are replaced by consumers with lamps that have a different efficacy, the luminaire will no longer perform at the rated level. Luminaire design can protect against this possibility to a significant degree—a factor that is worth considering in the design of lighting efficiency policies and programs. Luminaires with integral LED lighting (also called inseparable solid-state lighting) do not have replaceable lamps, so their efficacies are known with a high degree of certainty over their lifetimes. The socket type used in the luminaire may also ensure that lamp replacement cannot result in large efficiency losses. For example, ANSI Standard ANSI_ANSLG C81.62-2009, GU-24 bases are not allowed to be used for incandescent lamp technologies; therefore, the efficacies of replacement lamps will be at least at the fluorescent or LED levels.

7.4 Cost-Effectiveness of Efficient Lighting Technology

7.4.1 Life-Cycle Cost

While lighting provides obvious benefits to consumers, it also imposes costs to society, only some of which are borne by the consumer. While the costs of production are typi-cally included in the price paid by the consumer, environmental and social costs are typically not included in the price. Therefore, lighting policies are increasingly incorporat-ing, to the degree possible, total life-cycle costs.

The environmental costs of lighting depend on lamp technology. The U.S. Department of Energy has produced a three-part study of the life-cycle environmental and resource costs in the manufacturing, transport, use, and disposal of LED lighting products in rela-tion to comparable traditional lighting technologies.† Figure 7.8, from part 3 of that series [25], compares the life-cycle impacts of different CFLs and LED lamps manufactured in 2012, and those expected to be in the market by 2017, relative to those of general-service incandescent lamps. The life-cycle environmental impacts of the current generation of LEDs are approximately comparable to the impacts of CFLs, but the impacts of the next generation of LED technology are expected to be significantly smaller.

7.4.2 Current Cost of Ownership of Different Lamp Technologies

In contrast to policymakers, typical lighting consumers generally consider only those costs associated with the ownership of lighting: the purchase price and possibly, if relevant informa-tion is available, the costs associated with operation, maintenance, and disposal. This section compares the current cost of ownership of different general-service lighting technologies.

High-efficacy lighting technologies, (e.g., fluorescent or LED) lamps generally have longer lifetimes (Table 7.1) and higher prices than lower-efficacy (incandescent) options.

---

* The LER system was developed by the National Electrical Manufacturers Association. The method for measuring LER was published as a series of three standards for fluorescent lighting (LE5), for commercial and residential downlights (LE5A), and for HID industrial luminaires (LE5B).
† These reports are available online at http://www1.eere.energy.gov/buildings/ssl/tech_reports.html.

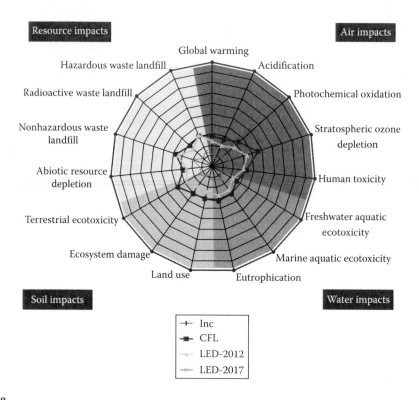

FIGURE 7.8

Life-cycle impacts of different general-service lamp technologies normalized to the impact of an incandescent lamp. (From Tuenge, J. et al., Life-cycle assessment of energy and environmental impacts of LED lighting products, Part 3: LED environmental testing, U.S. Department of Energy, Solid State Lighting Program, 2013, http://apps1.eere.energy.gov/buildings/publications/pdfs/ssl/lca_webcast_03-28-2013.pdf, accessed March 23, 2013.)

Determining the cost-effectiveness of a given technology, by balancing the higher initial price of a lamp against its energy cost savings and longer service life, can be a complex task for the consumer. Different technologies also often have different typical lumen outputs, further complicating the comparison. As a simplifying approach to determine the cost-effectiveness of a particular lighting technology, one can compare the total cost *per year* and *per unit light output* of owning and operating a particular lamp. Additional technical considerations, such as the need for a particular light distribution, CRI, or color temperature, must also be taken into account when choosing a lighting technology, but this approach can be used to compare the cost-effectiveness of a set of lighting options that have been prescreened to meet the technical needs of a particular application.

To determine the total cost of owning and operating a lamp, one first estimates the number of hours for which the lamp will be used in the course of a year. Since lamp lifetimes are typically measured in hours, this roughly determines the lamp's lifetime in years (although the calculation can be more complicated for technologies, like fluorescent lamps, whose lifetimes are shortened by short-term and intermittent use). The annual energy cost is then calculated by multiplying the lamp wattage by the annual hours of use and by the price per watt-hour of electricity. The lifetime operating cost is determined by summing the annual energy cost over the lifetime of the lamp, often multiplied by a discount factor that accounts for the reduced value of future savings compared to up-front costs (e.g., owing to financing costs). The total cost of ownership is then the sum of the price, the

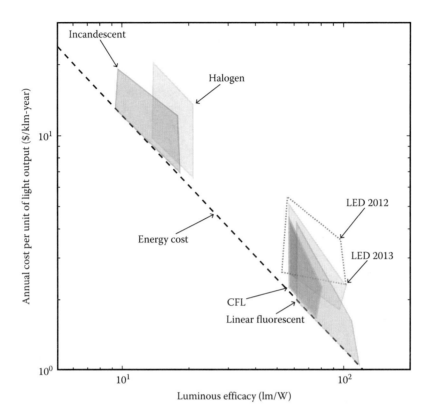

FIGURE 7.9
Relative cost-effectiveness of different general-service lighting technologies.

lifetime operating cost, and any end-of-life disposal costs. This value can then be divided by the lamp's lifetime in years and the lamp's total light output in kilolumens to produce a value that can be used to compare the cost of different lighting technologies on an equal basis. Cost per kilolumen-year is a particularly convenient set of units, because it typically yields values in the order of $1–$10 to be used for comparison (Figure 7.9).

For the purposes of the figure, we have assumed 1095 annual hours of use, corresponding to 3 h/day, which is an industry-standard value for computing lamp lifetimes and is fairly typical in residential applications. As shown in the figure, linear fluorescent lamps are the most cost-effective and efficacious general-service lighting technology in 2013, yielding a significant savings in both energy and cost over incandescent technologies and a small savings compared to CFL and LED lamps. LED lamps are evolving rapidly, however, and have seen extremely rapid improvements in both cost and efficacy in recent years, as illustrated by the dotted region in the figure. They are expected over the coming decade to surpass all competing technologies in both efficacy and cost. Zero discounting of future energy savings has been assumed in this figure, since the particular choice of discount rate is highly dependent on the individual consumer. Zero disposal cost is also assumed, which is typical in most applications. The impact of discounting would be to slightly reduce the cost-effectiveness of longer-lifetime technologies relative to shorter-lifetime technologies, so that, for example, LEDs would become slightly less cost-effective relative to CFLs. For typical discount rates, however, the overall ranking of the technologies by cost-effectiveness shown in the figure would not change.

7.5 Policy Approaches to Improve Lighting Energy Efficiency

Many governments have implemented policies to promote the adoption of energy-efficient lighting because of the large potential for energy, economic, and environmental savings. The primary policy instruments that have been applied to lighting are the following:

- Minimum energy performance standards (MEPSs)
- Labeling and certification
- Economic and fiscal incentives
- Bulk purchasing and procurement specifications
- Building codes

The primary policy instruments are briefly reviewed in the following text. For more information on how lighting policies have been applied in different countries, see en.lighten [26,27] and Collaborative Labeling & Appliance Standards Program [28].

7.5.1 Minimum Energy Performance Standards

Many nations and states have set MEPSs, also known as efficacy standards, for lighting. Conventionally, such standards have been set for relatively narrow classes of lamps that prevent different light source technologies from competing against each other (e.g., 4 ft fluorescent lamps, CFLs, or general-service incandescent lamps). When standards operate on broader categories of lighting equipment, like luminaires or general-service lamps, larger gains are possible.

Recently, a significant number of nations have adopted such standards on general-service lamps that have the effect of eliminating conventional incandescent lamps from the market. The International Energy Agency's 4E program examined the early impact on the lighting market of such these standards in nine countries and the European Union [29]. IEA concluded that the standards were proving successful in Australia, Korea, and the United Kingdom, with the average efficacy of lamp sales having risen by up to 50% in 3 years, with no significant effect observable at the time in the other jurisdictions, although it is important to note that the requirements had not yet gone into force in some jurisdictions. Minimum efficacy standards for luminaires could have a similar effect, necessitating widely used luminaires to use high efficacy lamps, incorporate lighting controls, or both.

Also relatively new are standards for LED lamps, which in some cases also include quality standards [26]. For example, the International Electrotechnical Commission (IEC)'s IEC/PAS 62612 standard contains performance requirements for *self-ballasted* LED lamps (LEDs lamps with integrated drivers) for general-service lighting. Quality standards can be important to avoid the LED market being spoiled by flooding with low-quality products that undermine consumer confidence.

7.5.2 Labeling and Certification

Labeling and certification programs educate the public about the efficiency of lighting products and provide assurance that products perform as advertised. Information programs include mandatory and voluntary product labeling and certification.

Lighting facts per bulb	
▬▬▬▬▬▬▬▬▬▬▬▬▬▬▬	
Brightness	820 lm
Estimated yearly energy cost $7.23 Based on 3 h/day, 11¢/kWh Cost depends on rates and use	
Life Based on 3 h/day	1.4 years
Light appearance Warm	Cool
2700 K	
Energy used	60 W

(a)   (b)

FIGURE 7.10
(a) EU energy label for a high-efficiency lamp. (b) Lighting Facts label developed by the U.S. FTC, required for use on general-service lamps sold in the United States.

Certification programs like U.S. Energy Star and Brazil's Selo Procel brand efficient products with an easily recognized symbol ensuring consumer that the product is efficient [30,31]. Energy-efficiency labels, on the other hand, provide information on the efficiency of the actual efficiency of the product or how that efficiency compares with other products.

There has been much discussion about appropriate labeling protocols for lighting, because residential customers in particular are unfamiliar with the lumen metric, and have generally used wattage as an indicator of brightness. With the increased using of high efficiency, low wattage lamps, consumers need clearly conveyed information on brightness and efficiency. Comparative labels, like the European Union's EU Energy Label (Figure 7.10a), clearly indicate the performance of the labeled product relative to other products in the market. The label assigns a letter grade, with high grades corresponding to high efficacy. The consumer does not need to understand efficacy units to understand the performance of the lamps relative to other products in the market. While the Federal Trade Commission's (FTC) Light Facts label (Figure 7.10b) provides information on light output (lumens) and energy consumption (watts), to determine the efficacy, the consumer would need to know to divide the two and then would not know how this product performs relative to other products in the market. The FTC label does provide easily understood information on color temperature, however.

7.5.3 Economic and Fiscal Incentives

Many jurisdictions offer fiscal incentives for the adoption of high-efficiency lighting. Rebates are the most common incentive for efficient lighting, which are often implemented as part of utility demand-side management programs. In recent years, rebates have begun to focus on lighting controls as well as on efficient lamps. Low interest loans for commercial and residential lighting systems upgrades are also widely available. Tax incentives are also widely offered at the state and federal levels for lighting system upgrades. The Database of State Incentives for Renewables and Efficiency

(DSIRE) compiles information online on the panoply of economic and fiscal incentives available to residential and commercial customers in U.S. states.

7.5.4 Bulk Purchasing and Procurement Specifications

Many governments (national, regional, and municipal) use bulk purchasing and procurement specifications to stimulate production and therefore reduce the cost of energy-efficient lighting, with the intention of inducing a shift toward more efficient products in the larger on-governmental market (e.g., in Vietnam and the United States, see Energy Sector Management Assistance Program [32] and Department of Energy [33]).

7.5.5 Building Codes

Energy codes are a subset of building codes, which govern the design and construction of residential and commercial structures. Energy codes generally include lighting provisions. Code provisions are generally applicable only to new construction and to renovations that affect some threshold fraction of a building's floor area. Therefore, incorporation of code provisions into the building stock depends on construction and renovation rates, which can result in relatively slow penetration rates, depending on economic conditions. On the other hand, building codes have the advantage of being able to address lighting efficiency at the application level, accommodating different lighting needs in different spaces.

Lighting provisions in building codes may be prescriptive or performance based. The prescriptive approach stipulates minimum efficacy requirements for lighting products used in different spaces. The more flexible, performance-based approach specifies the maximum allowable lighting power density in watts/square meter (square foot) for different spaces. The lighting designer can then choose any equipment that together meets the overall requirement for the space.

Lighting controls have been increasingly incorporated in building code requirements. This includes increasing the building space required to incorporate lighting controls, incorporation of increasingly sophisticated controls, and the layering of multiple types of controls. See, for example, updates to ASHRAE/IESNA 90.1 [34,35]. In some cases, lighting controls have been used as an alternative compliance path, allowing the use of less efficient lamps. For example, the State of California's Title 24 provisions require that all permanently installed outdoor luminaires using lamps rated over 100 W either have a lamp efficacy of at least 60 lm/W or that they be controlled by a motion sensor [36].

Because all of these policy instruments described earlier rely on accurate ratings of product efficiency, a sound underlying national infrastructure of product testing, monitoring, and enforcement is needed for lighting policies to be effective.

Governments may also support product development to spur efficiency innovations. The L-Prize is creative and highly successful example, in which the U.S. Department of Energy offered a $10 M cash prize for the development of a 90 lm/W solid-state lamp to replace the standard A19 60-W incandescent. This represented an enormous advancement in LED efficiency when the prize was announced in May 2008. The L-Prize resulted in a fierce competition, which Philips won in August 2011. But the benefits of the competition continued after the prize was won. Today there are eight products in the market that are certified by DOE's Lighting Facts program to meet or exceed the 90 lm/W requirements, with the best omnidirectional A-lamps achieving almost 100 lm/W.

7.6 Conclusions

With the advent of LED lighting, efficiency improvements can now be achieved without the degradation of light quality. Indeed, new LED lighting technologies can improve light quality by offering color-tuning capacity as well as all of the conventional standard and automated controls. Not only aesthetically pleasing, this feature could offer a potent tool to provide light that is health enhancing, by better mimicking the color dynamics of the diurnal cycle of natural light, essential for well-being.

While at this point falling prices and improved light quality alone may be enough to induce the shift to an LED lighting future, well-crafted lighting policies can continue to drive energy and environmental savings. Energy efficiency standards, coupled with quality standards, can ensure that low-efficiency and low quality LEDs are removed from the market. Bulk purchasing requirements, building codes, and luminaire standards can be used to drive the adoption of lighting controls and earlier adoption of high-efficiency lamps.

In conclusion, global lighting is poised for great increases in energy efficiency and reductions in total life-cycle costs, as a result of solid-state lighting products entering the market and policies aimed at removing low-efficiency incandescent lamps from the market. This is especially true in the residential market where new LED lamps are more than six times as efficient as the incumbent incandescent lamps. Increased use of lighting controls is also likely to significantly reduce energy use. Further opportunities to reduce environmental impacts are available by integrating PV power into lighting systems. Taken together, these opportunities suggest a bright future for lighting.

Glossary

Color rendering index (CRI) is a measure of the degree of color shift objects undergo when illuminated by the light source as compared with the color of those same objects when illuminated by a reference source of comparable color temperature. [RP-16-10] CRI is used to quantify how well a light source renders color across the color spectrum. For color temperatures below 5000 K (which applies to most indoor lighting), the reference source is a blackbody radiator, which is most nearly approximated by an incandescent lamp. For color temperatures above 5000 K, the reference source is a standard daylight condition of the same color temperature. Therefore, effectively, this metric quantifies how closely other light source types mimic an incandescent light source or a daylight source.

Correlated color temperature (CCT) is the absolute temperature of a blackbody radiator, expressed in degrees Kelvin, whose chromaticity (color appearance) most nearly resembles that of the light source [adapted from RP-16-10]. In lighting design, this metric is used to describe how warm or cool a light source appears. Light with a CCT at or above 4000 K, with a greater proportion of its flux at shorter (blue–violet) wavelengths, is considered cool. Light with a CCT at or below 3000 K, with more flux at longer (red–yellow) wavelengths, is considered warm.

Efficacy is the ratio of the lumens to the power (watts) drawn by a lighting system. The concept may be applied to the lamp alone (lamp efficacy), to the lamp-and-ballast system (system efficacy) for fluorescent or HID systems, to the lamp-and-driver system for LED lamps (system efficacy), or to the luminaire (luminaire efficacy). The distinction between these is which luminous flux is included and where the power input is measured.

Lumen (lm) is the SI unit for measuring luminous flux.

Lumen maintenance is the lumen output at a given time in the life of the lamp and expressed as a percentage of the initial lumen output of the light source.

Luminaire is a complete lighting unit consisting of the light fixture, in combination with its light source(s), any necessary ballasts or drivers, and all parts connecting the power source to the light source, including any integrated controls.

Luminous Flux is the rate of flow of electromagnetic radiation emitted by a light source in all directions, weighted by the spectral sensitivity of the human visual system [37].

References

1. en.lighten. 2013. Lighten initiative, United Nations Environment Program. Accessed June 5, 2013. Available online: http://www.enlighten-initiative.org/portal/Home/tabid/56373/Default.aspx.

2. en.lighten. 2013. en.lighten initiative. Country lighting assessments. Accessed April 2013. Available online: http://www.enlighten-initiative.org/portal/CountrySupport/CountryLightingAssessments/tabid/104272/Default.aspx.

3. IEA. 2006. *Light's Labour's Lost.* International Energy Agency, Paris, France. Accessed September 13, 2013. Available online: http://www.iea.org/publications/freepublications/publication/light2006.pdf.

4. DOE. April 2012. Solid-state lighting research and development: Multi-year Program Plan. U.S. Department of Energy, Energy Efficiency & Renewable Energy Building Technologies Program, Washington, DC.

5. DOE. January 2010. 2010 U.S. lighting market characterization. U.S. Department of Energy, Energy Efficiency & Renewable Energy Building Technologies Program, Washington, DC. Accessed September 13, 2013. Available online: http://apps1.eere.energy.gov/buildings/publications/pdfs/ssl/2010-lmc-final-jan-2012.pdf.

6. Navigant Research, Inc. 2013. Energy efficient lighting for commercial markets—LED lighting adoption and global outlook for LED, fluorescent, halogen and HID lamps and luminaires in commercial buildings: Market analysis and forecasts (Executive Summary). Navigant Consulting. Accessed September 13, 2013. Available online: http://www.navigantresearch.com/research/energy-efficient-lighting-for-commercial-markets.

7. Navigant Consulting, Inc. January 2012. Energy savings potential of solid-state lighting in general illumination applications. U.S. Department of Energy, Solid State Lighting Program. Navigant Consulting, Inc. Washington, DC. Accessed September 13, 2013. Available online: http://apps1.eere.energy.gov/buildings/publications/pdfs/ssl/ssl_energy-savings-report_10-30.pdf.

8. McNamara, G. 2012. PubSpectra—Open data access fluorescence spectra. Selected works. Accessed September 13, 2013. Available online: http://works.bepress.com/gmcnamara/9.

9. Roberts, J.E. December 2012. Light and dark and human health. *J R Astron Soc Can* 11–14 (Special report). Available online: http://www.rasc.ca/sites/default/files/LPA_Special_Issue_LR.pdf

10. Wielgus, A.R. and Roberts, J.E. 2012. Invited review: Retinal photodamage by endogenous and xenobiotic agents. *Photochem Photobiol* 88: 1320–1345.

11. Eastman, C.I. and Martin, S.K. 1999. How to use light and dark to produce circadian adaptation to night shift work. *Ann Med* 31: 87–98.

12. Roberts, J.E. 2000. Light and immunomodulation. *N Y Acad Sci* 917: 435–445.

13. Wehr, T.A. et al. 2001. Evidence for a biological dawn and dusk in the human circadian timing system. *J Physiol* 535(Pt 3): 937–951.

14. Mercier, K. October 2012. Maximizing health and sleep in the elderly. *LD+A*, Illuminating Engineering Society, pp. 42–47.

15. International Dark Sky Association. 2009. The BUG system—A new way to control stray light from outdoor luminaires. Specifier Bulletin for Dark Sky Applications, International Dark-Sky Association, 2(1), Tucson, AZ. Accessed September 21, 2014. Available online: http://www.aal.net/content/resources/files/BUG_rating.pdf.

16. IES. 2011. Model lighting ordinance (MLO) with user's guide. Illuminating Engineering Society. Accessed June 15, 2011. Available online: http://www.ies.org/PDF/MLO/MLO_FINAL_June2011.pdf.

17. Philips. 2011. The ABCs of electronic fluorescent ballasts. A guide to fluorescent ballasts. Philips Advance. Philips Lighting Electronics, N.A., Rosemont, IL. Accessed September 13, 2013. Available online: http://www.siongboon.com/projects/2010-08-22 electronic ballast/ABC of electronic fluorescent ballast.pdf.

18. DOE. 2013. Energy efficiency of LEDs, building technologies program solid-state lighting technology fact sheet. U.S. Department of Energy. National Academies Press, Washington, DC. Accessed September 13, 2013. Available online: http://www1.eere.energy.gov/buildings/ssl/factsheets.html.

19. DOE. January 2006. 2005 High intensity discharge lighting technology workshop report. U.S. Department of Energy, Energy Efficiency & Renewable Energy Building Technologies Program R&D, Washington, DC.

20. Scholand, M. 2012. Max Tech and beyond: High-intensity discharge lamps. Lawrence Berkeley National Laboratory, Berkeley, CA.

21. NAS. 2013. Assessment of advanced solid state lighting. National Academy of Sciences, Committee on Assessment of Solid State Lighting; Board on Energy and Environmental Systems; Division on Engineering and Physical Sciences; National Research Council. Accessed September 13, 2013. Available online: http://www.nap.edu/openbook.php?record_id=18279.

22. Williams, A. et al. 2012. A meta-analysis of energy savings from lighting controls in commercial buildings. *LEUKOS* 8(3): 161–180. Accessed September 13, 2013. Available online: http://www.ies.org/leukos/samples/1_Jan12.pdf.

23. DOE. 2013. GSFL-IRL preliminary analysis technical support document, Appendix 9A. Lighting controls, U.S. Department of Energy. Accessed February 28, 2013. Available online: http://www.regulations.gov/#!documentDetail;D=EERE-2011-BT-STD-0006-0022.

24. NEMA. 2008. Procedure for determining target efficacy ratings for commercial, industrial, and residential luminaires. National Electrical Manufacturers Association, NEMA Standards Publication LE 6, Rosslyn, VA.

25. Tuenge, J. et al. 2013. Life-cycle assessment of energy and environmental impacts of LED lighting products, Part 3: LED environmental testing. U.S. Department of Energy, Solid State Lighting Program. Accessed March 23, 2013. Available online: http://apps1.eere.energy.gov/buildings/publications/pdfs/ssl/lca_webcast_03-28-2013.pdf.

26. en.lighten. 2013. en.lighten initiative. Achieving the global transition to energy efficient lighting toolkit. Available online: http://www.thegef.org/gef/sites/thegef.org/files/publication/Complete EnlightenToolkit_1.pdf.

27. en.lighten. 2013. en.lighten initiative. Global policy map. Accessed September 13, 2013. Available online: http://www.enlighten-initiative.org/portal/CountrySupport/GlobalPolicyMap/tabid/104292/Default.aspx.

28. CLASP. 2013. Collaborative labeling and appliance standards project. Accessed September 13, 2013. Available online: http://www.clasponline.org/.

29. IEA-4E. July 2011. Impact of 'phase-out' regulations on lighting markets, Benchmarking report. International Energy Agency 4E, Efficient Electric End-Use Equipment. Accessed September 13, 2013. Available online: http://mappingandbenchmarking.iea-4e.org/shared_files/231/download.

30. EPA and DOE. 2013. Find ENERGY STAR products. Energy Star Program, U.S. Environmental Protection Agency and U.S. Department of Energy. Accessed September 13, 2013. Available online: http://www.energystar.gov/index.cfm?c=products.pr_find_es_products.
31. PROCEL. 2013. Selo PROCEL: Apresentação. Programa Nacional de Conservação de Energia Elétrica, Eletrobras. Accessed September 13, 2013. Available online: http://www.eletrobras.com/elb/procel/main.asp?TeamID0{95F19022-F8BB-4991-862A-1C116F13AB71}.
32. ESMAP. 2012. Case study Vietnam compact fluorescent lamp Program. Energy Sector Management Assistance Program. Accessed September 13, 2013. Available online: http://www.org//sites/.org/files/18.%20Vietnam_CFL_Case_Study.pdf.
33. DOE. 2013. Energy-efficient product procurement. U.S. Department of Energy, Federal Energy Management Program. Accessed September 13, 2013. Available online: http://www1.eere.energy.gov/femp/technologies/procuring_eeproducts.html.
34. Lighting Controls Association. 2013. ASHRAE Releases 90.1-2010—Part 2: Lighting controls. Accessed September 13, 2013. Available online: http://lightingcontrolsassociation.org/ashrae-releases-90-1-2010-part-2-lighting-controls/.
35. DiLaura, D. et al. 2011. *The Lighting Handbook: Reference and Application*, 10th edn. Illuminating Engineering Society of North America, New York.
36. CEC. December 2008. 2008 Building energy efficiency standards for residential and non-residential buildings. California Energy Commission. CEC-400-2008-001-CMF. Accessed September 13, 2013. Available online: http://www.energy.ca.gov/2008publications/CEC-400-2008-001/CEC-400-2008-001-CMF.PDF.
37. ALG. 2013. Advanced lighting guidelines. Sources & auxiliaries. Accessed April 2013. Available online: http://algonline.com/index.php?sources-auxiliares.

8

Energy-Efficient Technologies: Major Appliances and Space Conditioning Equipment

Eric Kleinert, James E. McMahon, Greg Rosenquist, James Lutz,
Alex Lekov, Peter Biermayer, and Stephen Meyers

CONTENTS

8.1 Introduction

The annual electricity consumption in the U.S. residential sector in 2012 was 3.76 billion kW h/day. By the end of 2013, the total was 3.77 billion kW h/day. In 2014, the U.S. residential sector used approximately 3.73 billion kW h/day. The average annual residential electricity usage per customer in 2012 was 10,834 kW h. By the end of 2013, it was approximately 10,786 kW h and in 2014, it was 10,612 kW h (EIA short-term outlook, October 2013). That is roughly an annual cost of $1200.00 in 2012, $1300.00 in 2013, and $1320.00 in 2014.

The U.S. Energy Information Administration (EIA) uses the British thermal unit (Btu) as a common energy unit to compare or add up energy consumption across the different energy sources that produce electricity. The United States uses several kinds of energy (petroleum, coal, natural gas, propane, nuclear, hydroelectric, wind, wood, biomass waste, geothermal, and solar). It is much easier to convert the various types of energy into Btus and come up with a total for electricity consumption for the various sectors and products.[*]

In April, 2013, the EIA came out with a scenario for the next 30 years of electrical consumption in the residential sector indicating that the United States will be consuming more electricity. This does not mean that our products are consuming more electricity; it means that there are going to be more homes, people,[†] and electrical products over the next 30 years (see Table 8.1).

Reducing the energy consumption of residential appliances and space conditioning equipment depends on replacing older equipment with the much more efficient models available now and on continuing to design even more energy-efficient appliances. National energy efficiency standards for appliances have driven efficiency improvements over the last 20 years, and appliances have become significantly more efficient as a result. Further improvement of appliance efficiency represents a significant untapped technological opportunity.

---

[*] Electricity 1 kilowatt hour (kW h) = 3412 Btu (but on average, it takes about three times the Btu of primary energy to generate the electricity [EIA]).

[†] U.S. population will kept increasing, see http://www.census.gov/popclock/.

TABLE 8.1

Residential Sector Key Indicators and Consumption (Quadrillion, Btu, Unless Otherwise Noted)

Report Annual Energy Outlook 2013
Scenario ref2013 Reference Case
Datekey d102312a
Release Date April 2013

ref2013.d102312a	2010	2011	2012	2013	2014	2015	2016	2017	2018	2019	2020	2021	2022	2023	2024	2025	2026	2027	2028	2029	2030	2031	2032	2033	2034	2035	2036	2037	2038	2039	2040
4. Residential Sector Key Indicators and Consumption (quadrillion Btu, unless otherwise noted)																															
Key Indicators and Consumption																															
Key indicators																															
Households (millions)																															
Single-family	82.85	83.56	84.53	85.57	86.08	86.86	87.73	88.63	89.52	90.39	91.25	92.07	92.87	93.68	94.52	95.37	96.22	97.03	97.81	98.57	99.34	100.09	100.82	101.54	102.28	103.03	103.80	104.55	105.30	106.03	106.77
Multifamily	25.78	26.07	26.43	26.82	27.19	27.57	27.98	28.44	28.90	29.36	29.82	30.27	30.72	31.16	31.60	32.05	32.52	33.01	33.52	34.03	34.54	35.05	35.55	36.06	36.55	37.05	37.55	38.04	38.54	39.04	39.53
Mobile homes	6.60	6.54	6.50	6.46	6.40	6.36	6.35	6.36	6.39	6.42	6.45	6.48	6.51	6.54	6.57	6.60	6.63	6.66	6.69	6.72	6.75	6.77	6.80	6.83	6.85	6.88	6.91	6.93	6.96	6.99	7.02
Total	115.23	116.17	117.46	118.86	119.67	120.78	122.05	123.42	124.80	126.18	127.52	128.83	130.10	131.38	132.69	134.02	135.37	136.71	138.02	139.32	140.63	141.91	143.17	144.42	145.69	146.96	148.25	149.53	150.80	152.06	153.32
Average house square footage	1653	1659	1665	1671	1676	1682	1687	1691	1696	1700	1704	1708	1712	1716	1720	1724	1728	1731	1734	1737	1740	1743	1746	1748	1751	1754	1757	1759	1762	1764	1767
Energy intensity (million Btu per household)																															
Delivered energy consumption	99.2	97.2	91.6	94.7	92.2	91.1	89.9	88.8	87.8	87.0	86.0	85.1	84.4	83.7	83.0	82.5	81.9	81.3	80.7	80.2	79.7	79.2	78.7	78.2	77.8	77.3	76.9	76.5	76.1	75.8	75.5
Total energy consumption	189.0	185.0	174.1	175.4	171.9	169.2	166.6	164.9	163.9	163.3	161.7	160.5	159.6	158.8	158.0	157.4	156.5	155.9	155.2	154.6	154.0	153.4	152.8	152.3	151.9	151.4	151.0	150.7	150.5	150.5	150.6
(thousand Btu per square foot)																															
Delivered energy consumption	60.0	58.6	55.0	56.6	55.0	54.1	53.3	52.5	51.8	51.2	50.4	49.8	49.3	48.8	48.3	47.8	47.4	47.0	46.6	46.2	45.8	45.4	45.1	44.7	44.4	44.1	43.7	43.5	43.2	43.0	42.7
Total energy consumption	114.3	111.5	104.5	104.9	102.6	100.6	98.8	97.5	96.6	96.1	94.9	93.9	93.2	92.5	91.9	91.3	90.6	90.1	89.5	89.0	88.5	88.0	87.6	87.1	86.7	86.3	86.0	85.7	85.4	85.3	85.2
Delivered energy consumption by fuel																															
Purchased electricity																															
Space heating	0.30	0.27	0.25	0.29	0.28	0.28	0.28	0.29	0.29	0.29	0.29	0.30	0.30	0.30	0.30	0.30	0.31	0.31	0.31	0.31	0.31	0.32	0.32	0.32	0.32	0.32	0.32	0.32	0.32	0.32	0.32
Space cooling	0.92	0.93	0.93	0.80	0.89	0.89	0.90	0.91	0.92	0.94	0.95	0.97	0.98	1.00	1.02	1.04	1.06	1.08	1.10	1.12	1.14	1.15	1.17	1.19	1.21	1.23	1.25	1.27	1.28	1.30	1.32
Water heating	0.45	0.45	0.46	0.47	0.48	0.48	0.49	0.49	0.49	0.50	0.50	0.51	0.51	0.52	0.52	0.52	0.53	0.53	0.53	0.53	0.53	0.53	0.53	0.53	0.53	0.54	0.54	0.54	0.54	0.55	0.55
Refrigeration	0.38	0.38	0.38	0.38	0.38	0.37	0.37	0.38	0.38	0.38	0.38	0.38	0.38	0.39	0.39	0.39	0.39	0.40	0.40	0.41	0.41	0.41	0.42	0.42	0.43	0.43	0.43	0.44	0.44	0.45	0.45

(Continued)

TABLE 8.1 (Continued)

Residential Sector Key Indicators and Consumption (Quadrillion, Btu, Unless Otherwise Noted)

ref2013.d102312a	2010	2011	2012	2013	2014	2015	2016	2017	2018	2019	2020	2021	2022	2023	2024	2025	2026	2027	2028	2029	2030	2031	2032	2033	2034	2035	2036	2037	2038	2039	2040
Cooking	0.11	0.11	0.11	0.11	0.11	0.12	0.12	0.12	0.12	0.12	0.12	0.13	0.13	0.13	0.13	0.13	0.13	0.14	0.14	0.14	0.14	0.14	0.15	0.15	0.15	0.15	0.15	0.15	0.16	0.16	0.16
Clothes dryers	0.20	0.20	0.20	0.20	0.21	0.21	0.21	0.21	0.21	0.21	0.22	0.22	0.22	0.22	0.22	0.23	0.23	0.23	0.23	0.24	0.24	0.24	0.24	0.25	0.25	0.25	0.25	0.25	0.26	0.26	0.26
Freezers	0.08	0.08	0.08	0.08	0.08	0.08	0.08	0.08	0.08	0.08	0.08	0.08	0.08	0.08	0.08	0.08	0.08	0.08	0.08	0.08	0.08	0.08	0.08	0.08	0.08	0.08	0.08	0.08	0.08	0.08	0.09
Lighting	0.65	0.63	0.63	0.55	0.52	0.50	0.50	0.49	0.49	0.49	0.45	0.43	0.42	0.41	0.40	0.40	0.39	0.39	0.39	0.38	0.38	0.38	0.37	0.37	0.37	0.37	0.37	0.37	0.38	0.38	0.38
Clothes washers[a]	0.03	0.03	0.03	0.03	0.03	0.03	0.03	0.03	0.03	0.03	0.03	0.03	0.03	0.03	0.03	0.02	0.02	0.02	0.02	0.02	0.02	0.02	0.02	0.02	0.02	0.02	0.02	0.02	0.03	0.03	0.03
Dishwashers[a]	0.10	0.10	0.10	0.10	0.10	0.10	0.10	0.10	0.10	0.10	0.10	0.10	0.10	0.10	0.10	0.12	0.11	0.11	0.11	0.11	0.11	0.11	0.11	0.12	0.12	0.12	0.12	0.12	0.12	0.13	0.13
Televisions and related equipment[b]	0.32	0.32	0.32	0.33	0.33	0.33	0.34	0.34	0.34	0.35	0.35	0.36	0.36	0.37	0.37	0.37	0.38	0.38	0.39	0.39	0.40	0.41	0.41	0.42	0.42	0.43	0.43	0.44	0.44	0.45	0.45
Computers and related equipment[c]	0.16	0.16	0.15	0.14	0.14	0.13	0.13	0.13	0.13	0.13	0.13	0.13	0.13	0.13	0.13	0.12	0.12	0.12	0.12	0.12	0.12	0.12	0.12	0.12	0.12	0.12	0.12	0.12	0.12	0.13	0.13
Furnace fans and boiler circulation pumps	0.13	0.13	0.13	0.14	0.14	0.14	0.14	0.14	0.14	0.14	0.14	0.14	0.14	0.14	0.14	0.14	0.14	0.14	0.14	0.14	0.14	0.14	0.14	0.14	0.14	0.14	0.14	0.14	0.14	0.14	0.14
Other uses[d]	1.11	1.07	0.95	0.99	0.97	0.98	1.00	1.01	1.03	1.06	1.08	1.11	1.13	1.16	1.18	1.21	1.23	1.26	1.28	1.30	1.33	1.36	1.38	1.41	1.43	1.46	1.48	1.52	1.55	1.59	1.62
Delivered energy	**4.93**	**4.86**	**4.72**	**4.64**	**4.65**	**4.66**	**4.68**	**4.72**	**4.77**	**4.83**	**4.84**	**4.88**	**4.92**	**4.97**	**5.02**	**5.08**	**5.13**	**5.19**	**5.25**	**5.30**	**5.36**	**5.42**	**5.48**	**5.54**	**5.61**	**5.67**	**5.73**	**5.80**	**5.88**	**5.95**	**6.03**
Natural gas																															
Space heating	3.32	3.25	2.89	3.32	3.16	3.15	3.13	3.10	3.06	3.04	3.02	3.00	2.98	2.95	2.94	2.92	2.91	2.89	2.88	2.87	2.85	2.84	2.82	2.81	2.79	2.77	2.75	2.73	2.70	2.68	2.67
Space cooling	0.00	0.00	0.00	0.00	0.00	0.00	0.00	0.00	0.00	0.00	0.00	0.00	0.00	0.00	0.00	0.00	0.00	0.00	0.00	0.00	0.00	0.00	0.00	0.00	0.00	0.00	0.00	0.00	0.00	0.00	0.00
Water heating	1.30	1.30	1.33	1.33	1.34	1.33	1.32	1.32	1.32	1.32	1.33	1.33	1.33	1.33	1.33	1.33	1.33	1.33	1.32	1.32	1.31	1.30	1.29	1.28	1.28	1.27	1.27	1.26	1.26	1.26	1.26
Cooking	0.22	0.22	0.22	0.22	0.22	0.22	0.22	0.22	0.22	0.22	0.22	0.22	0.22	0.22	0.22	0.22	0.22	0.22	0.23	0.23	0.23	0.23	0.23	0.23	0.23	0.23	0.23	0.23	0.23	0.23	0.24
Clothes dryers	0.06	0.06	0.06	0.06	0.06	0.06	0.06	0.06	0.06	0.06	0.06	0.06	0.06	0.06	0.06	0.06	0.07	0.07	0.07	0.07	0.07	0.07	0.07	0.07	0.07	0.07	0.07	0.07	0.07	0.07	0.07
Delivered energy	**4.89**	**4.83**	**4.49**	**4.93**	**4.78**	**4.76**	**4.73**	**4.70**	**4.66**	**4.64**	**4.62**	**4.61**	**4.59**	**4.57**	**4.55**	**4.54**	**4.52**	**4.51**	**4.49**	**4.47**	**4.46**	**4.44**	**4.41**	**4.39**	**4.37**	**4.34**	**4.31**	**4.29**	**4.27**	**4.25**	**4.23**
Distillate fuel oil																															
Space heating	0.49	0.50	0.49	0.51	0.51	0.50	0.49	0.48	0.47	0.46	0.45	0.44	0.43	0.42	0.41	0.40	0.39	0.38	0.37	0.36	0.36	0.35	0.34	0.33	0.33	0.32	0.31	0.31	0.30	0.29	0.29
Water heating	0.10	0.09	0.08	0.08	0.08	0.07	0.07	0.06	0.06	0.06	0.06	0.06	0.05	0.05	0.05	0.05	0.05	0.05	0.04	0.04	0.04	0.04	0.04	0.04	0.04	0.04	0.04	0.04	0.03	0.03	0.03
Delivered energy	**0.58**	**0.59**	**0.57**	**0.63**	**0.59**	**0.58**	**0.56**	**0.55**	**0.53**	**0.52**	**0.51**	**0.49**	**0.48**	**0.47**	**0.46**	**0.45**	**0.44**	**0.43**	**0.42**	**0.41**	**0.40**	**0.39**	**0.38**	**0.37**	**0.36**	**0.36**	**0.35**	**0.34**	**0.33**	**0.33**	**0.32**
Propane																															
Space heating	0.28	0.27	0.25	0.28	0.27	0.27	0.26	0.26	0.26	0.25	0.25	0.25	0.25	0.24	0.24	0.24	0.24	0.23	0.23	0.23	0.23	0.23	0.22	0.22	0.22	0.22	0.22	0.22	0.21	0.21	0.21
Water heating	0.07	0.07	0.07	0.06	0.06	0.06	0.06	0.05	0.05	0.05	0.05	0.05	0.05	0.05	0.05	0.05	0.05	0.05	0.05	0.05	0.05	0.04	0.04	0.04	0.04	0.04	0.04	0.04	0.04	0.04	0.04
Cooking	0.03	0.03	0.03	0.03	0.03	0.03	0.03	0.03	0.03	0.03	0.03	0.03	0.03	0.03	0.03	0.03	0.03	0.03	0.03	0.03	0.03	0.03	0.03	0.03	0.03	0.03	0.03	0.03	0.03	0.03	0.02
Other uses[e]	0.15	0.16	0.17	0.17	0.18	0.18	0.18	0.19	0.19	0.19	0.19	0.20	0.20	0.20	0.20	0.21	0.21	0.21	0.21	0.22	0.22	0.22	0.23	0.23	0.23	0.23	0.24	0.24	0.24	0.25	0.25
Delivered energy	**0.53**	**0.53**	**0.51**	**0.54**	**0.53**	**0.53**	**0.53**	**0.53**	**0.53**	**0.53**	**0.52**	**0.52**	**0.52**	**0.52**	**0.52**	**0.52**	**0.52**	**0.52**	**0.52**	**0.52**	**0.52**	**0.52**	**0.52**	**0.52**	**0.52**	**0.52**	**0.52**	**0.52**	**0.52**	**0.52**	**0.52**
Marketed renewables (wood)[f]	0.44	0.45	0.42	0.48	0.44	0.43	0.43	0.43	0.43	0.43	0.44	0.44	0.44	0.44	0.44	0.44	0.44	0.44	0.44	0.45	0.45	0.45	0.45	0.45	0.45	0.45	0.45	0.45	0.45	0.45	0.45
Other fuels[g]	0.04	0.02	0.02	0.03	0.02	0.02	0.02	0.02	0.02	0.02	0.02	0.02	0.02	0.02	0.02	0.02	0.02	0.02	0.02	0.02	0.02	0.02	0.02	0.02	0.02	0.02	0.02	0.02	0.02	0.02	0.02
Delivered energy consumption by end use																															
Space heating	4.86	4.76	4.32	4.94	4.69	4.66	4.62	4.58	4.54	4.50	4.47	4.44	4.41	4.38	4.35	4.32	4.30	4.28	4.26	4.24	4.22	4.19	4.17	4.15	4.12	4.09	4.07	4.04	4.01	3.98	3.96

(Continued)

TABLE 8.1 (Continued)

Residential Sector Key Indicators and Consumption (Quadrillion, Btu, Unless Otherwise Noted)

ref2013.d102312a	2010	2011	2012	2013	2014	2015	2016	2017	2018	2019	2020	2021	2022	2023	2024	2025	2026	2027	2028	2029	2030	2031	2032	2033	2034	2035	2036	2037	2038	2039	2040
Space cooling	0.92	0.93	0.93	0.80	0.89	0.89	0.90	0.91	0.92	0.94	0.95	0.97	0.98	1.00	1.02	1.04	1.06	1.08	1.10	1.12	1.14	1.15	1.17	1.19	1.21	1.23	1.25	1.27	1.28	1.30	1.32
Water heating	1.91	1.91	1.94	1.95	1.96	1.94	1.93	1.93	1.93	1.93	1.94	1.94	1.95	1.95	1.95	1.95	1.95	1.95	1.94	1.94	1.93	1.92	1.91	1.90	1.89	1.89	1.88	1.88	1.88	1.88	1.89
Refrigeration	0.38	0.38	0.38	0.38	0.38	0.37	0.37	0.38	0.38	0.38	0.38	0.38	0.38	0.39	0.39	0.39	0.39	0.40	0.40	0.41	0.41	0.41	0.42	0.42	0.43	0.43	0.43	0.44	0.44	0.45	0.45
Cooking	0.36	0.36	0.36	0.36	0.36	0.36	0.36	0.37	0.37	0.37	0.37	0.37	0.38	0.38	0.38	0.38	0.38	0.39	0.39	0.39	0.40	0.40	0.40	0.40	0.40	0.41	0.41	0.41	0.41	0.42	0.42
Clothes dryers	0.25	0.25	0.26	0.26	0.27	0.27	0.27	0.27	0.27	0.28	0.28	0.28	0.28	0.29	0.29	0.29	0.29	0.30	0.30	0.30	0.30	0.31	0.31	0.31	0.32	0.32	0.32	0.32	0.33	0.33	0.33
Freezers	0.08	0.08	0.08	0.08	0.08	0.08	0.08	0.08	0.08	0.08	0.08	0.08	0.08	0.08	0.08	0.08	0.08	0.08	0.08	0.08	0.08	0.08	0.08	0.08	0.08	0.08	0.08	0.08	0.08	0.08	0.09
Lighting	0.65	0.63	0.63	0.55	0.52	0.50	0.50	0.49	0.49	0.49	0.45	0.43	0.42	0.41	0.40	0.40	0.39	0.39	0.39	0.38	0.38	0.38	0.37	0.37	0.37	0.37	0.37	0.37	0.38	0.38	0.38
Clothes washers[a]	0.03	0.03	0.03	0.03	0.03	0.03	0.03	0.03	0.03	0.03	0.03	0.03	0.03	0.03	0.03	0.02	0.02	0.02	0.02	0.02	0.02	0.02	0.02	0.02	0.02	0.02	0.02	0.02	0.03	0.03	0.03
Dishwashers[a]	0.10	0.10	0.10	0.10	0.10	0.10	0.10	0.10	0.10	0.10	0.10	0.10	0.10	0.10	0.10	0.10	0.11	0.11	0.11	0.11	0.11	0.11	0.11	0.12	0.12	0.12	0.12	0.12	0.13	0.13	0.13
Televisions and related equipment[b]	0.32	0.32	0.32	0.33	0.33	0.33	0.34	0.34	0.34	0.35	0.35	0.36	0.36	0.37	0.37	0.37	0.38	0.38	0.39	0.39	0.40	0.41	0.41	0.42	0.42	0.43	0.43	0.44	0.44	0.45	0.45
Computers and related equipment[c]	0.16	0.16	0.15	0.14	0.14	0.13	0.13	0.13	0.13	0.13	0.13	0.13	0.13	0.13	0.13	0.12	0.12	0.12	0.12	0.12	0.12	0.12	0.12	0.12	0.12	0.12	0.12	0.12	0.12	0.13	0.13
Furnace fans and boiler circulation pumps	0.13	0.13	0.13	0.14	0.14	0.14	0.14	0.14	0.14	0.14	0.14	0.14	0.14	0.14	0.14	0.14	0.14	0.14	0.14	0.14	0.14	0.14	0.14	0.14	0.14	0.14	0.14	0.14	0.14	0.14	0.14
Other uses[h]	1.26	1.23	1.12	1.17	1.14	1.16	1.18	1.20	1.22	1.25	1.28	1.30	1.33	1.36	1.38	1.41	1.44	1.47	1.50	1.52	1.55	1.58	1.61	1.64	1.66	1.69	1.72	1.76	1.79	1.83	1.87
Delivered energy	11.41	11.28	10.75	11.24	11.02	10.99	10.96	10.95	10.95	10.97	10.95	10.96	10.97	10.99	11.01	11.04	11.07	11.11	11.14	11.17	11.20	11.23	11.26	11.29	11.32	11.35	11.39	11.43	11.47	11.52	11.57
Electricity related losses	10.35	10.20	9.69	9.59	9.54	9.44	9.36	9.40	9.50	9.63	9.66	9.71	9.79	9.87	9.95	10.04	10.11	10.20	10.28	10.36	10.45	10.53	10.61	10.70	10.80	10.90	11.00	11.10	11.23	11.36	11.50
Total energy consumption by end use																															
Space heating	5.49	5.33	4.83	5.55	5.26	5.23	5.19	5.15	5.11	5.08	5.05	5.03	5.00	4.97	4.95	4.93	4.90	4.88	4.86	4.84	4.83	4.81	4.78	4.76	4.74	4.71	4.68	4.65	4.62	4.60	4.57
Space cooling	2.84	2.88	2.83	2.46	2.71	2.70	2.69	2.72	2.76	2.81	2.86	2.90	2.94	2.99	3.04	3.10	3.15	3.20	3.24	3.29	3.35	3.40	3.45	3.50	3.55	3.60	3.65	3.69	3.74	3.79	3.84
Water heating	2.85	2.85	2.89	2.93	2.94	2.92	2.90	2.90	2.91	2.93	2.95	2.96	2.97	2.98	2.98	2.99	2.99	2.99	2.98	2.97	2.97	2.95	2.94	2.93	2.92	2.92	2.92	2.92	2.92	2.93	2.94
Refrigeration	1.16	1.16	1.15	1.16	1.15	1.13	1.12	1.12	1.13	1.13	1.14	1.14	1.15	1.15	1.16	1.16	1.17	1.18	1.19	1.20	1.21	1.22	1.23	1.23	1.24	1.25	1.26	1.27	1.28	1.30	1.31
Cooking	0.58	0.59	0.59	0.59	0.59	0.60	0.60	0.60	0.61	0.61	0.62	0.62	0.63	0.63	0.64	0.65	0.65	0.66	0.66	0.67	0.67	0.68	0.68	0.69	0.69	0.70	0.70	0.71	0.71	0.72	0.72
Clothes dryers	0.66	0.66	0.67	0.68	0.69	0.69	0.68	0.69	0.69	0.70	0.71	0.71	0.72	0.73	0.73	0.74	0.74	0.75	0.76	0.76	0.77	0.77	0.78	0.79	0.79	0.80	0.80	0.81	0.82	0.82	0.83
Freezers	0.25	0.26	0.25	0.26	0.26	0.25	0.25	0.25	0.25	0.25	0.25	0.25	0.25	0.25	0.25	0.25	0.25	0.25	0.25	0.25	0.25	0.24	0.24	0.24	0.24	0.24	0.24	0.25	0.25	0.25	0.25
Lighting	2.02	1.97	1.93	1.69	1.58	1.52	1.49	1.47	1.47	1.48	1.35	1.29	1.25	1.23	1.21	1.19	1.17	1.16	1.14	1.13	1.11	1.10	1.10	1.09	1.09	1.09	1.09	1.09	1.09	1.10	1.10
Clothes washers[a]	0.10	0.10	0.10	0.10	0.10	0.10	0.09	0.09	0.09	0.08	0.08	0.08	0.08	0.08	0.08	0.07	0.07	0.07	0.07	0.07	0.07	0.07	0.07	0.07	0.07	0.07	0.07	0.07	0.07	0.07	0.07
Dishwashers[a]	0.32	0.32	0.32	0.32	0.32	0.31	0.31	0.31	0.31	0.31	0.31	0.31	0.31	0.31	0.31	0.31	0.31	0.32	0.32	0.32	0.33	0.33	0.34	0.34	0.35	0.35	0.35	0.36	0.36	0.37	0.37
Televisions and related equipment[b]	0.98	0.98	0.99	1.01	1.01	1.01	1.01	1.01	1.02	1.04	1.05	1.06	1.08	1.09	1.10	1.12	1.13	1.14	1.15	1.17	1.18	1.19	1.21	1.22	1.23	1.25	1.26	1.27	1.29	1.30	1.32
Computers and related equipment[c]	0.49	0.49	0.45	0.43	0.41	0.40	0.40	0.39	0.39	0.39	0.39	0.39	0.38	0.38	0.37	0.37	0.37	0.37	0.36	0.36	0.36	0.36	0.36	0.36	0.36	0.36	0.36	0.36	0.36	0.36	0.36
Furnace fans and boiler circulation pumps	0.42	0.42	0.39	0.43	0.42	0.42	0.42	0.41	0.42	0.42	0.42	0.42	0.42	0.42	0.42	0.42	0.42	0.42	0.42	0.42	0.42	0.42	0.42	0.42	0.42	0.42	0.42	0.41	0.41	0.41	0.41
Other uses[h]	3.60	3.48	3.06	3.22	3.13	3.15	3.17	3.22	3.28	3.36	3.44	3.51	3.59	3.66	3.73	3.80	3.86	3.93	4.00	4.07	4.14	4.22	4.28	4.35	4.42	4.49	4.57	4.66	4.76	4.86	4.97
Total	21.76	21.48	20.43	20.83	20.56	20.42	20.32	20.35	20.44	20.59	20.62	20.66	20.76	20.85	20.96	21.08	21.18	21.31	21.42	21.53	21.65	21.76	21.88	21.99	22.12	22.25	22.38	22.53	22.69	22.88	23.08

(Continued)

TABLE 8.1 (Continued)

Residential Sector Key Indicators and Consumption (Quadrillion, Btu, Unless Otherwise Noted)

ref2013.d102312a	2010	2011	2012	2013	2014	2015	2016	2017	2018	2019	2020	2021	2022	2023	2024	2025	2026	2027	2028	2029	2030	2031	2032	2033	2034	2035	2036	2037	2038	2039	2040
Nonmarketed renewables[i]																															
Geothermal heat pumps	0.01	0.01	0.01	0.01	0.01	0.02	0.02	0.02	0.02	0.02	0.02	0.02	0.02	0.02	0.02	0.02	0.02	0.02	0.02	0.02	0.02	0.02	0.02	0.02	0.03	0.03	0.03	0.03	0.03	0.03	0.03
Solar hot water heating	0.01	0.01	0.01	0.01	0.02	0.02	0.02	0.02	0.02	0.02	0.02	0.02	0.02	0.02	0.02	0.02	0.02	0.02	0.02	0.02	0.02	0.02	0.02	0.02	0.02	0.02	0.02	0.02	0.02	0.02	0.02
Solar photovoltaic	0.01	0.02	0.04	0.07	0.09	0.11	0.14	0.14	0.14	0.14	0.14	0.15	0.15	0.15	0.15	0.15	0.16	0.16	0.16	0.16	0.17	0.17	0.17	0.18	0.18	0.18	0.19	0.19	0.20	0.21	0.21
Wind	0.00	0.00	0.00	0.00	0.00	0.01	0.01	0.01	0.01	0.01	0.01	0.01	0.01	0.01	0.01	0.01	0.01	0.01	0.01	0.01	0.01	0.01	0.01	0.01	0.01	0.01	0.01	0.01	0.01	0.01	0.01
Total	0.03	0.04	0.07	0.10	0.12	0.16	0.19	0.19	0.19	0.19	0.20	0.20	0.20	0.20	0.20	0.21	0.21	0.21	0.21	0.22	0.22	0.22	0.23	0.23	0.24	0.24	0.25	0.25	0.26	0.26	0.27
Heating degree Days																															
New England	5944	6138	5796	6490	6215	6200	6186	6172	6158	6144	6131	6117	6103	6089	6075	6062	6048	6034	6020	6006	5992	5978	5964	5950	5936	5922	5907	5893	5879	5865	5850
Middle Atlantic	5453	5413	5038	5779	5459	5443	5427	5411	5394	5378	5362	5346	5330	5314	5298	5281	5265	5249	5233	5217	5201	5185	5169	5153	5137	5121	5105	5089	5073	5058	5042
East North Central	6209	6187	5462	6358	6137	6127	6116	6105	6094	6084	6073	6062	6051	6040	6030	6019	6008	5997	5986	5976	5965	5954	5943	5932	5922	5911	5900	5889	5878	5867	5856
West North Central	6585	6646	5633	6619	6372	6360	6348	6336	6323	6310	6297	6284	6271	6257	6244	6230	6216	6202	6189	6175	6161	6147	6133	6119	6105	6091	6077	6063	6049	6035	6020
South Atlantic	3183	2555	2337	2800	2699	2693	2687	2680	2674	2667	2660	2654	2647	2640	2634	2627	2621	2615	2608	2602	2596	2590	2584	2578	2572	2566	2560	2554	2549	2543	2538
East South Central	4003	3397	2934	3599	3438	3435	3431	3428	3424	3421	3417	3414	3410	3407	3403	3400	3396	3393	3389	3386	3382	3378	3375	3371	3367	3364	3360	3356	3352	3349	3345
West South Central	2503	2203	1829	2280	2084	2076	2068	2060	2052	2044	2036	2028	2020	2012	2004	1996	1988	1980	1972	1964	1956	1948	1940	1932	1924	1916	1908	1900	1892	1884	1876
Mountain	4882	5054	4624	5021	4666	4647	4628	4608	4587	4566	4545	4522	4500	4476	4453	4430	4406	4383	4359	4336	4312	4288	4264	4240	4216	4192	4168	4144	4119	4095	4071
Pacific	3202	3411	3170	3239	3112	3109	3106	3104	3101	3097	3094	3091	3087	3083	3079	3076	3072	3068	3064	3061	3057	3053	3050	3046	3043	3039	3036	3033	3029	3026	3022
United States	4388	4240	3811	4355	4146	4131	4116	4101	4085	4070	4054	4039	4024	4008	3993	3978	3963	3947	3932	3918	3903	3888	3873	3858	3844	3829	3814	3799	3785	3770	3756
Cooling degree Days																															
New England	655	607	604	465	559	564	569	573	578	583	588	592	597	602	607	611	616	621	625	630	635	640	644	649	654	659	664	668	673	678	683
Middle Atlantic	997	887	889	688	834	841	848	854	861	868	875	882	889	896	902	909	916	923	930	937	944	950	957	964	971	978	984	991	998	1005	1011
East North Central	978	898	1011	746	793	795	797	799	801	803	805	807	809	811	813	815	817	819	821	822	824	826	828	830	832	834	836	838	840	842	844
West North Central	1123	1116	1241	957	987	988	990	991	992	994	995	997	998	1000	1002	1003	1005	1007	1008	1010	1012	1014	1015	1017	1019	1021	1022	1024	1026	1028	1030
South Atlantic	2289	2357	2240	2041	2177	2185	2193	2202	2210	2219	2228	2236	2245	2254	2262	2271	2279	2288	2296	2305	2313	2322	2330	2339	2347	2356	2364	2372	2380	2389	2397
East South Central	1999	1811	1817	1597	1740	1746	1753	1760	1766	1773	1779	1786	1792	1799	1805	1812	1818	1825	1831	1838	1845	1851	1858	1864	1871	1877	1884	1890	1897	1903	1910
West South Central	2755	3194	2881	2512	2772	2784	2797	2810	2822	2835	2847	2860	2873	2885	2898	2911	2923	2936	2948	2961	2974	2986	2999	3011	3024	3037	3049	3062	3074	3087	3099
Mountain	1490	1396	1522	1420	1628	1639	1650	1662	1674	1686	1698	1711	1725	1739	1752	1766	1780	1794	1808	1822	1837	1851	1866	1880	1895	1910	1925	1940	1955	1970	1985
Pacific	746	809	888	844	898	901	903	906	908	910	913	915	918	923	923	925	928	930	933	935	938	940	942	945	947	950	952	954	957	959	961
United States	1498	1528	1518	1319	1444	1453	1462	1471	1480	1489	1499	1508	1517	1526	1535	1545	1554	1563	1572	1582	1591	1600	1610	1619	1628	1638	1647	1657	1666	1676	1685

Source: 2010 and 2011 consumption based on: U.S. Energy Information Administration (EIA), Annual Energy Review 2011, DOE/EIA-0384(2011) (Washington, DC; September 2012); 2010 and 2011 degree days based on state-level data from the National Oceanic and Atmospheric Administration's Climatic Data Center and Climate Prediction Center.
Projections: EIA, AEO2013 National Energy Modeling System run ref2013.d102312a.
Note: Totals may not equal sum of components due to independent rounding. Data for 2010 and 2011 are model results and may differ slightly from official EIA data reports. Btu = British thermal unit. - - = Not applicable.
a Does not include water heating portion of load.
b Includes televisions, set-top boxes, and video game consoles.
c Includes desktop and laptop computers, monitors, printers, speakers, networking equipment, and uninterruptible power supplies.
d Includes small electric devices, heating elements, and motors not listed above. Electric vehicles are included in the transportation sector.
e Includes such appliances as outdoor grills and mosquito traps.
f Includes wood used for primary and secondary heating in wood stoves or fireplaces as reported in the Residential Energy Consumption Survey 2005.
g Includes kerosene and coal.
h Includes all others not listed above.
i Consumption determined by using the fossil fuel equivalent of 9756 Btu/kW h.

8.2 Description of Major Appliances and Space Conditioning Equipment

8.2.1 Refrigerator-Freezers and Freezers

Refrigerators, refrigerator-freezers, and freezers keep food cold by transferring heat from the air in the appliance cabinet to the outside of the cabinet. A refrigerator is a well-insulated cabinet used to store food at 34°F (1.1°C) or above, a refrigerator-freezer is a refrigerator with an attached freezer compartment that stores food below 0°F (–17.7°C), and a standalone freezer is a refrigerated cabinet to store and freeze foods at –12°F (–24.4°C) or below. Almost all refrigerators are fueled by electricity. The refrigeration system includes an evaporator, a condenser, a metering device, and a compressor. The system uses a vapor compression cycle, in which the refrigerant changes phase (from high pressure vapor to high pressure liquid and back to low pressure vapor) while circulating in a closed loop system. The refrigerant absorbs or discharges heat as it changes phase. Although most refrigerants and insulating materials once contained chlorofluorocarbons (CFCs), all U.S. models sold after January 1, 1996, are CFC-free. Under the Montreal Protocol, in 1996, refrigerator-freezer manufacturers are using R134a (1,1,1,2-tetrafluoroethane) as the replacement for R-12 (dichlorodifluoromethane) refrigerant. R134a is also sold under other names such as: Dymel 134a, Forane 134a, Genetron 134a, HFA-134a, HFC-134a, R-134a, Suva 134a, and Norflurane.

There are over 170 million refrigerators and refrigerator-freezers currently in use today in U.S. homes with over 60 million refrigerators over 10 years old and are costing consumers billions in excessive energy costs.

8.2.2 Water Heaters

A water heater is an appliance that is used to heat potable water for use outside the heater upon demand. Water heaters supply water to sinks, bathtubs and showers, dishwashers, and clothes washing machines. Most water heaters in the United States are storage water heaters, which continuously maintain a tank of water at a thermostatically controlled temperature. The most common storage water heaters consist of a cylindrical steel tank that is lined with glass in order to prevent corrosion. Most hot water tanks manufactured today are insulated with polyurethane foam and wrapped in a steel jacket. Although some use oil, almost all storage water heaters are fueled by natural gas (or LPG) or electricity.

Rather than storing water at a controlled temperature, instantaneous water heaters (tankless) heat water as it is being drawn through the water heater. Both gas-fired and electric instantaneous water heaters (tankless) are available. Instantaneous water heaters are quite popular in Europe and Asia. Although they are not commonly used in the United States, their presence does seem to be increasing.

Like refrigerators, water heaters are present in almost all U.S. households. Approximately 54% of households have gas-fired water heaters, and approximately 38% have electric water heaters. Hot water use varies significantly from household to household, mostly due to differences in household size and occupant behavior.

The Department of Energy (DOE) final rule effective on April 16, 2015, mandates will require higher energy factor (EF) ratings on virtually all residential gas, electric, oil, and tankless gas water heaters. The EF is the ratio of useful energy output from the water heater to the total amount of energy delivered to the water heater. The higher the EF, the more efficient is the water heater (see Table 8.2).

TABLE 8.2

2015 Energy Conservation Standards for Residential Water Heaters

Product Classes Affected by Change	Rated Storage Volume/Inputs Affected by Change	New EF Requirements
Electric	≥20 and ≤55 gal, ≤12 kW input	$0.960 - (0.0003 \times V)$
	>55 and ≤120 gal, ≤12 kW input	$2.057 - (0.00113 \times V)$
Gas fired	≥20 and ≤55 gal, ≤75,000 Btu/h	$0.675 - (0.0015 \times V)$
	>55 and ≤100 gal, ≤75,000 Btu/h	$0.8012 - (0.00078 \times V)$
Oil fired	≤50 gal, ≤105,000 Btu/h	$0.68 - (0.0019 \times V)$
Instantaneous electric[a]	≤2 gal, ≤12 kW input	$0.93 - (0.00132 \times V)$
Instantaneous gas fired	≤2 gal, ≤200,000 Btu/h	$0.82 - (0.0019 \times V)$

[a] No change.

8.2.3 Furnaces and Boilers

Furnaces and boilers are major household appliances used to provide central space heating. Both fuel-burning and electric furnaces and boilers are available. A typical gas furnace installation is composed of the following basic components: (1) a cabinet or casing; (2) heat exchangers; (3) a system for obtaining air for combustion; (4) a combustion system including burners and controls; (5) a venting system for exhausting combustion products; (6) a circulating air blower and motor; and (7) an air filter and other accessories. (Furnaces that burn oil and liquid petroleum gas (LPG) are also available, though not as common.) In an electric furnace, the casing, air filter, and blower are very similar to those used in a gas furnace. Rather than receiving heat from fuel-fired heat exchangers, however, the air in an electric furnace receives heat from electric heating elements. Controls include electric overload protection, contactor, limit switches, and a fan switch. Furnaces provide heated air through a system of ducts leading to spaces where heat is desired. In a better system, hot water or steam is piped to terminal heating units placed throughout the household. The boiler itself is typically a pressurized heat exchanger of cast iron, steel, or copper in which water is heated.

According to the U.S. Census Bureau in its report *Annual 2012 Characteristics of New Housing* report, 59% of new family homes completed in 2012 used natural or LPG gas for heating, followed by 39% that use electricity for heating, 1% that use oil, and 2% that use other forms.

8.2.4 Central Air Conditioning, Room Air Conditioners, and Ductless Minisplit Air Conditioners

A central air conditioning (AC) system is an appliance designed to provide cool air to an enclosed space. Typically, central AC systems consist of an indoor unit and an outdoor unit (split system). Central air conditioning (AC) units are also available in a package unit. A packaged unit air conditioning system has all of the components contained in one unit. This type of air conditioning unit can be installed on a rooftop with duct work added. A package unit air conditioner operates the same as a central unit. The outdoor condenser unit contains a compressor, condenser coil (outdoor heat exchanger coil), condenser fan, and condenser fan motor; the indoor unit consists of an evaporator coil (indoor heat exchanger coil) and a refrigerant flow control device (a capillary tube, thermostatic expansion valve, or orifice) residing either in a forced-air furnace or an air handler. Refrigerant tubing connects the two units. A central AC system provides conditioned air by drawing warm air from the living area space and blowing it through the evaporator coil; as it is passing through the evaporator coil, the air gives up its heat content to the refrigerant. The conditioned air is then delivered back to the living area space (via a ducted system) by the blower residing in the furnace or air handler. The compressor takes the vaporized refrigerant aiming out of the evaporator and raises it to a temperature exceeding that of the outside air. The refrigerant then passes on to the condenser unit (outside coil), where the condenser coil rejects the heat from the refrigerant to the cooler outside air, and condenses. The liquid refrigerant passes through the flow control device, and its pressure and temperature are reduced. The refrigerant reenters the evaporator coil, where the refrigeration cycle is repeated.

Unlike the two-unit, central AC system, a room air conditioner is contained within one cabinet and is mounted in a window or a wall so that part of the unit is outside the building and part is within the occupied space. The two sides of the cabinet are typically separated by an insulated divider wall in order to reduce heat transfer. The components in the outdoor portion of the cabinet are the compressor, condenser coil, condenser fan, fan motor, and capillary tube. The components in the indoor portion of the cabinet are the evaporator coil and evaporator fan. The fan motor drives both the condenser and evaporator fans. A room AC provides conditioned air in the same manner described for a central AC system but without air ducts.

Like conventional central air conditioners, minisplit air conditioners use an outside compressor/condenser unit and an indoor evaporator coil/ductless air handler unit. The difference is that each room or zone to be cooled has its own ductless air handler. Each indoor ductless air handler unit is connected to the outdoor condensing unit via a conduit carrying the electrical power, refrigerant lines, and condensate lines. The primary advantage is that by providing dedicated units to each occupied space or zones, it is easier to meet the varying comfort needs of different rooms of the residence. By avoiding the use of ductwork, minisplit AC also avoid energy losses associated with central AC. Some minisplit air conditioner condensers are designed to handle up to five ductless air handlers at one time.

Approximately 89% of U.S. households had an AC system in 2009; room ACs is used in 6% of households, and 5% without AC. According to the U.S. Census Bureau in its report *Annual 2012 Characteristics of New Housing* report, 89% of new family homes completed in 2012, followed by 11% completed without AC installed.

AC manufacturers are beginning to offer HFC-410A refrigerant in AC systems as an alternative to HCFC-22 (R-22) refrigerant units. The EPA has established the phase-out of the HCFC-22 refrigerant with no production or importing beginning in 2020. However, manufacturers of AC equipment must phase out the use of HCFC-22 refrigerant in new AC equipment by January 1, 2010. In general, existing R-22 systems will probably be converted to R-407C, an alternative refrigerant; however, new AC equipment will be designed to operate on R-410A.

8.2.5 Heat Pumps

Unlike air conditioners, which provide only space cooling, heat pumps use the same equipment to provide both space heating and cooling. A heat pump draws heat from the outside air into a building during the heating season and removes heat from a building to the outside during the cooling season. An air source heat pump contains the same components and operates in the same way as a central AC system but is able to operate by reversing the refrigerant directional flow as well, in order to provide space heating. In providing space heat, the indoor coil acts as the condenser while the outdoor coil acts as the evaporator. When the outside air temperature drops below 2°C (35.6°F) during the heating season, the available heat content of the outside air significantly decreases; in this case, a heat pump will utilize supplementary electric-resistance backup heat. A ground-source heat pump operates on the same principle as air-source equipment except that heat is rejected or extracted from the ground instead of the air. Since ground temperatures do not vary over the course of a day or a year as much as the ambient air temperature, more stable operating temperatures are achieved. The ground loop for a ground-source heat pump is a closed system that uses a pressurized, sealed piping system filled with a water/antifreeze mixture. The indoor mechanical equipment of a ground-source system includes a fan coil unit with an indoor coil, a compressor, and a circulation pump for the ground loop. Almost all heat pumps are powered by electricity.

Heat pumps were used in approximately 11% of U.S. households in 2001. The energy consumption of heat pumps varies according to the same user characteristics discussed earlier for AC systems. According to the U.S. Census Bureau in its report *Annual 2012 Characteristics of New Housing* report, 38% of new family homes completed in 2012 have heat pump units installed.

The EPA has established the phase-out of the HCFC-22 refrigerant for heat pumps with no production or importing beginning in 2020. However, manufacturers of heat pump equipment must phase out the use of HCFC-22 refrigerant in new heat pump equipment by January 1, 2010. In general, existing R-22 systems will probably be converted to R-407C, an alternative refrigerant; however, new heat pump equipment will be designed to operate on R-410A.

8.2.6 Clothes Washers

A clothes washer is an appliance that is designed to clean fabrics by using water, detergent, and mechanical agitation. The clothes are washed, rinsed, and spun within the perforated basket that is contained within a water-retaining tub. Top-loading washers move clothes up and down, and back and forth, typically about a vertical axis. Front-loading machines move clothes around a horizontal axis. Electricity is used to power an electric motor that agitates and spins the clothes, as well as a pump that is used to circulate and drain the water in the washer tub. Some washer models use a separate water heater element to heat the water used in the washer.

Approximately 84% of households had clothes washers in 2009. Most of the clothes washers sold in the United States are top-loading, vertical-axis machines. The majority of energy used for clothes washing (85%–90%) is used to heat the water. User behavior significantly affects the energy consumption of clothes washers. The user can adjust the amount of water used by the machine to the size of the load, and thereby save water and energy. Choosing to wash with cold water rather than hot water reduces energy consumption by the water heater. Similarly, rinsing with cold water rather than warm can reduce energy consumption. Energy consumption depends on how frequently the washer is used. The DOE's test procedure assumes clothes washers are used 300 times a year on average. According to the U.S. Energy Star program, if every washer purchased in the United States in 2013 earned the Energy Star rating, we would save about $250 million in electricity, water usage, and gas every year thereafter.

8.2.7 Clothes Dryers

A clothes dryer is an appliance that is designed to dry fabrics by tumbling them in a cabinet like drum with forced-air circulation. The source of heated air may be powered either by electricity or natural gas. The motors that rotate the drum and drive the fan are powered by electricity. Approximately 79% of U.S. households have automatic clothes dryers and 21% do not use a clothes dryer in 2009. An automatic dryer in U.S. homes consists of 80% electric dryers, 19% natural gas dryers, and 1% propane gas (LPG) dryers.

8.2.8 Dishwashers

A dishwasher is an appliance that is designed to wash and dry kitchenware by using water and detergent. Typically, in North America, hot water is supplied to the dishwasher by an external water heater. In addition, an internal electric heater further raises the water temperature within the dishwasher. Electric motors pump water through spray arms impinging on the kitchenware in a series of wash and rinse cycles. An optional drying function is also enabled by electric heaters and sometimes a fan. In recent years, some dishwashers incorporate soil sensors that determine when the dishes are clean and the washing cycle can be stopped. Approximately 69% of U.S. households had dishwashers in 2011.

8.2.9 Cooktops and Ovens

A cooktop is a horizontal surface on which food is cooked or heated from below; a conventional oven is an insulated, cabinet like appliance in which food is surrounded by heated air. When a cooktop and an oven are combined in a single unit, the appliance is referred to as a range. Both gas and electric ranges are available. Cooktops and ovens are present in almost all households. Almost 60% of households use electric cooktops and ovens, and the remaining 40% of households use gas cooktops and ovens.

In a microwave oven, nonionizing microwaves directed into the oven cabinet cavity cause the water molecules inside the food to vibrate. Movement of the water molecules heats the food from the inside out. The fraction of households with microwave ovens has increased dramatically in recent years. Appliance manufacturers have sold over 5 million microwave ovens in 2013.

8.3 Current Production

Table 8.3 shows the number of various appliances that was shipped by manufacturers in 2012 and 2013. Shipments have been increasing for most of the major appliances and air conditioners.

8.4 Efficient Designs

State and federal standards requiring increased efficiency for residential appliances, utility programs, and labels (such as Energy Star) have improved appliance efficiency dramatically since the late 1970s (Meyers et al., 2004). For example, the annual energy consumption (according to the DOE test procedure) of a new refrigerator in 2003 was less than half the consumption in 1980. Because of the slow turnover rate of appliances, however, the older, less efficient equipment remains in use for a long time. Promising design options for further improving the efficiency of residential appliances are discussed next.

The DOE has set standards and test procedures for appliances and air conditioners and has required manufacturers to comply with their rulings. See website for further details on the appliances and air conditioners listed in this chapter at: http://www1.eere.energy.gov/buildings/appliance_standards/standards_test_procedures.html.

TABLE 8.3

Shipments of Major Appliances and Space Conditioning Equipment in the United States

	YTD-2012	YTD-2013	%Chg
Major home appliances and space conditioning equipment (thousands of units)			
All major appliances	36,981.10	37,971.10	2.70
Cooking—total	9,014.60	9,166.70	1.70
Electric cooking—total	2,386.60	2,555.60	7.10
Electric ranges	1,902.70	2,009.00	5.60
Electric ovens	316.5	365.9	15.60
Surface cooking units	167.4	180.7	8.00
Gas cooking—total	1,412.80	1,521.50	7.70
Gas ranges	1,228.30	1,304.20	6.20
Gas ovens	17.1	17.7	3.70
Surface cooking units	167.5	199.5	19.20
Microwave ovens	5,215.20	5,089.70	−2.40
Home laundry—total	7,320.90	8,258.00	12.80
Automatic washers	4,104.40	4,591.90	11.90
Dryers—total	5,215.20	3,666.00	14.00
Electric	2,615.80	2,963.50	13.30
Gas	600.7	702.5	17.00
Kitchen cleanup—total	6,621.40	7,253.00	9.50
Disposers	3,419.80	3,725.00	8.90
Dishwashers—total	3,181.30	3,505.00	10.20
Built in	3,153.70	3,480.20	10.40
Portable	27.6	24.8	−10.10
Compactors	20.3	23	12.90
Food preservation—total	5,976.30	6,090.00	1.90
Refrigerators 6.5 and over	4,897.90	5,134.10	4.80
Freezers—total	1,078.50	955.9	−11.40
Chest	714.1	581.2	−18.60
Upright	364.4	374.6	2.80
Gas water heaters—total	3,186.50	3,535.50	11.00
Electric water heaters—total	3,077.30	3,330.70	8.20
Gas furnaces—total	1,787.30	2,093.90	17.20
Oil furnaces—total	25,567.00	24,593.00	−10.80
Air conditioners and heat pumps—total	5,019.00	5,540.40	10.40
Air conditioners only—total	3,526.60	3,807.50	8.00
Heat pumps only—total	1,492.30	1,732.90	16.10
Home comfort—total	8,047.90	7,206.50	−10.50
Room air conditioners	7,224.70	6,436.10	−10.90
Dehumidifiers	823.2	770.4	−6.40

Source: Association of Home Appliance Manufacturers (AHAM), Washington, DC.

Notes: Figures (in units) include shipments for the U.S. market, whether imported or domestically produced. Export shipments are not included; Industry figures are estimates derived from the best available figures supplied by a sample of AC manufacturers and are subject to revision.

8.4.1 Refrigerators and Freezers

Relative to the 2001 U.S. federal efficiency standard, achieving a 15% energy use reduction for refrigerator-freezers is possible with the use of a high-efficiency compressor, high-efficiency motors for the evaporator and condenser fans, and adaptive defrost control. Models at this level of efficiency account for a modest market share in the United States. Achieving a 25% energy use reduction generally would require a reduction in load transmitted through the unit's walls and doors, which might require the use of vacuum panels.

8.4.2 Improved Fan Motors

The evaporator and condenser fans of large refrigerators are powered by motors. The most common motor used for this purpose is a shaded-pole motor. Large efficiency gains are possible in refrigerators and freezers by switching to electronically commutated motors (ECMs), also known as brushless permanent-magnet motors, which typically demand less than half as much power as shaded-pole motors.

8.4.3 Vacuum Insulation Panels

The use of vacuum insulation panels (VIPs) can significantly reduce heat gain in a refrigerated cabinet and thereby decrease the amount of energy necessary to maintain a refrigerator or freezer at a low temperature. When using VIP, a partial vacuum is created within the walls of the insulation panels. Because air is conductive, the amount of heat transfer from the outside air to the refrigerated cabinet is reduced as the amount of air within the panels is reduced. Evacuated panels are filled with low-conductivity powder, fiber, or aerogel in order to prevent collapse. Energy savings associated with the use of vacuum panel insulation range from 10% to 20%. Vacuum panel technology still faces issues regarding cost and reliability before it can come into widespread use in refrigeration applications (Malone and Weir, 2001).

8.4.4 Water Heaters

8.4.4.1 Gas-Fired Storage Water Heater

The current models of gas-fired storage water heaters have a central flue that remains open when the water heater is not firing. This leads to large off-cycle standby losses. It should be possible to dramatically reduce off-cycle losses with relatively inexpensive technical modifications to the water heater. Energy savings derived from models with these modifications are expected to be about 25% compared to the 2004 U.S. standards.

The amount of heat extracted from the fuel used to fire a gas appliance can be increased by condensing the water vapor in the flue gases. In a condensing storage water heater, the flue is lengthened by coiling it around inside the tank. The flue exit is located near the bottom of the tank where the water is coolest. Because the flue gases are relatively cool, a plastic venting system may be used. A drain must be installed in condensing systems. Energy savings associated with the use of a gas-fired condensing water heater are approximately 40% compared to the 2004 efficiency standard. At this time, the high cost of the water heater results in a payback time that exceeds the typical lifetime of a water heater, but it is reasonable to assume that the cost could be reduced to the point that the water heater would be cost effective. In applications with heavy hot water use, such as laundromats or hotels, they may already be cost effective. Currently, a few companies produce condensing storage water heaters for commercial markets and they are sometimes sold as combined water heater/space heating systems for residential use.

8.4.4.2 Heat Pump Water Heaters

Heat pumps used with water heaters capture heat from the surrounding air or recycle waste heat from AC systems and then transfer the heat to the water in the storage tank. In this way, less energy is used to bring the water to the desired temperature. The heat pump can be a separate unit that can be attached to a standard electric water heater. Water is circulated out of the water heater storage tank, through the heat pump, and back to the storage tank. The pump is small enough to sit on the top of a water heater but could be anywhere nearby. Alternatively, the heat pump can be directly integrated into the water heater. Research indicates that this technology uses 60%–70% less energy than conventional electric resistance water heaters. Field and lab tests have been completed for several prototypes. A few models are currently available for sale.

8.4.4.3 Solar Water Heaters

Technological improvements in the last decade have improved the quality and performance of both passive and active solar water heaters. Research indicates that, in general, solar water heaters use 60% less energy than conventional electric resistance water heaters. There are several types of solar water heaters commercially available today.

8.4.5 Furnaces

8.4.5.1 Condensing Furnaces

The efficiency of a conventional gas furnace can be increased by using an additional heat exchanger to capture the heat of the flue gases before they are expelled to the outside. The secondary heat exchanger is typically located at the outlet of the circulating air blower, upstream of the primary heat exchanger. A floor drain is required for the condensate. A condensing furnace has efficiency up to 96% AFUE, well above the 80% AFUE rating of a standard noncondensing gas furnace. Condensing furnaces have been on the market since the 1980s and now constitute one-third of all gas furnace sales. They are particularly popular in colder areas of the United States, where the cost of heating is high. An early technical problem, corrosion of the secondary heat exchanger, has been resolved by the industry.

8.4.5.2 Integrated Water Heaters and Furnaces

Traditionally, water heating and space heating have required two separate appliances— a hot water heater and a furnace. Combining a water heater and a furnace into a single system can potentially provide both space heating and hot water at a lower overall cost. Integrated water and space heating is most cost effective when installed in new buildings, because gas connections are necessary for only one appliance rather than two.

Combination of space- and water-heating appliances fall into two major classes: (1) boiler/ tankless-coil combination units, and (2) water-heater/fancoil combination units. A great majority of boiler/tankless-coil combination units are fired with oil, whereas most water-heater/fancoil combination units are fired with natural gas. In the latter units, the primary design function is domestic water heating. Domestic hot water is circulated through a heating coil of an air-handling system for space heating. Usually the water heater is a tank-type gas-fired water heater, but instantaneous gas-fired water heaters can be used as well.

The efficiencies of these integrated systems are determined largely by the hot water heating component of the system. Compared to a system using a standard water heater or boiler, an integrated system using a condensing water heater or condensing boiler can reduce energy consumption by as much as 25%.

8.4.6 Central Air Conditioning, Room Air Conditioners, and Ductless Minisplit Air Conditioners

8.4.6.1 Electric Variable-Speed Air Conditioning

Variable-speed central air conditioners use ECMs, which are more efficient than the induction motors used in a single-speed system. In addition, the speed of the ECM can be varied to match system capacity more precisely to a building load. Cycling losses, which are associated with a system that is continually turned off and on in order to meet building load conditions, are thus reduced. Unlike induction motors, ECMs retain their efficiency at low speeds; consequently, energy use is also reduced at low-load conditions. A variable-speed AC system uses approximately 40% less energy than a standard single-speed AC system. Although these AC systems are now available from major manufacturers, they account for a small fraction of sales.

8.4.6.2 Electric Two-Speed Air Conditioning

Two-speed induction motors are not as efficient as variable-speed ECMs, but they are less expensive. Like variable-speed air conditioners, two-speed air conditioners reduce cycling losses. When two-speed induction motors are used to drive compressor and fans, the system can operate at two distinct capacities. Cycling losses are reduced, because the air conditioner can operate at a low speed to meet low building loads. In some models, two-speed compressors are coupled with variable-speed indoor blowers to improve system efficiency further. A two-speed AC system reduces energy consumption by approximately one-third. Although these AC systems are available from several major manufacturers, they account for a small fraction of sales.

8.4.6.3 Room Air Conditioners

The most efficient room air conditioners have relatively large evaporator and condenser heat-exchanger coils, high-efficiency rotary compressors, and permanent split-capacitor fan motors. Compared to standard room ACs, highly efficient room ACs reduce energy consumption by approximately 25%. Such room ACs are available from several manufacturers. Units with Energy Star designation, which must exceed federal minimum efficiency standards by 10%, accounted for 35% of sales in 2004.

8.4.6.4 Ductless Minisplit Air Conditioners

Ductless minisplit air conditioners are available in numerous mix-and-match capacities and configurations. This type of AC system does not require bulky ductwork or complicated installation or expensive modifications to be installed in a home or office. Multiple indoor units (zones) with varying capacities can be connected to one condenser unit. These units are available in heat and/or cool. These high-efficiency air conditioners can save up to 20% energy usage.

8.4.7 Heat Pumps

8.4.7.1 Variable-Speed and Two-Speed Heat Pumps

Like central air conditioners, heat pumps can be made more efficient by the use of two-speed and variable-speed motors (see the earlier discussion of efficient central air conditioners). Both two-speed and variable speed air-source heat pumps are available; variable-speed ground-source heat pumps are not commercially available at this time. Compared to standard models, two-speed air-source heat pumps reduce energy consumption by approximately 27%, variable-speed air-source heat pumps reduce energy consumption by 35%, and two-speed ground-source heat pumps reduce energy consumption by 46%. Variable-speed and two-speed air-source heat pumps are made by the same companies that make variable speed and two-speed AC systems. Several manufacturers produce efficient air-source and two-speed ground-source heat pumps, but they account for a small fraction of all heat pump sales.

8.4.7.2 Gas-Fired Heat Pumps

Currently, all residential heat pumps are electric, but researchers have been developing gas heat pumps. The Gas Research Institute (GRI) and a private corporation jointly developed a natural gas, engine-driven, variable-speed heat pump in which the compressor is driven by an internal combustion spark-ignition engine and heat is recovered in the space-heating mode. This engine-driven heat pump was put on the market in 1994, but was withdrawn in the late 1990s due to lack of a maintenance infrastructure. In addition, the DOE has been funding the development of a gas-fired ammonia-water absorption-cycle heat pump. Gas-driven heat pumps have the potential to reduce heat pump energy consumption by approximately 35%–45%.

8.4.8 Distribution Systems

When assessing the efficiency of a space conditioning system, it is important to consider the efficiency of the distribution system as well as the appliance. It is not uncommon for air ducts to have distribution losses of 20%–40% due to conduction as well as leakage. Better insulation as well as more careful duct sealing can reduce these losses. In general, the most effective strategy for reducing distribution losses is to include the distribution system in the conditioned space so that any losses due to conduction or air leakage go directly into the space to be conditioned. This requires careful attention by the architect in the design of new buildings and is typically expensive as a retrofit measure.

8.4.9 Clothes Washers

8.4.9.1 Horizontal-Axis Washers

Although horizontal-axis clothes washers dominate the European market, the vast majority of clothes washers sold in the United States are top-loading, vertical-axis machines. However, this is expected to change by 2007 when more stringent minimum efficiency regulations on clothes washers will take effect. Horizontal-axis washers, in which the tub spins around a horizontal axis, use much less water than their vertical-axis counterparts, and less hot water is therefore required from water heaters. As mentioned earlier, the majority of energy used for clothes washing is used for heating water, so a significant amount of energy can be saved by using horizontal-axis washers. Research has indicated that horizontal-axis washers are more than twice as efficient as vertical-axis washers of comparable size (U.S. Department of Energy, 2000).

8.4.9.2 High-Spin-Speed Washers

Clothes washers can be designed so that less energy is required to dry clothes after they have been washed. Extracting water from clothes mechanically in a clothes washer uses approximately 70 times less energy than extracting the water with thermal energy in electric clothes dryer. Thus, by increasing the speed of a washer's spin cycle, one can reduce the energy required to dry clothes.

Because gas clothes dryers require so much less energy than electric dryers, based on energy consumption including that consumed at the electric power station, energy savings are much more significant when a high-spin-speed washer is used with an electric dryer. In a vertical-axis clothes washer, an increase in spin speed from 550 to 850 rpm reduces moisture retention from 65% to 41%. In an electric dryer, this reduces the energy consumption by more than 40%. In a horizontal-axis washer, an increase in spin speed from 550 to 750 rpm reduces moisture retention from 65% to 47%; in an electric dryer, energy consumption is reduced by more than 30%. High-spin-speed washers have been common on European horizontal axis machines for some time, and are becoming more common in the United States.

8.4.10 Clothes Dryers

8.4.10.1 Microwave Dryers

In conventional clothes dryers, hot air passes over wet clothes and vaporizes the surface water. During the later stages of drying, the surface dries out and heat from the hot air must be transferred to the interior, where the remaining moisture resides. In contrast, in microwave drying, water molecules in the interior of a fabric absorb electromagnetic energy at microwave wavelengths, thereby heating the water and allowing it to vaporize. Several U.S. appliance manufacturers have experimented with microwave clothes dryers, and a few small companies have built demonstration machines. Issues of arcing on metal objects and possible fire within the dryer would need to be resolved before microwave clothes dryers are likely to be sold.

8.4.10.2 Heat Pump Dryers

A heat pump dryer is essentially clothes dryer and an air conditioner packaged as one appliance. In a heat pump dryer, exhaust heat energy is recovered by recirculating all the exhaust air back to the dryer; the moisture in the recycled air is removed by a refrigeration–dehumidification system. A drain is required to remove the condensate; because washers and dryers are usually located side by side, a drain is generally easily accessible. Heat pump dryers can be 50%–60% more efficient than conventional electric dryers. Introduced in Europe in 1999, heat pump dryers are available both in the United States and Europe.

TABLE 8.4

Cost Effectiveness of Efficient Designs for Selected Appliances and Space Conditioning Equipment

Product	Baseline Technology	Efficient Technology	CCE[a]
Refrigerator	484 kW h/year	426 kW h/year	4.9 cents/kW h
Room air conditioner	9.85 EER	10.11 EER	5.2 cents/kW h
Electric water heater	92 EF	Heat pump	3.9 cents/kW h
Clothes washer[b]	3.23 kW h/cycle	Horizontal axis and high spin speed (1.87 kW h/cycle)	5.0 cents/kW h
Gas furnace[c]	80% AFUE	Condensing furnace (90% AFUE)	$9.30/MBtu

[a] The CCE may be compared to expected residential energy prices. In the United States, long-term forecasts (as of 2005) place average prices in the range of 8.5–9.0 cents/kW h for electricity and $7.7–$8.4/MBtu for natural gas.

[b] Includes the energy for water heating and drying associated with clothes washing. Refers to electric water heater and dryer. Baseline refers to the 1994 DOE standard level. Additional consumer benefits derive from reduction in water usage.

[c] The CCE value is a U.S. national average. In cold climates, the amount of saved energy is higher, so the CCE is lower by a third or more.

8.5 Cost Effectiveness of Energy-Efficient Designs

The cost effectiveness of energy-efficient designs for major appliances and space conditioning equipment depends on consumer energy prices and also on per-unit costs when new technologies are manufactured on a large scale. Therefore, the cost effectiveness of such designs varies.

A U.S. study, conducted in 2004 for the National Commission on Energy Policy (NCEP), reviewed the literature on energy-efficient designs for major appliances and space conditioning equipment and estimated their costs and energy savings relative to baseline designs (Rosenquist et al., 2004). For a number of products, it determined the efficiency level that provides users with the lowest life-cycle cost over the appliance lifetime. Table 8.4 lists the selected efficiency levels (expressed in terms of annual energy consumption in some cases), along with their associated cost of conserved energy (CCE) for consumers. The CCE in terms of $/kW h or MBtu (gas) is an expression of the extra first cost incurred to save a unit of energy. Calculation of CCE requires application of a present worth factor (PWF) to spread the initial incremental cost over the lifetime of the equipment. The PWF uses a discount rate to effectively amortize costs over time. The CCEs for electric appliances in Table 8.4 are all below the average U.S. residential electricity price.

8.6 Conclusion

Residential appliances consume significant amounts of electricity and natural gas. This chapter described the basic engineering principles of the major appliances and space conditioning equipment as well as promising energy-efficient designs.

Significant potential energy savings are possible beyond the models typically sold in the marketplace today. Among the various end-uses, energy savings range from 10% to 50%. Many of these efficient appliances appear to be cost effective at currently projected manufacturing costs, with simple payback times that are shorter than the typical appliance lifetimes of 10–20 years. If costs of efficiency improvements decrease, or if future energy prices increase, more of the potential energy savings that are already technically possible will become economically attractive. In addition, future research is likely to identify additional technological opportunities to save energy.

Acknowledgments

This work was supported by the Assistant Secretary for Energy Efficiency and Renewable Energy, Office of Buildings Technology, Office of Code and Standards, of the U.S. DOE, under Contract No. DE-AC03-76SF00098. The opinions expressed in this chapter are solely those of the authors and do not necessarily represent those of the Ernest Orlando Lawrence Berkeley National Laboratory, or the U.S. DOE.

Author Note

Eric Kleinert an experienced professional in the major appliance, air-conditioning, refrigeration, and service industry for 45 years. He is accomplished on all sides of the industry: owning and operating successful sales, service and parts companies; teaching vocational students preventive, diagnostic services and techniques; and, the author of the leading major appliance and HVAC repair textbooks on the market, Troubleshooting and Repairing Major Appliances, McGraw-Hill and HVAC and Refrigeration Preventive Maintenance, McGraw-Hill.

References

Energy Information Administration (EIA), 2005a. *Household Energy Consumption and Expenditures 2001*. U.S. Department of Energy, Washington, DC.

Energy Information Administration (EIA), 2005b. *Annual Energy Outlook 2005*. U.S. Department of Energy, Washington, DC.

Energy Information Administration (EIA), 2013. *Annual Energy Outlook 2013*. U.S. Department of energy, Washington, DC.

Malone, B. and Weir, K., 2001. State of the art for VIP usage in refrigeration applications. *International Appliance Manufacturing*.

9

Heat Pumps

Herbert W. Stanford III

CONTENTS

9.1 Heat Pump Concept

Heat pumps are *reverse cycle* building heating/cooling units or systems that can extract heat *from a building* and reject that heat to the environment, providing cooling for a building, and can switch from providing cooling to providing heating by extracting heat from the environment and rejecting that heat *into a building*. This heat transfer cycle can be accomplished using a vapor-compression refrigeration cycle or an absorption refrigeration cycle; though by far, vapor-compression systems are more widely used (see Section 9.4 for more discussion of absorption cycle systems).

All refrigeration cycles hinge on one common physical characteristic: if a chemical compound (which we can call a *refrigerant*) changes phase from a liquid to a gas, a process called *evaporation*, the compound must absorb heat to do so. Likewise, if refrigerant changes phase back from a gas to a liquid, the process of *condensation*, the absorbed heat must be rejected. Thus, all refrigeration cycles depend on circulating a refrigerant between a heat *source* (from which heat is removed, thus producing cooling) and heat *sink* (somewhere to which the collected heat can be rejected).

The *vapor-compression refrigeration cycle*, wherein a chemical substance alternately changes from liquid to gas and from gas to liquid, actually consists of four distinct steps:

1. *Compression*: Low-pressure, low-temperature refrigerant gas is compressed, thus raising its pressure by expending mechanical energy. There is a corresponding increase in temperature along with the increased pressure.

2. *Condensation*: The high-pressure, high-temperature gas is cooled by air or water that serves as a *heat sink* and condenses into liquid form.

3. *Expansion*: The high-pressure, high-temperature liquid flows through an orifice in the expansion valve, thus reducing both liquid pressure and temperature.

4. *Evaporation*: The low-pressure, low-temperature liquid absorbs heat from an air or water that serves as a *heat source* and evaporates as a gas or vapor. The low-pressure, low-temperature vapor flows to the compressor and the process repeats.

As shown in Figure 9.1, the vapor-compression refrigeration system consists of four components that perform these four steps. The *compressor* raises the pressure of the initially low-pressure refrigerant gas. The *condenser* is a heat exchanger that cools the high-pressure gas so that it changes phase to liquid. The *expansion valve* (or other pressure-reducing device) controls the pressure ratio, and thus flow rate, between the high- and low-pressure regions of the system. The *evaporator* is a heat exchanger that heats the low-pressure liquid, causing it to change phase from liquid to vapor.

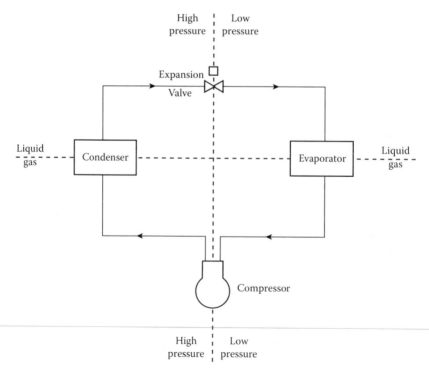

FIGURE 9.1
Schematic of basic vapor-compression refrigeration cycle.

For any heat pump utilizing the vapor-compression refrigeration cycle, there is a fifth element required: a *reversing valve*. This allows the condenser and evaporator heat exchangers to switch roles as the system switches from providing cooling to providing heating and vice versa by reversing the direction of refrigerant flow between these two components.

The amount of energy consumed by the compressor is dictated by the *lift* or pressure increase that the compressor must provide to raise the pressure/temperature of the low-pressure, low-temperature vapor that enters to the high-pressure, high-temperature vapor that leaves. On the low-pressure, low-temperature side of the system, refrigerant conditions depend on the temperature needed to provide required sensible cooling and dehumidification (typically about 45°F [7°C]), while the high-pressure/hot side refrigerant must be at conditions that allow it to condense at ambient temperatures of 105°F–125°F (51°C–65°C). If condensing temperatures are lowered to the heat sink's temperature condition, then the required lift is decreased and required energy input to the compressor also decreases.

The *type* of heat pump is typically defined by the environmental component that is used as a heat *source*, from which heat is extracted when providing heating, and heat *sink*, to which heat is rejected when providing cooling. The heat source/sink may be atmospheric air, directly or indirectly; the ground; or large water sources, such as lakes, rivers, or oceans. Subtypes of heat pumps are also sometimes defined on the basis of heat transfer media utilized: air, refrigerant, or water.

Vapor-compression cycle heat pumps have been available since the 1950s and factory-built heat pump units are widely used in residential and light commercial applications in mild climate regions, mostly in the South. However, since the early 2000s, the energy efficiency of these units, both air source and water source, has been improved dramatically:

Variable air volume supply air: Variable indoor airflow provided by the use of *electronically commutated motors* (ECMs). ECMs are DC motors that function using a built-in inverter and a magnet rotor and as a result are able to achieve greater efficiency in airflow systems than conventional *permanent split capacitor* (PSC) motors used in the past. Initially, silicon rectifier controllers were added to PSC motors to provide variable speed control. But, as PSC motor speed was reduced, efficiency suffered, falling from 65%–70% to as low as 12%. ECMs, on the other hand, maintain a high level (65%–75%) of efficiency over almost its full speed range. Additionally, unlike PSC motors, ECMs are not prone to overheating and do not require additional measures to offset the generation of heat.

By varying the indoor airflow rate in response to the imposed cooling and heating loads, indoor fan energy consumption is reduced significantly since no heat pump system is required to operate at peak capacity more than a few hours each year. Additionally, when providing cooling, variable airflow results in much better humidity control than provided by older, single-speed systems, a real boon for installations in the South.

Variable speed compressor(s): Variable compressor speed/load control provides a better match between the imposed load and the compressor capacity. Older, single-speed compressors cycled on and off as the imposed cooling/heating load was reduced, resulting in thermal losses at the beginning and end of each run cycle. The next level of improvement was to provide two-speed compressors and, for the last 10 years or so, fully variable speed compressors. Some very high-efficiency systems even use multiple variable speed compressor configurations in which one compressor is used during normal *part load* operation and the second is used only during *peak load* conditions.

    Variable heat sink/source medium flow: For both air-source and water-source heat pumps, variable sink/source flow is being applied to reduce energy consumption required for air or water transport.

    Demand defrost control: For air-source heat pumps, demand defrost control is incorporated to minimize energy losses associated with the buildup of frost or ice on the outdoor heat exchanger when the unit is providing heating.

And vapor-compression cycle heat pumps now typically utilize R-410A or R-134a refrigerant in lieu of the older R-22 refrigerant that is no longer available, significantly reducing the potential for atmospheric ozone depletion and warming in the event of a refrigerant leak. The refrigerant change has resulted in the heat transfer area of both indoor and outdoor heat exchanges being increased, improving both cooling/heating performance and energy consumption even further.

9.2 Air-Source Heat Pumps

Air-source heat pumps utilize atmospheric (outdoor) air as their ultimate heat source/sink, either extracting heat from outdoor air to provide heating or rejecting excess heat into it when cooling. Thus, the performance and seasonal efficiency of these heat pumps is significantly impacted by outdoor air temperature.

9.2.1 Premium Efficiency Air-Source Heat Pumps

Standard 90.1-2010, promulgated by the American Society of Heating, Refrigerating, and Air-Conditioning Engineers, Inc. (ASHRAE), forms the basis of most current energy conservation building codes in the United States. For air-source heat pumps, this standard defines efficiency on the basis of specific required minimum performance measurements:

    For cooling by units rated with a cooling output of less than 65,000 Btu/h (19 kW), *seasonal energy efficiency ratio* (SEER) is defined as the total cooling output of the heat pump during its normal cooling usage period (in Btu) divided by the total electric energy input during that same period (in watt hours).

    For cooling by units rated with a cooling output of 65,000 Btu/h (19 kW) or greater, *energy efficiency ratio* (EER) is defined as the ratio of net cooling capacity (in Btu/h) to the total rate of electrical input (in watts) under designated operating conditions.

    For heating by units rated with a cooling output of less than 65,000 Btu/h (19 kW), *heating seasonal performance factor* (HSPF) is defined as the total heating output of the heat pump during its normal heating usage period (in Btu) divided by the total electric energy input during that same period (in watt hours).

    For heating by units rated with a cooling output of 65,000 Btu/h (19 kW) or greater, heating efficiency is defined in terms of *coefficient of performance* (COP). COP is a dimensionless value defined as the net heating capacity of the heat pump (in Btu/h [kW]) divided by the energy input (in Btu/h [kW]) under designated operating conditions.

ASHRAE has established current minimum performance requirements for packaged (all components in one outdoor enclosure) and split system (separate indoor and outdoor components) air-source heat pumps at SEER = 13.0 and HSPF = 7.7 for heat pumps rated with a cooling output of less than 65,000 Btu/h (19 kW). All air-source heat pumps with a cooling output of less than 65,000 Btu/h (19 kW) available from larger equipment manufacturers will meet or exceed the minimum ASHRAE performance standards. Additionally, most manufacturers offer *high*-efficiency units that have SEER = 15–16 and HSPE = 8–9 and some offer *premium* efficiency units that have SEER = 19–20 and HSPE = 9–10, which comply generally with the EPA's Energy Star requirements.

For larger-capacity air-source heat pump units, the following table summarizes ASHRAE's current minimum requirements:

System Cooling Capacity (Btu/h)	Minimum Cooling Performance (EER)	Minimum Heating Performance (COP)
65,000–134,999 (19–39 kW)	11.0	3.3 at 47°F (9°C)/2.25 at 17°F (−8°C)
135,000–239,999 (40–70 kW)	10.6	3.2 at 47°F (9°C)/2.05 at 17°F (−8°C)
≥240,000 (≥70 kW)	9.5	

Again, though, manufacturers typically offer larger-capacity premium efficiency equipment with EER = 12–12.5 and COP (at 47°F [9°C]) = 3.5–4.5.

An additional important part of achieving high-efficiency air-source heat pump energy performance is to select each heat pump capacity as close as possible to the anticipated peak load. Since most heat pumps are installed in mild climate areas where cooling is more of a consideration than heating, units are typically selected on the basis of the peak imposed cooling load in order to minimize oversizing and loss of cooling efficiency. For example, a residence or commercial area may require 40,800 Btu/h (12 kW) or 3.4 tons of cooling. Because heat pump units are manufactured in specific capacity ranges, typically 2–6 tons at 1 ton increments, 7–1/2 tons, and 10 tons, contractors would normally recommend installation of a 4.0 tons heat pump for this example.

But the more efficient and less costly approach would be to install a 3.0 tons heat pump. Doing so would result in better match between imposed load and available capacity during part load operating periods, reducing energy consumption. The shortfall in capacity at peak load may result in indoor temperatures rising 1°F–2°F (1°C) above cooling set point temperature, but since the peak cooling period occurs for only a few hours each year, the negative comfort impact is minimal.

For applications located where heating requirements are more significant, sizing the heat pump unit on the basis of heating need rather than cooling need may result in improved heat pump performance and improved energy efficiency. While this selection may result in some oversizing relative to cooling and a small loss in cooling efficiency, the larger system will be able to provide compressor-based heating for longer periods, reducing the amount of low-efficiency supplement heating required, saving energy, and reducing operating costs.

Careful review of imposed heating and cooling loads is required to select an air-source heat pump unit and maximize energy efficiency. It is important that consumers and designers evaluate alternative unit selections and make the final decision on the basis of the lowest life-cycle cost, considering both first cost and anticipated energy costs of the anticipated life of the unit, typically 15 years.

9.2.2 Cold Climate Air-Source Heat Pumps

For all air-source heat pumps, and based on system thermodynamics, as the outdoor air temperature falls, the heat pump is able to extract less and less heat from the environment, while requiring more and more energy input to the compressor. Ultimately, the heat pump will be unable to extract the amount of heat needed to offset building losses and additional *supplemental* heating will be required. Typically, the supplemental heating capacity is provided by electric resistance heating coils that, by themselves, have a COP = 0.98. When the outdoor air is too cold for the refrigerant to extract any heat from it, there is no need to operate the compressor at all and all of the heating is done by the supplemental heater, at low efficiency and high expense.

However, since the 1990s, research to develop an air-source heat pump that would work in colder climates has been underway. One manufacturer now offers a cold-climate heat pump that features a two-speed, two-cylinder compressor for efficient operation, a backup booster compressor that allows the system to operate efficiently down to 15°F (−9°C), and a plate heat exchanger called an *economizer* that further extends the performance of the heat pump to well below 0°F (−18°C).

9.2.3 Dual Fuel Air-Source Heat Pumps

An alternative to the use of electric resistance supplemental heating, the dual fuel heat pump, has been in use since about 1980. This type of heat pump utilizes, today, a high-efficiency condensing furnace firing natural gas or LPG (propane) to provide heating when fossil fuel firing is more cost-effective than operating the unit compressor(s) and consuming electricity.

While the COP of a condensing furnace is essentially the same as for electric resistance heating, the much lower cost of natural gas or LPG relative to cost of electricity can result in much more economical heating. For example, with a typical residential electricity cost of $0.12/kWh, the cost of electric resistance heating will be $34 per 1,000,000 Btu ($0.116/kW). In December 2014, residential natural gas sold for $9.98 (U.S. average) per 1,000,000 Btu ($0.034/kW). Therefore, the heat pump system must maintain a minimum heating COP of 3.41, well above the ASHRAE minimum, to provide heating at a cost that is equal to or less than the cost of fossil fuel firing. Typically, the *tipping* point for switching from the heat pump's compressor-based heating cycle to fossil fuel–fired heating is when the outdoor temperature falls below about 30°F (−1°C).

9.3 Water-Source Heat Pumps

Air-source heat pumps units cannot easily be applied to larger, particularly multistoried, buildings that require a large number of independently controlled thermal zones. Larger apartment buildings, hotels, schools, office buildings, etc. simply will not have the outdoor space available for a large number of independent air-source heat pumps.

The alternative, then, is to go to water-source heat pumps. Typically, a water-source heat pump system consists of one or more individual single zone, water-cooled packaged units with a supply fan, indoor air coil, compressor, and water-to-refrigerant heat exchanger, connected to a common water supply system that acts as the heat source/sink for each individual heat pump unit.

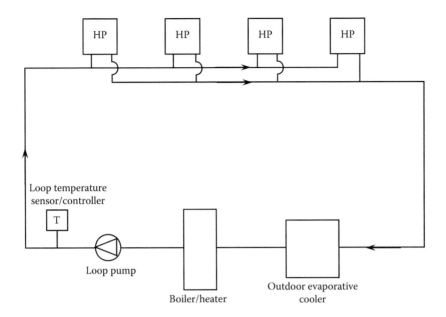

FIGURE 9.2
Schematic of basic closed-circuit water-source heat pump system.

9.3.1 Closed-Circuit Water-Source Heat Pump Systems

The most common, and very efficient, approach to using water-source heat pumps is to utilize a closed-circuit water heat exchanger as illustrated by Figure 9.2.

In this configuration, each packaged water-source heat pump refrigerant system reverses the direction of refrigerant flow from summer to winter with the following results: during summer, heat is removed from the space via the air coil (evaporator) and rejected to the central circulating water heat exchanger via the water-to-refrigerant heat exchanger (condenser). Excess heat rejected to this heat exchanger is, in turn, rejected to the outdoors through a closed-circuit evaporative cooler in order to maintain the heat exchanger temperature at or below 85°F (29°C).

During winter, heat is removed from the circulating water heat exchanger via the water-to-refrigerant heat exchanger (evaporator) and rejected to the space via the air coil (condenser). If enough heat is removed from the heat exchanger to lower its temperature to below 65°F (18°C), an auxiliary heater (typically a high-efficiency condensing hot water boiler) adds supplemental heat to the heat exchanger to maintain the required temperature.

ASHRAE stipulates minimum energy performance requirements only for closed heat exchanger water-source heat pump units of less than 135,000 Btu/h (40 kW) cooling capacity. The minimum requirements are a cooling EER = 12.0 and a heating COP = 4.2. Again, manufacturers offer *premium* performance units that can produce a cooling EER of 16 or higher and a heating COP of almost 5.0.

9.3.2 Closed-Circuit Geothermal Heat Pump Systems

Closed circuit geothermal heat pumps are a variation to the basic closed-circuit heat pump system, as discussed in the previous subsection, that are designed to take advantage of the earth (ground) or a large body of water as heat source/sink. The supplemental heater(s)

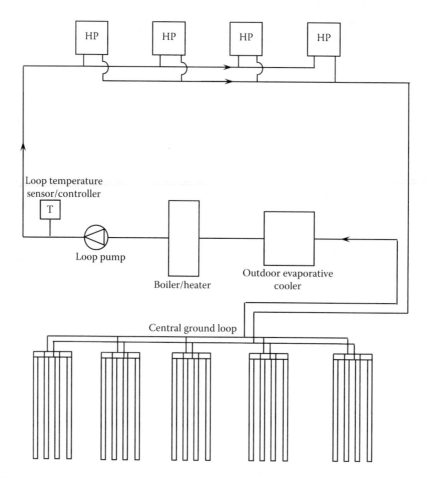

FIGURE 9.3
Schematic of basic closed-circuit geothermal heat pump system.

and evaporative cooler(s) that represent the heat source and sink in the basic closed-circuit heat pump system are augmented by a *ground-coupled heat exchanger* or water-coupled heat exchanger that is utilized as the primary system heat source and sink. The supplemental heater(s) and evaporative cooler(s) are maintained to augment the ground-coupled heat exchanger and/or to operate under emergency conditions.

Figure 9.3 illustrates the common ground-coupled geothermal heat pump system configuration.

The ground-coupled heat exchanger may be installed either horizontally directly in the ground or vertically through one or more wells, typically referred to as *boreholes*. The fluid could be either water or a refrigerant.

Geothermal heat pumps take advantage of the natural constant temperature of the ground 5–6 ft [1.5–1.8 m] below grade where the ground mean annual temperature remains between 45°F and 65°F [7°C–18°C], depending on location and water table. The ground temperature is warmer than the air above it in the winter and cooler than the air in the summer, resulting in the need for less compressor lift and more efficient heating and cooling.

The geothermal heat pump system has three major parts: the ground-coupled heat exchanger, the heat pump unit, and the air delivery system (ductwork and air outlets/inlets). The heat exchanger is a system of pipes that is buried in the shallow ground near the

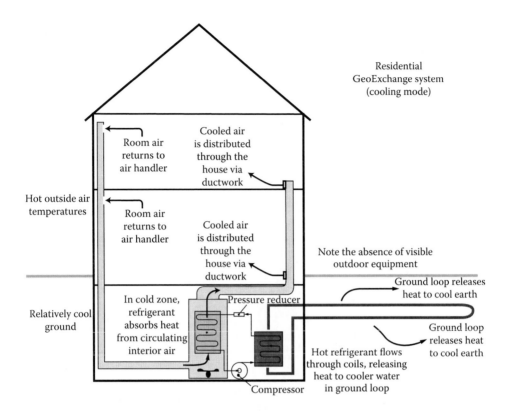

FIGURE 9.4
Example of a residential geothermal heat pump configuration in the summer months. (Courtesy of the Geothermal Heat Pump Consortium.)

building. A fluid (usually water or a mixture of water and antifreeze) circulates through the pipes to absorb or deposit heat within the ground (Figure 9.4).

In the winter, the heat pump removes heat from the heat exchanger and pumps it into the indoor air delivery system. In the summer, the process is reversed, and the heat pump moves heat from the indoor air into the heat exchanger. The heat removed from the indoor air during the summer can also be used to heat water (Figure 9.5).

The ground-coupled heat exchanger provides the means of transferring heat to the earth in summer and extracting heat from the earth in winter. Physically, the heat exchanger consists of several lengths of plastic pipe typically installed either in horizontal trenches or vertical boreholes that are backfilled to provide close contact with the earth.

Fluid inside the heat exchanger is pumped through a refrigerant heat exchanger in the geothermal heat pump. In the summer, it absorbs heat from the refrigerant and carries it to the ground through the ground-coupled heat exchanger piping to be rejected. In winter, it absorbs heat from the earth and transfers that heat to be extracted by the refrigerant.

Once installed, the heat exchanger remains out of sight beneath the surface. The ground heat exchanger consists of high-density polyethylene piping with an anticipated service life of at least 50 years.

Horizontal closed heat exchangers are the most cost-effective configuration when (1) adequate land is available, (2) soil conditions are such that trenches are easy to dig (e.g., no rock), and (3) trench depths are not defined by frost lines exceeding 36 in. [1 m] below grade. Trenching machines or backhoes are utilized to dig the trenches normally 5–6 ft

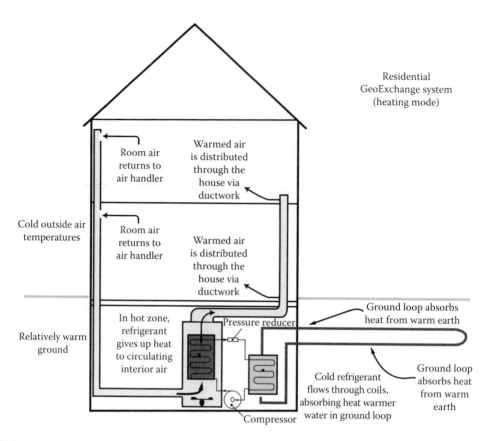

FIGURE 9.5
Example of a residential geothermal heat pump configuration in the winter months. (Courtesy of the Geothermal Heat Pump Consortium.)

below finished grade and then lay a series of parallel plastic pipes. Backfilling the trench requires that great care is taken to eliminate sharp rocks or debris that may damage the heat exchanger piping. Typically, compacted sand or a mix of sand and bentonite is backfilled for the first 6–12 in. [15–30 cm] to create good contract between the piping and earth. A typical horizontal heat exchanger will be about 400–600 ft [120–180 m] long per ton of required cooling capacity (Figure 9.6).

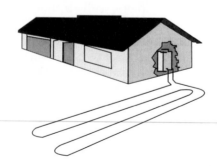

FIGURE 9.6
Example of a horizontal heat exchanger configuration in a residential installation. (Courtesy of the Geothermal Heat Pump Consortium.)

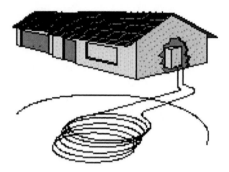

FIGURE 9.7
Example of a *slinky* heat exchanger configuration in a residential installation. (Courtesy of the Geothermal Heat Pump Consortium.)

The pipe may be curled into a slinky shape in order to fit more of it into shorter trenches. While this reduces the amount of land space needed, it may require more pipes (Figure 9.7).

For vertical heat exchangers, boreholes are drilled into the ground from 150 to 450 ft [45–140 m] deep. Each hole contains a single heat exchanger of pipe with a U-bend at the bottom. After the pipe is inserted, the hole is backfilled or grouted with a slurry of mixture of sand and bentonite. Each vertical pipe is then connected to a horizontal pipe, which is also concealed underground. The horizontal pipe then carries fluid in a closed system to and from each geothermal heat pump unit (Figure 9.8).

ASHRAE stipulates minimum performance requirements only for ground-coupled heat pump systems of less than 135,000 Btu/h [40 kW] cooling capacity, requiring a cooling EER = 13.4 and a heating COP = 3.6.

Despite their excellent energy efficiency, the application of ground-coupled heat pump systems has been relatively limited due to a number of *challenges* that significantly reduce their cost effectiveness, especially as conventional air-source and water-source heat pumps continue to improve their energy performance. These challenges include:

Land type and space for a ground-coupled heat exchanger: The soil in which a horizontal ground-coupled heat exchanger is installed must be reasonably finely textured, no rock, and soil that contains little clay. The soil in which boreholes and vertical heat

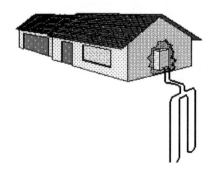

FIGURE 9.8
Example of a vertical heat exchanger configuration in a residential installation. (Courtesy of the Geothermal Heat Pump Consortium.)

exchangers are installed are less critical since the borehole is backfilled with a mix of sand and bentonite after the piping is installed to insure good heat transfer with the surrounding undisturbed earth.

Horizontal trench heat exchangers installed with two pipes per trench typically require 2000–4000 ft² (600–1200 m²) of ground surface area per ton of cooling capacity required. Obviously, that requirement limits the applicability of this type of heat exchanger. Vertical heat exchangers required only about 200–400 ft² (60–120 m²) of ground surface area per ton of cooling capacity required.

Costs of ground-coupled heat exchanger: Horizontal trenching, assuming no rock is present near the ground surface, is relatively inexpensive since it is accomplished by conventional backhoes or trenching machines fairly quickly. Drilling boreholes for vertical heat exchangers is more difficult and expensive than trenching. And, even with vertical heat exchangers, some horizontal trenching is required to connect each borehole piping heat exchanger to a common piping system.

Another cost element associated with vertical heat exchangers is the need to drill a test bore prior to designing the system in order to determine actual subsurface conditions. The soil thermal conductivity can vary as much as 400% depending on the type of soil and rock encountered, ground water hydrology, etc. This means that the number and depth of geothermal boreholes may vary significantly from site to site for the same system capacity.

Circuit pumping energy consumption: Geothermal heat pumps require a flow ranging from 2.5 to 3.5 gal/min (9.4–13.2 L/min) for each ton of cooling load. In older systems, the pumps to provide this flow operated at all times at a constant speed and, thus, significant pumping energy was consumed. Today's systems utilized variable flow pumping and the costs of pumping have been reduced.

System complexity: Geothermal heat pump systems have more components, more complex controls, and higher maintenance costs than conventional systems.

A more efficient alternative to the ground-coupled heat exchanger is a *water-coupled heat exchanger*. Large ponds, lakes, rivers, and even the ocean represent excellent heat sources/sinks for use with heat pumps. While water temperatures vary in these bodies of water summer to winter, the water temperature swing is always more moderate than for air, rarely falling below about 32°F [0°C] or rising above 85°F [29°C]. Even water that freezes on the surface is above freezing below the ice.

Water-to-water heat exchange is 20%–50% more efficient than water-to-ground heat exchange, significantly improving the efficiency of heat exchange of water-coupled systems over ground-coupled systems, and these systems typically have a cooling EER of 18–20 and a heating COP of 3–4.

Water-coupled heat exchangers may be configured with horizontal coiled piping submerged in the body of water. The alternative is to install a high-efficiency plate-and-frame heat exchanger with closed-circuit flow to the heat pump unit(s) on one side and flow from the body of water on the other. This second configuration, despite having somewhat higher pumping energy consumption, is preferred to avoid the problems with silting, drift, and mechanical damage that can occur to submerged piping loops.

While the applicability of water-coupled heat pump systems is limited simply because availability of usable water sources/sinks is more limited, the systems themselves are both less expensive and more energy efficient than ground-coupled systems and represent a very cost-effective approach when applicable.

9.3.3 Open-Circuit Geothermal Heat Pump Systems

Open-circuit geothermal heat pump systems, often called *groundwater heat pump systems*, use two or more groundwater wells for their heat source/sink, one or more wells from which groundwater is withdrawn and one or more wells into which the water is *injected* back into the ground after passing through the heat pump units. These systems are as efficient, sometimes even more so, than water-coupled heat pump systems.

However, these systems have a significant negative environmental impact. Since the water flow rate required for heat pump operation is relatively high, the amount of water withdrawn from an available aquifer amounts for a significant demand on the aquifer. Then, since the water reinjection point is typically far outside the *recharge zone* for the aquifer, simply pumping the water back into the ground really does not benefit the aquifer from which it was removed.

Since most aquifer levels in the United States have already fallen significantly over the past 50 years and continue to fall, the use of this heat pump system configuration and further stress on aquifers is not recommended.

9.3.4 Gas-Fired Engine-Driven Heat Pumps

To many, the use of electricity as a primary energy source is not desirable. The electrical power industry routinely releases huge quantities of CO_2 into the atmosphere, along with large quantities of SO_2 and NO_x. On a comparative basis, these poor greenhouse gas emissions make grid electricity a poor sustainable choice for any building heating and cooling system.

In the early 1990s, a U.S. manufacturer introduced a small (5 tons) natural-gas-fired engine-driven packaged air-source heat pump for the residential and light commercial failure. While the engine itself is no more efficient than grid electricity power production, its use of clean-burning natural gas or LPG (propane) resulted in significantly reduced greenhouse gas emissions and a lower operating cost due to the low cost of gas fuels compared to electricity.

But that heat pump was a dismal failure! Of over 4000 units sold, 100% of the units failed within the first 3 years of their operation and the manufacturer was forced to replace almost all of these systems with conventional units. The problem with this concept was not with the heat pump, but with the engine. In this case, the manufacturer attempted to use a modified engine designed for lawn tractors and small engine–generators designed for 150–500 h annual use in an application that required the engine to operate 4000–5000 h/year.

Theoretically, an engine-driven heat pump, coupled with recovered heat from the engine jacket and exhaust, can achieve a heating COP of 1.4–2.0. And, today, manufacturers have teamed with automotive engine manufacturers to offer gas-engine-driven packaged air-source heat pumps in the 8–20 tons cooling capacity range, primarily for commercial applications. These units have an advertised SEER of 18+ and heating COP of 1.5.

The development of smaller gas-engine-driven systems is reportedly underway in both Japan and at Oak Ridge National Laboratory, and domestic-scale units could be brought to market within a few years.

9.3.5 Heat Recovery Chiller/Heat Pump System

Large buildings are often cooled by chilled water systems that circulated water at 40°F–45°F [4°C–7°C] to cooling coils located in air-handling units located throughout the building. While small residential or light commercial buildings require cooling *or* heating, large buildings may require simultaneous cooling *and* heating. And, there is almost always a need to generate service hot water in large buildings.

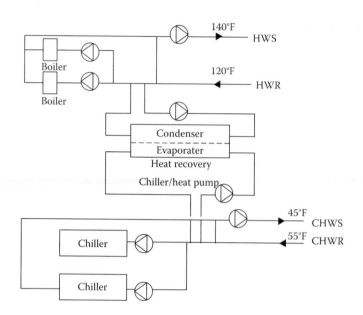

FIGURE 9.9
Schematic of HRC/heat pump system.

When providing cooling, the chilled water system rejects the excess collected heat to the outdoor air via cooling towers via condenser water at 85°F–95°F [29°C–35°C] temperature. These temperatures are typically too low to allow condenser water to be used for heating, so the *dedicated heat recovery chiller* (HRC) *heat pump* system is sometimes applied as illustrated by Figure 9.9.

Though not a truly *reverse cycle system*, this system is a heat pump configuration. A *lead* HRC is configured with a primary/secondary piping loop on the chilled water return piping upstream of the remaining cooling-only chillers. The HRC capacity is selected based on the need to provide heating during the summer. Thus, the HRC removes heat from the return chilled water, lowering its temperature, and rejects that heat to the hot water system at 130°F–140°F [54°C–60°C]. The effect is to reduce the imposed load on the cooling-only chillers by *precooling* the return chilled water while simultaneously producing hot water at sufficiently high temperature to be useful.

For this concept to be cost effective, there must be a need for heating during the summer cooling period. This heat may be needed for reheat associated with space temperature or humidity control, service hot water heating, pool heating, etc. To maximize the energy efficiency of this concept, though, there must be a need for chilled water cooling for as long as feasible. This means that the use of airside economizer cycles, required by most energy conservation building codes, needs to be carefully evaluated to determine when it is more efficient to provide mechanical cooling and capture the rejected heat versus using the outdoor air to provide *free* cooling and operating boilers to provide heat.

9.3.6 Variable Refrigerant Flow Heat Pump System

Since about 2000, the use of *variable refrigerant flow* (VRF) systems that achieve a better level of thermal zoning in buildings while using a common outdoor air unit has increased significantly. With VRF systems, up to 16 indoor evaporator units of varying capacity can be coupled with a single outdoor condensing unit. Currently widely applied in large buildings such as offices and hospitals outside the United States, especially in Japan, Europe, and Canada, these systems are just *catching on* in the United States.

These systems use multiple compressors, including inverter-driven variable speed units, and deliver excellent part-load performance and zoned temperature control, resulting in excellent occupant comfort. The basic difference between these systems and conventional HVAC systems is that they circulate refrigerant directly to multiple indoor units, rather than using water (as in a chiller) or air (as in a ducted DX system) to achieve heat transfer to the space.

VRF systems are extremely flexible, enabling a single condensing unit to be connected to a large number of indoor units of varying capacity and configuration. The exact number of indoor units varies according to the manufacturer, but one typical manufacturer allows connection of up to 16 indoor units to one condensing unit or up to 30 indoor units on a single refrigerant circuit supplied by 3 outdoor units.

Typically, each condensing unit uses two or three compressors, one of which is an inverter-driven variable speed compressor. Systems are commonly designed by combining multiple condensing units to achieve system capacities of up to several hundred tons.

Energy savings are due to several factors:

High part-load efficiency: Because VRF systems consist of multiple compressors, some of which are variable speed, the system's part load efficiency is excellent. A typical dual-compressor system can operate at 21 capacity steps. Since most HVAC systems spend most of their operating hours between 30% and 70% of their maximum capacity, a load range in which the COP of the system is very high and, thus, the seasonal energy efficiency of these systems is excellent.

Effective zone control: Indoor units can easily be turned off in locations needing no cooling, while the system retains highly efficient operation.

Heat pump operation: In buildings where simultaneous heating and cooling are needed, such as many office buildings, the system is configured as a three-pipe heat recovery system. In this configuration, refrigerant flow control is used to circulate refrigerant from the discharge of the evaporators in space being cooled to the evaporators of zones needing heat and vice versa. By using refrigerant to move heat between zones, a very high heating COP can be realized (perhaps as high as 4.0).

Because the energy savings of VRF systems are so application dependent, it is difficult to make definitive, general statements about their energy efficiency. Initial estimates of energy savings are in the range of 5%–15%, with higher savings in hot, humid climates, and lower savings in cold climates.

9.4 Advanced-Technology Heat Pumps

9.4.1 Absorption Cycle Heat Pumps

Absorption cycle heat pumps are essentially air-source heat pumps driven not by electricity but by a heat source such as natural gas, propane, solar-heated water, or geothermal-heated water. Because natural gas is the most common heat source for absorption heat pumps, they are also referred to as *gas-fired heat pumps*.

Residential absorption heat pumps use an ammonia–water absorption cycle to provide heating and cooling. As in a vapor-compression cycle heat pump, the refrigerant (in this case, ammonia) is condensed in one heat exchanger to release its heat; its pressure is then reduced and the refrigerant is evaporated to absorb heat.

The absorption cycle heat pump has no compressor. Evaporated ammonia is absorbed into water, and a relatively low-power pump then pumps the ammonia–water solution to a heat exchanger at a slightly higher pressure where heat is added to essentially boil the ammonia out of the water.

The *generator absorber heat exchanger* boosts the efficiency of the unit by recovering the heat that is released when the ammonia is absorbed into the water. Other innovations include high-efficiency vapor separation, variable ammonia flow rates, and low-emissions, variable-capacity combustion of the natural gas.

Absorption cycle cooling only is now commercially available in the 5+ ton capacity range and absorption cycle heat pumps are under development.

9.4.2 Solar-Assisted Heat Pumps

Solar-assisted (sometimes called *solar-augmented*) heat pumps utilize rooftop solar heat collectors configured to augment air-source or ground-source geothermal heat exchangers. These systems are generally series configured with solar panels that are arranged to act as an additional heat source for the heat pump.

For air-source or water-source heat pumps, solar panels can provide higher temperature heat during the winter, reducing the need for supplemental heating by electric coils or boilers. This configuration can also reduce surface land requirements for ground-coupled heat pumps, expanding their range of applicability.

Bibliography

American Society of Heating, Refrigerating, and Air-Conditioning Engineers. *Commercial/Institutional Ground-Source Heat Pump Engineering Manual*, American Society of Heating, Refrigerating, and Air-Conditioning Engineers, Atlanta, GA, 1995.

American Society of Heating, Refrigerating, and Air-Conditioning Engineers. *ASHRAE Handbook: HVAC Applications*, American Society of Heating, Refrigerating, and Air-Conditioning Engineers, Atlanta, GA, 2011 (Chapter 34).

American Society of Heating, Refrigerating, and Air-Conditioning Engineers. *ASHRAE Handbook: HVAC Systems and Equipment*, American Society of Heating, Refrigerating, and Air-Conditioning Engineers, Atlanta, GA, 2012 (Chapters 7 and 49).

ANSI/ASHRAE/IES Standard 90.1-2010. *Energy Standard for Buildings Except Low-Rise Residential Buildings*, American Society of Heating, Refrigerating, and Air-Conditioning Engineers, Atlanta, GA, 2012.

Staffell I., Brett D., Brandon N., and Hawkes, A. A review of domestic heat pumps, *Energy Environmental Science*, 2012, 5: 9291–9306.

Stanford, H.W., III. *Analysis and Design of Heating, Ventilating, and Air-Conditioning Systems*, Prentice-Hall, Englewood Cliffs, NJ, 1988.

Stanford, H.W., III. *HVAC Water Chillers and Cooling Towers: Fundaments, Application, and Operation*, 2nd edn. CRC Press/Taylor & Francis Group, Boca Raton, FL, 2010.

10

Industrial Energy Efficiency and Energy Management

Craig B. Smith, Barney L. Capehart, and Wesley M. Rohrer, Jr.*

CONTENTS

---

* Deceased.

10.1 Introduction

The industrial sector in the United States is highly diverse—consisting of manufacturing, mining, agriculture, and construction activities—and consumes one-third of the nation's primary energy use, at an annual cost of around $200 billion.[1] The industrial sector encompasses more than three million establishments engaged in manufacturing, agriculture, forestry, construction, and mining. These industries require energy to light, heat, cool, and ventilate facilities (end uses characterized as energy needed for comfort). They also use energy to harvest crops, process livestock, drill and extract minerals, power various manufacturing processes, move equipment and material, raise steam, and generate electricity. Some industries require additional energy fuels for use as raw materials—or feedstocks—in their production processes. Many industries use by-product fuels to satisfy part or most of their energy requirements. In the more energy-intensive manufacturing and nonmanufacturing industries, energy used by processes dwarfs the energy demand for comfort.

The U.S. sector energy use for 2014 is shown in Figure 10.1, industrial energy use is shown in Figure 10.2, and industrial electricity use is shown in Figure 10.3.

Manufacturing companies, which use mechanical or chemical processes to transform materials or substances to new products, account for about 69% of the total industrial sector use. The "big three" in energy use are petroleum, chemicals, and paper; these manufacturers together consume almost one-half of all industrial energy. The "big six," which adds the primary metals group, the food and kindred products group, as well as the stone, clay, and glass group, together accounts for 88% of manufacturing energy use, and over 60% of all industrial sector energy consumption.[1]

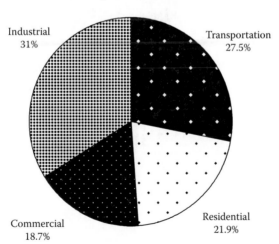

Total consumption: 98,324 trillion Btu

Industrial 31%

Transportation 27.5%

Commercial 18.7%

Residential 21.9%

FIGURE 10.1

The U.S. energy use by sector, 2014. (From U.S. DOE, Energy Information Agency, *Monthly Energy Review*, Washington, DC, March 27, 2014; U.S. Department of Energy, Office of Industrial Technologies, *Industrial Technology Program*, Washington, DC, 2005.)

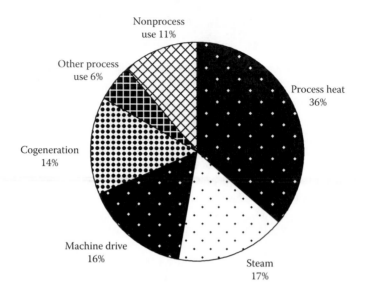

FIGURE 10.2
Manufacturing energy use 2006 (end-use basis). (From U.S. Department of Energy, Energy Information Agency, Manufacturing Energy Consumption Survey, Washington, DC, September 2012.)

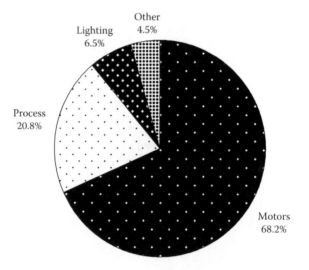

FIGURE 10.3
Manufacturing electrical energy use 2006 (end-use basis). (From U.S. Department of Energy, Energy Information Agency, Manufacturing Energy Consumption Survey, 2006, Washington, DC, 2002.)

According to the U.S. Energy Information Administration, energy efficiency in the manufacturing sector improved by 25% over the period 1980–1985.[2] During that time, manufacturing energy use declined 19%, and output increased 8%. These changes resulted in an overall improvement in energy efficiency of 25%. However, the "big five" did not match this overall improvement; although their energy use declined 2%, their output decreased by 5%, resulting in only a 17% improvement in energy efficiency during 1980–1985.

This 5-year record of improvement in energy efficiency of the manufacturing sector came to an end, with total energy use in the sector growing by 10% from 1986 to 1988, and overall manufacturing energy intensity stagnated during 1985–1994 due to falling and low energy prices, economic recession during part of this period, and a recovery in some of the more energy-intensive groups such as steel and aluminum production. However, industrial energy intensity again declined significantly during the late 1990s due to high capital investment, rapid industrial modernization, and explosive growth of "high-tech" industries that are not energy-intensive.

Since 1980, the overall value of industrial output increased through 2003, while the total energy consumed by the industrial sector has fallen overall throughout 2003.[3] This relationship is shown in Figure 10.4, where the consumption index for both primary and site energy is greater than the output index before 1980, and less afterward, with the gap consistently widening in the late 1980s. New energy-efficient technology and the changing production mix from the manufacture of energy-intensive products to less intensive products are responsible for this change.

Continuing this overall record of energy efficiency improvements in industry will require emphasis on energy management activities, as well as making capital investments in new plant processes and facilities improvements. Reducing the energy costs per unit of manufactured product is one way that the United States can become more competitive in the global industrial market. The U.S. Department of Energy has formally recognized these multiple benefits to the country by including the following statement in its 2004 Industrial Technology Program Report: "By developing and adopting more energy efficiency technologies, U.S. industry can boost its productivity and competitiveness while strengthening national energy security, improving the environment, and reducing emissions linked to global climate change."[4] Additionally, it is interesting to note that

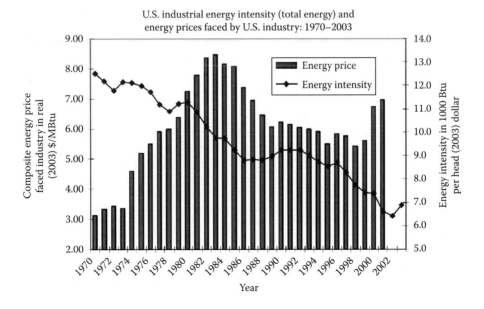

FIGURE 10.4
Industrial energy intensity 1980–2003. (From NAM, *Efficiency and Innovation in U.S. Manufacturing Energy Use*, National Association of Manufacturers, Washington, DC, 2005. With permission.)

Japan—one of the United States' major industrial competitors—has a law that says every industrial plant must have a full-time energy manager.[5]

Manufacturing energy intensity continued its gradual drop, as documented by the National Renewable Energy Laboratory report in 2009 showing the industrial use of Btu per dollar of gross domestic product fell by 20% from 1998 to 2007 (Ref. #7).

Several studies of industrial energy efficiency have been performed in recent years, and the results from the studies show that there is a readily achievable, cost-effective, 20% reduction in industrial consumption using good energy management practices and energy-efficient equipment. One highly credible study that has been done of the potential for industrial energy efficiency improvement in the United States is the "Energy Use, Loss and Opportunities Analysis: U.S. Manufacturing and Mining" report issued by the U.S. Department of Energy, Office of Industrial Technologies (OIT), in December 2004.[1] This report, together with the report on "Efficiency and Innovation in U.S. Manufacturing Energy Use" from the National Association of Manufacturers (NAM), conducted with the Alliance to Save Energy, makes a strong and very credible case for the achievement of a 20% savings in energy for the industrial sector.[3]

Earlier studies by the national laboratories in the United States produced a composite report, "Scenarios for a Clean Energy Future," in 2000 that also supported the achievement of a 20% reduction in industrial energy use. One element of this study was to examine data from the Industrial Assessment Center (IAC) Program (discussed in more detail later in this chapter) from 12,000 plant energy audits making 82,000 recommendations for actions to increase energy efficiency in the facilities audited.[6] Results from these real-world energy audits also supported the 20% reduction estimate.

Other groups, such as Lawrence Berkeley Laboratory, have performed studies of industrial energy efficiency, and their results from their studies show improvements of 20% or greater in industrial energy efficiency.[7] A recent study from the American Council for an Energy-Efficient Economy (ACEEE) showed the potential for a 40% reduction in electricity use for fan and pumping applications in industry.[8] Specific details and recommendations from these studies are presented in Section 10.5.

The most recent projections of the energy efficiency gains available in the manufacturing sector comes from the ACEEE, which in their January 2012 report (Ref. #29), "The Long Term Energy Efficiency Potential: What the Evidence Suggests," provided three different scenarios for energy consumption in the United States, and in the industrial sector, of between 40% and 58% reduction in energy use relative to their 2050 base case. The executive summary of their report is as follows:

> In 2010 the U.S. used just under 100 quads of total energy resources to power our economy. Using the Energy Information Administration's (EIA) Annual Energy Outlook we project that our total energy needs might rise to about 122 quads of energy by the year 2050. In this report we explore a set of energy efficiency scenarios that emphasizes a more productive investment pattern, one that can enable the U.S. economy to substantially lower overall energy expenditures should we choose to invest in and develop that larger opportunity. Building on the historical record of energy efficiency investments and their contribution to the nation's economic well being, we highlight three economy wide, long term scenarios to explore the potential contributions that more energy efficient behaviors and investments might play in reducing overall energy use by the year 2050. These three are
>
> 1. *Reference case*: A continuation of trends projected by EIA for the 2030–2035 period
> 2. *Advanced scenario*: It includes penetration of known advanced technologies

3. *Phoenix scenario*: In addition to advanced technologies also includes greater infrastructure improvements and some displacement of existing stock to make way for newer and more productive energy efficiency technologies, as well as configurations of the built environment that reduce energy requirements for mobility

For the industrial sector, energy intensity is projected to improve in the Reference Case by about 1%/year through 2050, but leading companies have been achieving continued improvements at more than double this rate. In our Advanced Scenario we project a 2%/year improvement rate for overall industrial energy intensity, which increases to 2.75% in our Phoenix Scenario. Future energy efficiency opportunities will come less from seeking out individual sources of waste and more from optimization of complex systems enabled by advances in information, communication and computational infrastructure. Most of the energy use in industry is in processes, not individual equipment, so improved processes represent the largest opportunity for energy intensity improvements. Current focus has been on process optimization, but we anticipate that even greater opportunities exist in the optimization of entire supply chains that may span many companies and supply chain integration that allows for efficient use of feedstocks and elimination of wasted production.

The really good news in the ACEEE study is that there is a tremendous potential for saving energy in the industrial sector, which is mainly the manufacturing sector. For their Advanced Scenario, which projects a 2%/year reduction in industrial energy intensity, this means that energy use in the industrial sector would fall to half its value in 2010 by 2045, a 35-year period.

10.2 Industrial Energy Management and Efficiency Improvement

10.2.1 Setting Up an Energy Management Program

The effectiveness of energy utilization varies with specific industrial operations because of the diversity of the products and the processes required to manufacture them. The organization of personnel and operations involved also varies. Consequently, an effective energy management program should be tailored for each company and its plant operations. There are some generalized guidelines, however, for initiating and implementing an energy management program. Many large companies have already instituted energy management programs and have realized substantial savings in fuel and electric costs. Smaller industries and plants, however, often lack the technical personnel and equipment to institute and carry out effective programs. In these situations, reliance on external consultants may be appropriate to initiate the program. Internal participation, however, is essential for success. A well planned, organized, and executed energy management program requires a strong commitment by top management.

Assistance also can be obtained from local utilities. Utility participation would include help in getting the customer started on an energy management program, technical guidance, or making information available. Most electric and gas utilities today have active programs that include training of customer personnel or provision of technical assistance. Table 10.1 summarizes the elements of an effective energy management program. These are discussed further in more detail.

TABLE 10.1

Elements of an Energy Management Program

Phase 1: Management Commitment

1.1   Commitment by management to an energy management program

1.2   Assignment of an energy management coordinator

1.3   Creation of an energy management committee of major plant and department representatives

Phase 2: Audit and Analysis

2.1   Review of historical patterns of fuel and energy use

2.2   Facility walk-through survey

2.3   Preliminary analyses, review of drawings, data sheets, equipment specifications

2.4   Development of energy audit plans

2.5   Conduct facility energy audit, covering
       a. Processes
       b. Facilities and equipment

2.6   Calculation of annual energy use based on audit results

2.7   Comparison with historical records

2.8   Analysis and simulation step (engineering calculations, heat and mass balances, theoretical efficiency
       calculations, computer analysis, and simulation) to evaluate energy management options

2.9   Economic analysis of selected energy management options (life cycle costs, rate of return, benefit–cost ratio)

Phase 3: Implementation and Submeters

3.1   Establish energy effectiveness goals for the organization and individual plants

3.2   Determine capital investment requirements and priorities

3.3   Establish measurement and reporting procedures, install monitoring and recording instruments as required

3.4   Institute routine reporting procedures ("energy tracking" charts) for managers and publicize results

3.5   Promote continuing awareness and involvement of personnel

3.6   Provide for periodic review and evaluation of overall energy management program

---

10.2.1.1 Phase I: Management Commitment

A commitment by the directors of a company to initiate and support a program is essential. An energy coordinator is designated, and an energy management committee is formed. The committee should include personnel representing major company activities utilizing energy. A plan is formulated to set up the programs with a commitment of funds and personnel. Realistic overall goals and guidelines in energy savings should be established based on overall information in the company records, projected activities, and future fuel costs and supply. A formal organization as described earlier is not an absolute requirement for the program; smaller companies will simply give the energy management coordination task to a staff member.

10.2.1.1.1 Organizing for Energy Conservation Programs

The most important organizational step that will affect the success of an energy management (EM) program is the appointment of one person who has full responsibility for its operation. Preferably that person should report directly to the top management position and be given substantial authority in directing technical and financial resources within the bounds set by the level of management commitment. It is difficult to stress enough the importance of making the position of plant energy manager a full-time job. Any diversion of interest and attention to other aspects of the business is bound to badly affect the EM program. One reason is that the greatest opportunity for energy cost control and energy efficiency gains is in improved operational and maintenance practices. Implementing and sustaining good operational and maintenance procedures is an exceedingly demanding job and requires a constant attention and a dedication to detail that is rarely found in corporate business life. The energy manager should be energetic, enthusiastic, dedicated, and political.

The second step is the appointment of the plant EM committee. This should consist of one group of persons who are able to and have some motivation for cutting fuel and electric costs and a second group who have the technical knowledge or access to data needed for the program department managers or their assistants. Thus, the EM committees should include labor representatives, the maintenance department head, a manager of finance or data storage, some engineers, and a public relations person. The energy manager should keep up to date on the energy situation daily, convene the committee weekly, and present a definitive report to top management at least monthly and at other times when required by circumstance. It is suggested also that several subcommittees be broken out of the main committee to consider such important aspects as capital investments, employee education, operator-training programs, external public relations, and so on. The committee will define strategy, provide criticism, publish newsletters and press releases, carry out employee programs, argue for the acceptance of feasible measures before management, represent the program in the larger community, and be as supportive as possible to the energy coordinator. This group has the most to risk and the most to gain. They must defend their own individual interests against the group, but at the same time must cooperate in making the program successful and thus be eligible for rewards from top management for their good work and corporate success.

As the EM program progresses to the energy audit and beyond, it will be necessary to keep all employees informed as to its purposes, its goals, and how its operation will impact plant operations and employee routine, comfort, and job performance. The education should proceed through written and oral channels as best benefits the organizational structure. Newsletters, posters, and employee meetings have been used successfully.

In addition to general education about energy conservation, it may prove worthwhile to offer specialized courses for boiler, mechanical, and electrical equipment operators and other workers whose jobs can affect energy utilization in the plant. The syllabuses should be based on thermodynamic principles applied to the systems involved and given on an academic level consistent with the workers' backgrounds. Long-range attempts to upgrade job qualifications through such training can have very beneficial effects on performance. The courses can be given by community colleges, by private enterprises professional societies, or by in-house technical staff, if available.

The material presented here on organization is based on the presumption that a considerable management organization already exists and that sufficient technical and financial resources exist for support of the energy management program as outlined. Obviously, very small businesses cannot operate on this scale; however, we have found many small companies that have carried out effective energy management efforts.

10.2.1.1.2 *Setting Energy Conservation Goals*

It is entirely appropriate and perhaps even necessary to select an energy reduction goal for the first year of the program very early in the program. The purpose is to gain the advantage of the competitive spirit of those employees that can be aroused by a target goal. Unfortunately, the true potential for conservation and the investment costs required to achieve it are not known until the plant energy audit is completed and a detailed study made of the data. Furthermore, a wide variety of energy-use patterns exists even with a single industry.

However, looking at the experience of other industries that have set goals and met them can provide some useful guidance.

An excellent example of a long-term successful energy management program in a large industrial corporation is that of the 3M Company, headquartered in St. Paul, Minnesota.[9] 3M is a large, diversified manufacturing company with more than 50 major product lines; it makes some 50,000 products at over 50 different factory locations around the country. The corporate energy management objective is to use energy as efficiently as possible in all operations; the management believes that all companies have an obligation to conserve energy and all other natural resources.

Energy productivity at 3M improved 63% from 1973 to 2004. They saved over $70 million in 2004 because of their energy management programs and saved a total of over $1.5 billion in energy expenses from 1973 to 2004. From 1998 to 2004, they reduced their overall energy use by 27% in their worldwide operations.[10] Their program is staffed by three to six people who educate and motivate all levels of personnel on the benefits of energy management. The categories of programs implemented by 3M include conservation, maintenance procedures, utility operation optimization, efficient new designs, retrofits through energy surveys, and process changes.

Energy efficiency goals at 3M are set, and then the results are measured against a set standard to determine the success of the programs. The technologies that have resulted in the most dramatic improvement in energy efficiency include heat recovery systems, high-efficiency motors, variable-speed drives (VSDs), computerized facility management systems, steam-trap maintenance, combustion improvements, variable-air-volume systems, thermal insulation, cogeneration, waste-steam utilization, and process improvements. Integrated manufacturing techniques, better equipment utilization, and shifting to non-hazardous solvents have also resulted in major process improvements.

The energy management program at 3M has worked very well, but the company's management is not yet satisfied. They have historically set a goal of further improving energy efficiency at a rate of 2%–3%/year for the next 5 years. This goal has produced a 10%–15% reduction in energy use per pound of product or per square foot of building space. They expect to substantially reduce their emissions of waste gases and liquids, to increase the energy recovered from wastes, and to constantly increase the profitability of their operations. 3M continues to stress the extreme importance that efficient use of energy can have on their industrial productivity.

In 2010, 3M met their 20% goal of reducing energy use from 2005. Their new goal is a 15% reduction in energy use per pound of product by 2015.

10.2.1.2 Phase 2: Audit and Analysis

10.2.1.2.1 Energy Audit of Equipment and Facilities

Historical data for the facility should be collected, reviewed, and analyzed. The review should identify gross energy uses by fuel types, cyclic trends, fiscal year effects, dependence on sales or workload, and minimum energy-use ratios. Historical data are graphed in a form similar to the one shown in Chapter 5. Historical data assist in planning a detailed energy audit and alert the auditors as to the type of fuel and general equipment to expect. A brief facility walk-through is recommended to establish the plant layout, major energy uses, and primary processes or functions of the facility.

The energy audit is best performed by an experienced or at least trained team, since visual observation is the principal means of information gathering and operational assessment. A team would have from three to five members, each with a specific assignment for the audit. For example, one auditor would check the lighting, another the HAVC system, another the equipment and processes, another the building structure (floor space, volume, insulation, age, etc.), and another the occupancy use schedule, administration procedures, and employees' general awareness of energy management.

The objectives of the audit are to determine how, where, when, and how much energy is used in the facility. In addition, the audit helps to identify opportunities to improve the energy-use efficiency and its operations. Some of the problems encountered during energy audits are determining the rated power of equipment, determining the effective hours of use per year, and determining the effect of seasonal, climatic, or other variable conditions on energy use. Equipment ratings are often obscured by dust or grease (unreadable nameplates). Complex machinery many not have a single nameplate listing the total capacity, but several giving ratings for component equipment. The effect of load is also important because energy use in a machine operating at less than full load may be reduced along with a possible loss in operating efficiency.

The quantitative assessment of fuel and energy use is best determined by actual measurements under typical operational conditions using portable or installed meters and sensing devices. Such devices include light meters, ammeters, thermometers, air flowmeters, and recorders. In some situations, sophisticated techniques such as infrared scanning or thermography are useful. The degree of measurement and recording sophistication naturally depends on available funds and the potential savings anticipated. For most situations, however, nameplate and catalog information are sufficient to estimate power demand. Useful information can be obtained from operating personnel and their supervisors particularly as it relates to usage patterns throughout the day.

The first two columns of the form are self-explanatory. The third column is used for the rated capacity of the device (e.g., 5 kW). The sixth column is used if the device is operated at partial load. Usage hours (column 7) are based on all work shifts and are corrected to account for the actual operating time of the equipment. The last three columns are used to convert energy units to a common basis (e.g., MJ or Btu).

Data recorded in the field are reduced easily by the use of specialized software or spread sheets that provide uniform results and summaries in a form suitable for review or for further analysis. Computer analysis also provides easy modification of the results to reflect specific management reporting requirements or to present desired comparisons for different energy use, types of equipment, and so on.

10.2.1.2.2 In-Plant Metering

Submetering reduces the work and time required for an energy audit; indeed, it does much more than that. Because meters are tools for assessing production control and for measuring equipment efficiency, they can contribute directly to energy conservation and cost containment. Furthermore, submetering offers the most effective way of evaluating the worth of an energy efficiency measure. Too many managers accept a vendor's estimate of fuel savings after buying a recuperator. They may scan the fuel bills for a month or two after the purchase to get an indication of savings—usually in vain—and then relax and accept the promised benefit without ever having any real indication that it exists. It may well be that, in fact, it does not yet exist. The equipment may not be adjusted correctly or it may be operated incorrectly, and there is no way of knowing without directly metering the fuel input. It is estimated that at least 2.5% waste is recoverable by in-plant metering.

Oil meters are just as effective as gas meters used in the same way and are even less expensive on an energy-flow basis. Electric meters are particularly helpful in monitoring the continued use of machines or lighting during shutdown periods and for evaluating the efficacy of lubricants and the machinability of feedstock. The use of in-plant metering can have its dark side too. The depressing part is the requirement for making periodic readings.

It does not stop even there. Someone must analyze the readings so that something can be done about them. If full use is to be made of the information contained in meter readings, it must be incorporated into the energy information portion of the management information system. At the very least, each subreading must be examined chronologically to detect malfunctions or losses of efficiency. Better still, a derived quantity such as average energy per unit of production should be examined.

10.2.1.2.3 A Special Case: Energy Audit of a Process

In some manufacturing and process industries, it is of interest to determine the energy content of a product. This can be done by a variation of the energy audit techniques described earlier. Since this approach resembles classical financial accounting, it is sometimes called energy accounting. In this procedure, the energy content of the raw materials is determined in a consistent set of energy units. Then, the energy required for conversion to a product is accounted for in the same units. The same is done for energy in the waste streams and the by-products. Finally, the net energy content per unit produced is used as a basis for establishing efficiency goals.

In this approach, all materials used in the product or used to produce it are determined. Input raw materials used in any specific period are normally available from plant records. Approximations of specific energy content for some materials can be found in the literature or can be obtained from the U.S. Department of Commerce or other sources. The energy content of a material includes that due to extraction and refinement as well as an inherent heating value it would have as a fuel prior to processing. Consequently, nonfuel-type ores in the ground are assigned zero energy, and petroleum products are assigned their alternate value as a fuel prior to processing in addition to the refinement energy. The energy of an input metal stock would include the energy due to extraction, ore refinement to metal, and any milling operations.

Conversion energy is an important aspect of the energy audit, since it is under direct control of plant management. All utilities and fuels coming into the plant are accounted for. They are converted to consistent energy units (joules or Btu) using the actual data available on the fuels or using approximate conversions.

Electrical energy is assigned the actual fuel energy required to produce the electricity. This accounts for power conversion efficiencies. A suggested approach is to assume (unless actual values are available from your utility) that 10.8 MJ (10,200 Btu) is used to produce 3.6 MJ (1 kW h), giving a fuel conversion efficiency of 3.6 + 10.8 = 0.33 or 33%.

The energy content of process steam includes the total fuel and electrical energy required to operate the boiler as well as line and other losses. Some complexities are introduced when a plant produces both power and steam, since it is necessary to allocate the fuel used to the steam and power produced. One suggested way to make this allocation is to assume that there is a large efficient boiler feeding steam to a totally condensing vacuum turbine. Then, one must determine the amount of extra boiler fuel that would be required to permit the extraction of steam at whatever pressure while maintaining the constant load on the generator. The extra fuel is considered the energy content of the steam being extracted.

Waste disposal energy is that energy required to dispose of or treat the waste products. This includes all the energy required to bring the waste to a satisfactory disposal state. In a case where waste is handled by a contractor or some other utility service, it would include the cost of transportation and treatment energy.

If the plant has by-products or coproducts, then energy credit is allocated to them. A number of criteria can be used. If the by-product must be treated to be utilized or recycled (such as scrap), then the credit would be based on the raw material less the energy expended to treat the by-product for recycle. If the by-product is to be sold, the relative value ratio of the by-product to the primary product can be used to allocate the energy.

10.2.1.2.4 Analysis of Audit Results and Identification of Energy Management Opportunities

Often, the energy audit will identify immediate energy management opportunities, such as unoccupied areas that have been inadvertently illuminated 24 h/day, equipment operating needlessly, and so on. Corrective housekeeping and maintenance action can be instituted to achieve short-term savings with little or no capital investment.

An analysis of the audit data is required for a more critical investigation of fuel waste and identification of the potential for conservation. This includes a detailed energy balance of each process, activity, or facility. Process modification and alternatives in equipment design should be formulated, based on technical feasibility and economic and environmental impact. Economic studies to determine payback, return on investment, and net savings are essential before making capital investments.

10.2.1.3 Phase 3: Implementation and Submeters

At this point, goals for saving energy can be established more firmly, and priorities set on the modification and alterations to equipment and the process: effective measurement and monitoring procedures are essential in evaluating progress in the energy management program. Routine reporting procedures between management and operations should be established to accumulate information on plant performance and to inform plant supervisors of the effectiveness of their operation. Time-tracking charts of energy use and costs can be helpful. Involving employees and recognizing their contributions facilitates the achievement of objectives. Finally, the program must be continually reviewed and analyzed with regard to established goals and procedures.

10.2.2 Energy Audit Report

Energy audits do not save money and energy for companies unless the recommendations are implemented. Audit reports should be designed to encourage implementation. The goal in writing an audit report should not be the report itself; rather, it should be to achieve implementation of the report recommendations and thus achieve increased energy efficiency and energy cost savings for the customer. In this section, the authors discuss their experience with writing industrial energy audit reports and suggest some ways to make the reports more successful in terms of achieving a high rate of the recommendations.[11]

- *Present information visually*: The authors present their client's energy-use data visually with graphs showing the annual energy and demand usage by month. These graphs give a picture of use patterns. Any discrepancies in use show up clearly.
- *Make calculation sections helpful*: The methodology and calculations used to develop specific energy management opportunity recommendations are useful in an audit report. Including the methodology and calculations gives technical personnel the ability to check the accuracy of one's assumptions and one's work. However, not every reader wants to wade through pages describing the methodology and showing the calculations. Therefore, the authors provide this information in a technical supplement to the audit report. Because this section is clearly labeled as the technical supplement, other readers are put on notice as to the purpose of this section.

- *Use commonly understood units*: When preparing the report, be sure to use units that the client will understand. Discussing energy savings in terms of Btu (British thermal units) may be meaningful only to the engineers and more technical readers. For management and operating personnel, kilowatt-hours (for electricity) or therms (for natural gas) are better units, because most energy bills use these units.

- *Explain your assumptions*: A major problem with many reports is the failure to explain the assumptions underlying the calculations. For example, when the authors use operating hours in a calculation, it is always carefully shown how the number was figured, for example, "Your facility operates from 7:30 a.m. to 8:00 p.m., 5 days a week, 51 weeks/year. Therefore, we will use 3188 h in our calculations."

When basic assumptions and calculations are shown, the reader can make adjustments if those facts change. In the example given, if the facility decided to operate 24 h/day, the reader would know where and how to make changes in operating hours, because that calculation had been clearly labeled.

The authors use one section of their report to list the standard assumptions and calculations. Thus, explanations for each of the recommendations do not have to be repeated. Some of the standard assumptions/calculations included in this section are operating hours, average cost of electricity, demand rate, off-peak cost of electricity, and the calculation of the fraction of air-conditioning load attributable to lighting.

- *Be accurate and consistent*: The integrity of a report is grounded in its accuracy. This does not just mean correctness of calculations. Clearly, inaccurate calculations will destroy a report's credibility, but other problems can also undermine the value of a report. Use the same terminology so that the reader is not confused. Make sure that the same values are used throughout the report. Do not use two different load factors for the same piece of equipment in different recommendations. This, for example, could happen if one calculated the loss of energy due to leaks from a compressor in one recommendation and the energy savings due to replacing the compressor motor with a high-efficiency motor in another recommendation.

- *Proofread the report carefully*: Typographical and spelling errors devalue an otherwise good product. With computer spell checkers, there is very little excuse for misspelled words. Nontechnical readers are likely to notice this type of error, and they will wonder if the technical calculations are similarly flawed.

10.2.2.1 Report Sections

The authors have found that the following report format meets their clients' needs and fits the authors' definition of a user-friendly report.

Executive summary: The audit report should start with an executive summary that basically lists the recommended energy conservation measures and shows the implementation cost and dollar savings amount. This section is intended for the readers who want to see only the bottom line. Although the executive summary can be as simple as a short table, the authors add some brief text to explain the recommendations and sometimes include other special information needed to implement the recommendations. They also copy the executive summary on colored paper so that it stands out from the rest of the report.

Energy management plan: Following the executive summary, some information is provided to the decision makers on how to set up an energy management program in their facility. The authors view this section as one that encourages implementation of the report, so every attempt is made to try to make it as helpful as possible.

Energy action plan: In this subsection, the authors describe the steps that a company should consider in order to start implementing the report's recommendations.

Energy financing options: The authors also include a short discussion of the ways that a company can pay for the recommendations. This section covers the traditional use of company capital, loans for small businesses, utility incentive programs, and the shared savings approach of the energy service companies.

Maintenance recommendations: The authors do not usually make formal maintenance recommendations in the technical supplement, because the savings are not often easy to quantify. However, in this section of the report, energy savings maintenance checklists are provided for lighting, heating/ventilation/air conditioning, and boilers.

The technical supplement: The technical supplement is the part of the report that contains the specific information about the facility and the audit recommendations. The authors' technical supplement has two main sections: one includes the report's assumptions and general calculations and the other describes the recommendations in detail, including the calculations and methodology. The authors sometimes include a third section that describes measures that were analyzed and determined not to be cost effective, or that have payback times beyond the client's planning horizon.

Standard calculations and assumptions: This section was briefly described earlier when the importance of explaining assumptions was discussed. Here, the reader is provided with the basis for understanding many of the authors' calculations and assumptions. Included is a short description of the facility: square footage (both air-conditioned and unconditioned areas); materials of construction; type and level of insulation; etc. If the authors are breaking the facility down into subareas, those areas are described, and each area is assigned a number that is then used throughout the recommendation section.

Standard values calculated in this section include operating hours, average cost of electricity, demand rate, off-peak cost of electricity, and the calculation of the fraction of air-conditioning load attributable to lighting. When a value is calculated in this section, the variable is labeled with an identifier that remains consistent throughout the rest of the report.

Audit recommendations: This section contains a discussion of each of the energy management opportunities the authors have determined to be cost effective. Each energy management recommendation (or EMR) that was capsulized in the executive summary is described in depth in this section.

Again, the authors try to make the EMRs user-friendly. To do this, the narrative discussion is placed at the beginning of a recommendation, and the technical calculations are left for the very end. In this manner, the authors allow the readers to decide for themselves whether they want to wade through the calculations.

Each EMR starts with a table that summarizes the energy, demand and cost savings, implementation cost, and simple payback period. Then follows a short narrative section that provides some brief background information about the recommended measure and explains how it should be implemented at the facility in question. If the authors are recommending installation of more than one item (lights, motors, air-conditioning units, etc.), a table is often used to break down the savings by unit or by area.

The final section of each EMR is the calculation section. Here the authors explain the methodology that was used to arrive at the report's savings estimates. The equations are provided, and it is shown how the calculations are performed so that the clients can see what has been done. If they want to change the report's assumptions, they can. If some of the data the authors have used are incorrect, they can replace it with the correct data and recalculate the results. However, by placing the calculations away from the rest of the discussion rather than intermingling it, the authors do not scare off the readers who need to know the other information.

Appendix: The authors use an appendix for lengthy data tables. For example, there is a motor efficiency table that is used in several of the authors' EMRs. Instead of repeating it in each EMR, it is printed in the appendix. The authors also include a table showing the facility's monthly energy-use history and a table listing the major energy-using equipment. Similar to the calculation section of the EMRs, the appendix allows the authors to provide backup information without cluttering up the main body of the report.

10.3 Improving Industrial Energy Audits

10.3.1 Preventing Overestimation of Energy Savings in Audits

A frequent criticism of energy audits is that they overestimate the savings potential available to the facility. This section addresses several problem areas that can result in overly optimistic savings projections and suggests ways to prevent mistakes.[12] This possibility of overestimation concerns many of the people and organizations that are involved in some part of this energy audit process. It concerns utilities that do not want to pay incentives for demand-side management programs if the facilities will not realize the expected results in energy or demand savings. Overestimates also make clients unhappy when their energy bills do not decrease as much as promised. The problem multiplies when a shared savings program is undertaken by the facility and an energy service company. Here, the difference between the audit projections and the actual metered and measured savings may be so significantly different that either there are no savings for the facility or the energy service company makes no profit.

More problems are likely with the accuracy of the energy audits for industrial and manufacturing facilities than for smaller commercial facilities or even large buildings because the equipment and operation of industrial facilities are more complex. However, many of the same problems discussed here in terms of industrial and manufacturing facilities can occur in audits of large commercial facilities and office buildings. Based on the authors' auditing experience for industrial and manufacturing facilities over the last 5 years, it is possible to identify a number of areas where problems are likely to occur, and a number of these are presented and discussed. In addition, the authors have developed a few methods and approaches to dealing with these potential problems, and a few ways have been found to initiate energy audit analyses that lead the authors to improved results. One of these approaches is to collect data on the energy-using equipment in an industrial or manufacturing facility and then to perform both an energy and a demand balance to help ensure that reasonable estimates of energy uses—and therefore, energy savings—are available for this equipment.

In addition, unfortunately, some analysts use the average cost of electricity to calculate energy savings. This can give a false picture of the actual savings and may result in overly optimistic savings predictions. This section also discusses how to calculate the correct values from the electricity bills and when to use these values. Finally, this section discusses several common energy savings measures that are frequently recommended by energy auditors. Some of these may not actually save as much energy or demand as expected, except in limited circumstances. Others have good energy-saving potential but must be implemented carefully to avoid increasing energy use rather than decreasing it.

10.3.2 Calculating Energy and Demand Balances

The energy and demand balances for a facility are an accounting of the energy flows and power used in the facility. These balances allow the energy analyst to track the energy and power inputs and outputs (uses) and see whether they match. A careful energy analyst should perform an energy-and-demand balance on a facility before developing and analyzing any EMRs.[13] In this way, the analyst can determine what the largest energy users are in a facility, can find out whether all—or almost all—energy uses have been identified, and can see whether more savings have been identified than are actually achievable. Making energy-use recommendations without utilizing the energy and demand balances is similar to making budget-cutting recommendations without knowing exactly where the money is currently being spent.

When the authors perform an energy survey (audit), all of the major energy-using equipment in the facility is inventoried. Then, the authors list the equipment and estimate its energy consumption and demand using the data gathered at the facility, such as nameplate ratings of the equipment and operating hours. The energy balance is developed by major equipment category, such as lighting, motors, HVAC, and air compressors. There is also a category called *miscellaneous* to account for loads that were not individually surveyed, such as copiers, electric typewriters, computers, and other plug loads. The authors typically allocate 10% of the actual energy use and demand to the *miscellaneous* category in the demand and energy balances. (For an office building instead of a manufacturing facility, this miscellaneous load might be 15%–20%.) Then, the energy and demand for each of the other categories is calculated.

10.3.2.1 Lighting

The first major category analyzed is lighting, because this is usually the category in which the authors have the most confidence for knowing the actual demand and hours of use. Thus, they believe that the energy and demand estimates for the lighting system are the most accurate and can then be subtracted from the total actual use to let the authors continue to build up the energy and demand balance for the facility. The authors record the types of lamps and number of lamps used in each area of the facility and ask the maintenance person to show them the replacement lamps and ballasts used. With this lamp and ballast wattage data, together with a good estimate of the hours that the lights are on in the various areas, they can construct what they believe to be a fairly accurate description of the energy and demand for the lighting system.

10.3.2.2 Air Conditioning

There is generally no other "easy" or "accurate" category to work on, so the authors proceed to either air conditioning or motors. In most facilities, there will be some air conditioning, even if it is just for the offices that are usually part of the industrial or manufacturing facility. Many facilities—particularly in the hot and humid southeast—are fully air conditioned. Electronics, printing, medical plastics and devices, and many assembly plants are common ones seen to be fully air conditioned. Boats, metal products, wood products, and plastic pipe-manufacturing facilities are most often not air conditioned. Air-conditioning system nameplate data are usually available and readable on many units, and efficiency ratings can be found from published ARI data[14] or from the manufacturers of the equipment. The biggest problem with air conditioning is to get runtime data that will allow the author(s) of the report to determine the number of full-load equivalent operating hours for the air-conditioning compressors or chillers. From the authors' experience in north and north-central Florida, about 2200–2400 h is used per year of compressor runtime for facilities that have air conditioning that responds to outdoor temperature. Process cooling requirements are much different and would typically have much larger numbers of full-load equivalent operating hours. With the equipment size, the efficiency data, and the full-load equivalent operating hours, it is possible to construct a description of the energy and demand for the air-conditioning system.

10.3.2.3 Motors

Turning next to motors, the authors begin looking at one of the most difficult categories to deal with in the absence of fully metered and measured load factors on each motor in the facility. In a 1-day plant visit, it is usually impossible to get actual data on the load factors for more than a few motors. Even then, that data are only good for the 1 day that it was taken. Very few energy-auditing organizations can afford the time and effort to make long-term measurements of the load factor on each motor in an industrial or manufacturing facility. Thus, estimating motor load factors becomes a critical part of the energy and demand balance, and also a critical part of the accuracy of the actual energy audit analysis. Motor nameplate data show the horsepower rating, the manufacturer, and sometimes the efficiency. If not, the efficiency can usually be obtained from the manufacturer, or from standard references such as the *Energy-Efficient Motor Systems Handbook*,[15] or from software databases such as MotorMaster produced by the Washington State Energy Office.[16] The authors inventory all motors over 1 hp and sometimes try to look at the smaller ones if there is enough time.

Motor runtime is another parameter that is very difficult to obtain. When the motor is used in an application where it is constantly on is an easy case. Ventilating fans, circulating pumps, and some process-drive motors are often in this class because they run for a known constant period of time each year. In other cases, facility operating personnel must help provide estimates of motor runtimes. With data on the horsepower, efficiency, load factor, and runtimes of motors, it is possible to construct a detailed table of motor energy and demands to use in the report's balances. Motor load factors will be discussed further in a later section of this chapter.

10.3.2.4 Air Compressors

Air compressors are a special case of motor use with most of the same problems. Some help is available in this category because some air compressors have instruments showing the load factor and some have runtime indicators for hours of use. Most industrial and manufacturing facilities will have several air compressors, and this may lead to some questions as to which air compressors are actually used and how many hours they are used. If the air compressors at a facility are priority-scheduled, it may turn out that one or more of the compressors are operated continuously, and one or two smaller compressors are cycled or unloaded to modulate the need for compressed air. In this case, the load factors on the larger compressors may be unity. Using these data on the horsepower, efficiency, load factor, and runtimes of the compressors, the authors develop a detailed table of compressor energy use and demand for the report's energy and demand balances.

10.3.2.5 Other Process Equipment

Specialized process equipment must be analyzed on an individual basis because it will vary tremendously depending on the type of industry or manufacturing facility involved. Much of this equipment will utilize electric motors and will be covered in the motor category. Other electrically powered equipment, such as drying ovens, cooking ovens, welders, and laser and plasma cutters, are nonmotor electric uses and must be treated separately. Equipment nameplate ratings and hours of use are necessary to compute the energy and demand for these items. Process chillers are another special class that are somewhat different from the comfort air-conditioning equipment, because the operating hours and loads are driven by the process requirements and not the weather patterns and temperatures.

10.3.2.6 Checking the Results

After the complete energy and demand balances are constructed for the facility, the authors check to see if the cumulative energy/demand for these categories plus the miscellaneous category is substantially larger or smaller than the actual energy usage and demand over the year. If it is, and it is certain that all of the major energy uses have been identified, the authors know that a mistake was made somewhere in their assumptions. As mentioned earlier, one area that has typically been difficult to accurately survey is the energy use by motors. Measuring the actual load factors is difficult on a 1-day walk-through audit visit, so the authors use the energy balance data to help estimate the likely load factors for the motors. This is done by adjusting the load factor estimates on a number of the motors to arrive at a satisfactory level of the energy and demand from the electric motors. Unless this is done, it is likely that the energy used by the motors will be overestimated and thus overestimate the energy savings from replacing standard motors with high-efficiency motors.

As an example, the authors performed an energy audit for one large manufacturing facility with a lot of motors. It was first assumed that the load factors for the motors were approximately 80%, based on what the facility personnel explained. Using this load factor gave a total energy use for the motors of over 16 million kW h/year and a demand of over 2800 kW. Because the annual energy use for the entire facility was just over 11 million kW h/year and the demand never exceeded 2250 kW, this load factor was clearly wrong. The authors adjusted the average motor load factor to 40% for most of the motors, which reduced the energy-use figure to 9 million kW h and the demand to just under 1600 kW. These values are much more reasonable with motors making up a large part of the electrical load of this facility.

After the energy/demand balances have been satisfactorily compiled, the authors use a graphics program to draw a pie chart showing the distribution of energy/demand between the various categories. This allows visual representation of which categories are responsible for the majority of the energy use. It also makes it possible to focus the energy savings analyses on the areas of largest energy use.

10.3.3 Problems with Energy Analysis Calculations

Over the course of performing 120 industrial energy audits, the authors have identified a number of problem areas. One lies with the method of calculating energy cost savings: whether to use the average cost of electricity or break the cost down into energy and demand cost components. Other problems include instances where the energy and demand savings associated with specific energy efficiency measures may not be fully realized or where more research should go into determining the actual savings potential.

10.3.3.1 On-Peak and Off-Peak Uses: Overestimating Savings by Using the Average Cost of Electricity

One criticism of energy auditors is that they sometimes overestimate the dollar savings available from various energy efficiency measures. One way overestimation can result is when the analyst uses only the average cost of electricity to compute the savings. Because the average cost of electricity includes a demand component, using this average cost to compute the savings for companies who operate on more than one shift can overstate the dollar savings. This is because the energy cost during the off-peak hours does not include a demand charge. A fairly obvious example of this type of problem occurs when the average cost of electricity is used to calculate savings from installing high-efficiency security lighting. In this instance, there is no on-peak electricity use, but the savings will be calculated as if all the electricity was used on-peak.

The same problem arises when an energy efficiency measure does not result in an expected—or implicitly expected—demand reduction. Using a cost of electricity that includes demand in this instance will again overstate the dollar savings. Examples of energy efficiency measures that fall into this category are occupancy sensors, photosensors, and adjustable-speed drives (ASDs). Although all of these measures can reduce the total amount of energy used by the equipment, there is no guarantee that the energy use will only occur during off-peak hours. While an occupancy sensor will save lighting kW h, it will not save any kW if the lights come on during the peak load period. Similarly, an ASD can save energy use for a motor, but if the motor needs its full-load capability—as an air-conditioning fan motor or chilled-water pump motor might—during the peak load period, the demand savings may not be there. The reduced use of the device or piece of equipment on peak load times may introduce a diversity factor that produces some demand savings. However, even this savings will be overestimated by using the average cost of electricity in most instances.

On the other hand, some measures can be expected to provide their full demand savings at the time of the facility's peak load. Replacing 40 W T12 fluorescent lamps with 32 W T8 lamps will provide a verifiable demand savings because the wattage reduction will be constant at all times and will specifically show up during the period of peak demand. Shifting loads to off-peak times should also produce verifiable demand savings. For example, putting a timer or energy management system control on a constant-load electric drying oven to ensure that it does not come on until the off-peak time will result in the full demand savings. Using high-efficiency motors also seems like it would also produce verifiable savings because of its reduced kW load, but in some instances, there are other factors that tend to negate these benefits. This topic is discussed later.

To help solve the problem of overestimating savings from using the average cost of electricity, the authors divide their energy savings calculations into a demand savings and an energy savings. In most instances, the energy savings for a particular piece of equipment is calculated by first determining the demand savings for that equipment and then multiplying by the total operating hours of the equipment. To calculate the annual cost savings (*CS*), the following formula is used:

$$CS = [\text{Demand savings} \times \text{Average monthly demand rate} \times 12 \text{ months/year}]$$

$$+ [\text{Energy savings} \times \text{Average cost of electricity without demand}]$$

If a recommended measure has no demand savings, then the energy cost savings is simply the energy savings times the average cost of electricity without demand (or off-peak cost of electricity). This procedure forces us to think carefully about which equipment is used on-peak and which is used off-peak.

To demonstrate the difference in savings estimates, consider replacing a standard 30 hp motor with a high-efficiency motor. The efficiency of a standard 30 hp motor is 0.901 and a high-efficiency motor is 0.931. Assume the motor has a load factor of 40% and operates 8760 h/year (three shifts). Assume also that the average cost of electricity is $0.068/kW h (including demand), the average demand cost is $3.79/kW/month, and the average cost of electricity without demand is $0.053/kW h. The equation for calculating the demand of a motor is

$$D = HP \times LF \times 0.746 \times \frac{1}{Eff}$$

The savings on demand (or demand reduction) from installing a high-efficiency motor is

$$DR = HP \times LF \times 0.746 \times \left(\frac{1}{Eff_S} - \frac{1}{Eff_H} \right)$$

$$= 30 \text{ hp} \times 0.40 \times 0.746 \text{ kW/hp} \times \left(\frac{1}{0.901} - \frac{1}{0.931} \right) = 0.32 \text{ kW}$$

The annual energy savings (*ES*) is

$$ES = DR \times H = 0.32 \text{ kW} \times 8760 \text{ h/year} = 2803.2 \text{ kW h/year}$$

Using the earlier average cost of electricity, the cost savings (*CS*₁) are calculated as

$$CS_1 = ES \times (\text{Average cost of electricity}) = 2803.2 \text{ kW h/year} \times 0.068/\text{kW h} = 190.62/\text{year}$$

Using the earlier recommended formula,

$$CS = [\text{Demand savings} \times \text{Average monthly demand rate} \times 12 \text{ months/year}]$$

$$\times [\text{Energy savings} \times \text{Average cost of electricity without demand}]$$

$$= (0.32 \text{ kW} \times 3.79/\text{month} \times 12 \text{ months/year}) + (2803.2 \text{ kW h/year} \times 0.053/\text{kW h})$$

$$= (14.55 + 148.57)/\text{year} = 163.12/\text{year}$$

In this example, using the average cost to calculate the energy cost savings overestimates the cost savings by $27.50/year, or 17%. Although the actual amount is small for one motor, if this error is repeated for all the motors for the entire facility as well as all other measures that reduce the demand component only during the on-peak hours, then the cumulative error in cost savings predictions can be substantial.

10.3.3.2 Motor Load Factors

Many in the energy-auditing business started off assuming that motors ran at full load or near full load, and based their energy consumption analysis and energy savings analysis on that premise. Most books and publications that give a formula for finding the electrical load of a motor do not even include a term for the motor load factor. However, since experience soon showed the authors that few motors actually run at full load or near full load, they were left in a quandary about what load factor to actually use in calculations, because good measurements on the actual motor load factor are rarely to be had. A classic paper by Hoshide shed some light on the distribution of motor load factors from his experience.[17] In this paper, Hoshide noted that only about one-fourth of all three-phase motors run with a load factor greater than 60%, with 50% of all motors running at load factors between 30% and 60%, and one-fourth running with load factors less than 30%. Thus, those auditors who had been assuming that a typical motor load factor was around 70% or 80% had been greatly overestimating the savings from high-efficiency motors, adjustable-speed drives, high-efficiency belts, and other motor-related improvements.

The energy and demand balances discussed earlier also confirm that overall motor loads in most facilities cannot be anywhere near 70%–80%. The authors' experience in manufacturing facilities has been that motor load factors are more correctly identified as being in the 30%–40% range. With these load factors, one obtains very different savings estimates and economic results than when one assumes that a motor is operating at a 70% or greater load factor, as shown in the example earlier.

One place where the motor load factor is critical—but often overlooked—is in the savings calculations for ASDs. Many motor and ASD manufacturers provide easy-to-use software that will determine savings with an ASD if you supply the load profile data. Usually a sample profile is included that shows calculations for a motor operating at full load for some period of time and at a fairly high overall load factor—for example, around 70%. If the motor has a load factor of only 50% or less to begin with, the savings estimates from a quick use of one of these programs may be greatly exaggerated. If the actual motor use profile with the load factor of 50% is use, one may find that the ASD will still save some energy and money, but often not as much as it looks like when the motor is assumed to run at the higher load factor. For example, a 20 hp motor may have been selected for use on a 15 hp load to ensure that there is a "safety factor." Thus, the maximum load factor for the motor would be only 75%. A typical fan or pump in an air-conditioning system that is responding to outside weather conditions may operate at its maximum load only about 10% of the time. Because that maximum load here is only 15 hp, the average load factor for the motor might be more like 40% and will not be even close to 75%.

10.3.3.3 High-Efficiency Motors

Another interesting problem area is associated with the use of high-efficiency motors. In Hoshide's paper mentioned earlier, he notes that, in general, high-efficiency motors run at a faster full-load speed than standard-efficiency motors. This means that when a standard motor is replaced by a high-efficiency motor, the new motor will run somewhat faster than the old motor in almost every instance. This is a problem for motors that drive centrifugal fans and pumps, because the higher operating speed means greater power use by the motor. Hoshide provides an example where he shows that a high-efficiency motor that should be saving about 5% energy and demand actually uses the same energy and demand as the old motor. This occurs because the increase in speed of the high-efficiency motor offsets the power savings by almost exactly the same 5% due to the cube law for centrifugal fans and pumps.

Few energy auditors ever monitor fans or pumps after replacing a standard motor with a high-efficiency motor; therefore, they have not realized that this effect has cancelled the expected energy and demand savings. Since Hoshide noted this feature of high-efficiency motors, the authors have been careful to make sure that their recommendations for replacing motors with centrifugal loads carry the notice that it will probably be necessary to adjust the drive pulleys or drive system so that the load is operated at the same speed to achieve the expected savings.

10.3.3.4 Motor Belts and Drives

The authors have developed some significant questions about the use of cogged and synchronous belts, and the associated estimates of energy savings. It seems fairly well accepted that cogged and synchronous belts do transmit more power from a motor to a load than if standard smooth V-belts are used. In some instances, this should certainly result in some energy savings. A constant-torque application like a conveyor drive may indeed save energy with a more efficient drive belt because the motor will be able to supply that torque with less effort. Consider also a feedback-controlled application, such as a thermostatically controlled ventilating fan or a level-controlled pump. In this case, the greater energy transmitted to the fan or pump should result in the task being accomplished faster than if less drive power were supplied, and some energy savings should exist. However, if a fan or a pump operates in a nonfeedback application—as is common for many motors—then there will not be any energy savings. For example, a large ventilating fan that operates at full load continuously without any temperature or other feedback may not use less energy with an efficient drive belt, because the fan may run faster as a result of the drive belt having less slip. Similarly, a pump that operates continuously to circulate water may not use less energy with an efficient drive belt. This is an area that needs some monitoring and metering studies to check the actual results.

Whether efficient drive belts result in any demand savings is another question. Because, in many cases, the motor is assumed to be supplying the same shaft horsepower with or without high-efficiency drive belts, a demand savings does not seem likely in these cases. It is possible that using an efficient belt on a motor with a constant-torque application that is controlled by an ASD might result in some demand savings. However, for the most common applications, the motor is still supplying the same load and thus would have the same power demand. For feedback-controlled applications, there might be a diversity factor involved so that the reduced operation times could result in some demand savings—but not the full value otherwise expected. Thus, using average cost electricity to quantify the savings expected from high-efficiency drive belts could well overestimate the value of the savings. Verification of the cases where demand savings are to be expected is another area where more study and data are needed.

10.3.3.5 Adjustable-Speed Drives

The authors would like to close this discussion with a return to ASDs because these are devices that offer a great potential for savings, but have far greater complexities than are often understood or appreciated. Fans and pumps form the largest class of applications where great energy savings is possible from the use of ASDs. This is a result again of the cube law for centrifugal fans and pumps where the power required to drive a fan or pump is specified by the cube of the ratio of the flow rates involved. According to the cube law, a reduction in flow to one-half the original value could now be supplied by a motor using only one-eighth of the original horsepower. Thus, whenever an air flow or liquid flow can be reduced, such as in a variable-air-volume system or with a chilled-water pump, there is a dramatic savings possible with an ASD. In practice, there are two major problems with determining and achieving the expected savings.

The first problem is the one briefly mentioned earlier, and that is determining the actual profile of the load involved. Simply using the standard profile in a piece of vendor's software is not likely to produce very realistic results. There are so many different conditions involved in fan and pump applications that taking actual measurements is the only way to get a good idea of the savings that will occur with an ASD. Recent papers have discussed the problems with estimating the loads on fans and pumps and have shown how the cube law itself does not always give a reasonable value.[18-20] The Industrial Energy Center at Virginia Polytechnic Institute and Virginia Power Company have developed an approach wherein they classify potential ASD applications into eight different groups and then estimate the potential savings from the analysis of each system and from measurements of that system's operation.[21] Using both an analytical approach and a few measurements allows them to get a reasonable estimate of the motor load profile and thus a reasonable estimate of the energy and demand savings possible.

The second problem is achieving the savings predicted for a particular fan or pump application. It is not sufficient to just identify the savings potential and then install an ASD on the fan or pump motor. In most applications, there is some kind of throttling or bypass action that results in almost the full horsepower still being required to drive the fan or pump most of the time. In these applications, the ASD will not save much, unless the system is altered to remove the throttling or bypass device and a feedback sensor is installed to tell the ASD what fraction of its speed to deliver. This means that in many air flow systems, the dampers or vanes must be removed so that the quantity of air can be controlled by the ASD changing the speed of the fan motor. In addition, some kind of feedback sensor must be installed to measure the temperature or pressure in the system to send a signal to the ASD or a PLC controller to alter the speed of the motor to meet the desired condition. The additional cost of the alterations to the system and the cost of the control system needed greatly change the economics of an ASD application compared to the case where only the purchase cost and installation cost of the actual ASD unit are considered.

For example, a dust collector system might originally be operated with a large 150 hp fan motor running continuously to pick up the dust from eight saws. However, because production follows existing orders for the product, sometimes only two, three, or four saws are in operation at a particular time. Thus, the load on the dust collector is much lower at these times than if all eight saws are in use. An ASD is a common recommendation in this case, but estimating the savings is not easy to begin with, and after the costs of altering the collection duct system and of adding a sophisticated control system to the ASD are considered, the bottom-line result is much different than the cost of the basic ASD with installation. Manual or automatic dampers must be added to each duct at a saw so that it can be shut off when the saw is not running. In addition, a PLC for the ASD must be added to the new system, together with sensors added to each damper so that the PLC will know how many saws are in operation and therefore what speed to tell the ASD for the fan to run to meet the dust collection load of that number of saws. Without these system changes and control additions, the ASD itself will not save any great amount of energy or money. Adding them in might well double the cost of the basic ASD and double the payback time that may have originally been envisioned.

Similarly, for a water or other liquid flow application, the system piping or valving must be altered to remove any throttling or bypass valves, and a feedback sensor must be installed to allow the ASD to know what speed to operate the pump motor. If several sensors are involved in the application, then a PLC may also be needed to control the ASD. For example, putting an ASD on a chilled-water pump for a facility is much more involved, and much more costly, than simply cutting the electric supply lines to the pump motor and inserting an ASD for the motor. Without the system alterations and without the feedback control system, the ASD cannot provide the savings expected.

10.3.4 General Rules

New energy auditors often do not have the experience to have engineering judgment about the accuracy of their analyses. That is, they cannot look at the result and immediately know that it is not within the correct range of likely answers. Because the authors' IAC program has a fairly steady turnover of students, they find the same type of errors cropping up over and over as draft audit reports are reviewed. To help new team members develop the engineering judgment that they will eventually gain through experience, the authors are developing "rules of thumb" for energy analyses. The rules of thumb are intended to provide a ballpark estimate of the expected results. For example, if the rule of thumb for the percentage for installing high-efficiency motors says that the savings range is 3%–5% of the energy use by the motors, then a student who comes up with a savings of 25% will immediately know that the calculations are wrong and will know to check the assumptions and data entry to see where the error lies. Without these rules of thumb, the burden for checking these results is shifted to the team leaders and program directors. Although this does not obviate the need for report review, it minimizes the likelihood that errors will occur. The authors suggest that other organizations who frequently utilize and train new energy auditors consider developing such rules of thumb for the major types of facilities or geographic areas that they audit.

Energy auditing is not an exact science, but a number of opportunities are available for improving the accuracy of the recommendations. Techniques that may be appropriate for small-scale energy audits can introduce significant errors into the analyses for large complex facilities. This chapter began by discussing how to perform an energy-and-demand balance for a company. This balance is an important step in doing an energy-use analysis, because it provides a check on the accuracy of some of the assumptions necessary to calculate savings potential. It also addressed several problem areas that can result in overly optimistic savings projections and suggested ways to prevent mistakes. Finally, several areas where additional research, analysis, and data collection are needed were identified. After this additional information is obtained, everyone can produce better and more accurate energy audit results.

10.3.4.1 Decision Tools for Improving Industrial Energy Audits: OIT Software Tools

The OIT—a program operated by the U.S. Department of Energy, Division of Energy Efficiency and Renewable Energy (EERE)—provides a series of computer software tools that can be obtained free from their website or by ordering a CD at no cost from them.

With the right know-how, these powerful tools can be used to help identify and analyze energy system savings opportunities in industrial and manufacturing plants. Although the tools are accessible at the U.S. DOE website for download, they also encourage users to attend a training workshop to enhance their knowledge and take full advantage of opportunities identified in the software programs. For some tools, advanced training is also available to help further increase expertise in their use.

10.3.4.2 Decision Tools for Industry: Order the Portfolio of Tools on CD

The Decision Tools for Industry CD contains the MotorMaster+ (MM+), Pump System Assessment Tool, Steam System Tool Suite, 3E Plus, and the new AIRMaster+ software packages described here. In addition, it includes MM+ training. The training walks the user through both the fundamentals and the advanced features of MM+ and provides examples for using the software to make motor purchase decisions. The CD can be ordered via e-mail from the EERE Information Center or by calling the EERE Information Center at 1-877-EERE-INF (877-337-3463).

DOE industry tools

- AIRMaster+
- Chilled Water System Analysis Tool (CWSAT)
- Combined Heat and Power Application Tool (CHP)
- Fan System Assessment Tool (FSAT)
- MotorMaster+ 4.0
- MotorMaster+ International
- NO_x and Energy Assessment Tool (NxEAT)
- Plant Energy Profiler for the Chemical Industry (ChemPEP Tool)
- Process Heating Assessment and Survey Tool (PHAST)
- Pumping System Assessment Tool 2004 (PSAT)
- Steam System Tool Suite

Other industry tools

- ASDMaster: Adjustable-Speed Drive Evaluation Methodology and Application

AIRMaster+: AIRMaster+ provides comprehensive information on assessing compressed-air systems, including modeling existing and future system upgrades, and evaluating savings and effectiveness of energy efficiency measures.

Chilled Water System Analysis Tool Version 2.0: Use the CWSAT to determine energy requirements of chilled-water distribution systems and to evaluate opportunities for energy and cost savings by applying improvement measures. Provide basic information about an existing configuration to calculate current energy consumption and then select proposed equipment or operational changes for comparison. The results of this analysis will help the user quantify the potential benefits of chilled-water system improvements.

Combined Heat and Power Application Tool: The CHP Application Tool helps industrial users evaluate the feasibility of CHP for heating systems such as fuel-fired furnaces, boilers, ovens, heaters, and heat exchangers. It allows the analysis of three typical system types: fluid heating, exhaust-gas heat recovery, and duct burner systems. Use the tool to estimate system costs and payback period, and to perform "what-if" analyses for various utility costs. The tool includes performance data and preliminary cost information for many commercially available gas turbines and default values that can be adapted to meet specific application requirements.

Fan System Assessment Tool: Use the FSAT to help quantify the potential benefits of optimizing fan system configurations that serve industrial processes. FSAT is simple and quick, and requires only basic information about the fans being surveyed and the motors that drive them. With FSAT, one can calculate the amount of energy used by one's fan system, determine system efficiency, and quantify the savings potential of an upgraded system.

MotorMaster+ 4.0: An energy-efficient motor selection and management tool, MotorMaster+ 4.0 software, includes a catalog of over 20,000 AC motors. This tool features motor inventory management tools, maintenance log tracking, efficiency analysis, savings evaluation, energy accounting, and environmental reporting capabilities.

MotorMaster+ International: MotorMaster+ International includes many of the capabilities and features of MotorMaster+; however, now it can help evaluate repair/replacement options on a broader range of motors, including those tested under the Institute of Electrical and Electronic Engineers standard and those tested using International Electrical Commission methodology. With this tool, analyses can be conducted in different currencies, and it will calculate efficiency benefits for utility rate schedules with demand charges, edit and modify motor rewind efficiency loss defaults, and determine "best available" motors. The tool can be modified to operate in English, Spanish, and French.

NO_x and Energy Assessment Tool: The NxEAT helps plants in the petroleum refining and chemical industries to assess and analyze NO_x emissions and application of energy efficiency improvements. Use the tool to inventory emissions from equipment that generates NO_x and then compare how various technology applications and efficiency measure affect overall costs and reduction of NO_x. Perform "what-if" analyses to optimize and select the most cost-effective methods for reducing NO_x from systems such as fired heaters, boilers, gas turbines, and reciprocating engines.

Plant Energy Profiler for the Chemical Industry: The ChemPEP Tool provides chemical plant managers with the information they need to identify savings and efficiency opportunities. The ChemPEP Tool enables energy managers to see overall plant energy use, identify major energy-using equipment and operations, summarize energy cost distributions, and pinpoint areas for more detailed analysis. The ChemPEP Tool provides plant energy information in an easy-to-understand graphical manner that can be very useful to managers.

Process Heating Assessment and Survey Tool: The PHAST provides an introduction to process heating methods and tools to improve thermal efficiency of heating equipment. Use the tool to survey process-heating equipment that uses fuel, steam, or electricity, and identify the most energy-intensive equipment. It can also help perform an energy (heat) balance on selected equipment (furnaces) to identify and reduce nonproductive energy use. Compare performance of the furnace under various operating conditions and test "what-if" scenarios.

Pumping System Assessment Tool 2004: The PSAT helps industrial users assess the efficiency of pumping system operations. The PSAT uses achievable pump performance data from Hydraulic Institute standards and motor performance data from the MotorMaster+ database to calculate potential energy and associated cost savings.

Steam System Tool Suite: In many industrial facilities, steam system improvements can save 10%–20% in fuel costs. To help tap into potential savings in typical industrial facilities, DOE offers a suite of tools for evaluating and identifying steam system improvements.

- *Steam System Assessment Tool (SSAT) Version 2.0.0*: The SSAT allows steam analysts to develop approximate models of real steam systems. Using these models, SSAT can be applied to quantify the magnitude—energy, cost, and emissions savings—of key potential steam improvement opportunities. SSAT contains the key features of typical steam systems. The enhanced and improved version includes features such as a steam demand savings project, a user-defined fuel model, a boiler stack loss worksheet for the SSAT fuels, a boiler flash steam recovery model, and improved steam trap models.

- *3E Plus, Version 3.2*: The program calculates the most economical thickness of industrial insulation for user input operating conditions. Calculations can be made using the built-in thermal performance relationships of generic insulation materials or supply conductivity data for other materials.

- *Steam Tool Specialist Qualification Training*: Industry professionals can earn recognition as Qualified Specialists in the use of the BestPractices Steam Tools. DOE offers an in-depth (2–1/2)-day training session for steam system specialists, including 2 days of classroom instruction and a written exam. Participants who complete the workshop and pass the written exam are recognized by DOE as Qualified Steam Tool Specialists. Specialists can assist industrial customers in using the BestPractices Steam Tools to evaluate their steam systems.

ASDMaster: Adjustable-Speed Drive Evaluation Methodology and Application. This Windows® software program helps plant or operations professionals determine the economic feasibility of an ASD application, predict how much electrical energy may be saved by using an ASD, and search a database of standard drives. The package includes two 3.5 in. diskettes, a user's manual, and a user's guide. Please order from EPRI. For more information, see the ASDMaster website.

10.3.4.3 Energy-Auditing Help from Industrial Assessment Centers

The IACs, sponsored by the U.S. Department of Energy, EERE Division Industrial Technologies Program (ITP), provide eligible small- and medium-sized manufacturers with no-cost energy assessments. Additionally, the IACs serve as a training ground for the next generation of energy-savvy engineers.

Teams composed mainly of engineering faculty and students from the centers, located at 24 universities around the country, conduct energy audits or industrial assessments and provide recommendations to manufacturers to help them identify opportunities to improve productivity, reduce waste, and save energy. Recommendations from industrial assessments have averaged about $55,000 in potential annual savings for each manufacturer.

As a result of performing these assessments, upper-class and graduate engineering students receive unique hands-on assessment training and gain knowledge of industrial process systems, plant systems, and energy systems, making them highly attractive to employers.

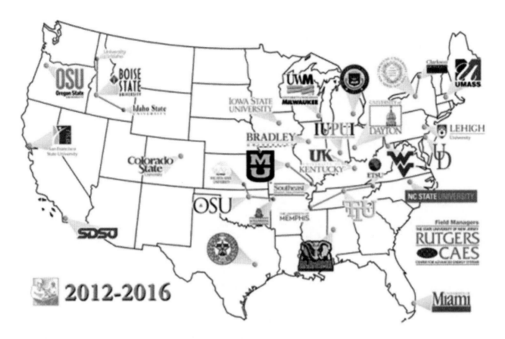

FIGURE 10.5
Industrial Assessment Center (IAC) locations, http://www1.eere.energy.gov/manufacturing/tech_assistance/iacs_locations, and service areas. (From U.S. Department of Energy, Office of Industrial Technologies, Washington, DC, 2005, http://www.energy.gov/eere/amo/industrial-assessment-centers-iacs.)

To be eligible for an IAC assessment, a manufacturing plant must meet the following criteria:

- Within Standard Industrial Codes 20-39
- Generally be located within 150 miles of a host campus
- Gross annual sales below $100 million
- Fewer than 500 employees at the plant site
- Annual energy bills more than $100,000 and less than $2.5 million
- No professional in-house staff to perform the assessment

Presently (2014), there are 26 schools across the country participating in the IAC Program. For additional information or to apply for an assessment, go to http://www.energy.gov/eere/amo/industrial-assessment-centers-iacs and click on one of the school names for contact information. A map of the IAC centers and their service areas is shown in Figure 10.5.

10.4 Industrial Electricity End Uses and Electrical Energy Management

10.4.1 Importance of Electricity in Industry

Electricity use in industry is primarily for electric drives, electrochemical processes, space heating, lighting, and refrigeration. Table 10.2 lists the relative importance of the use of electricity in the industrial sector. Timely data on industrial and manufacturing energy use are difficult to obtain, since its frequency of collection is only every 3 years, and then it takes EIA

TABLE 10.2

Manufacturing Electricity by End Use, 2006

	GW h	Percent Use
Machine drives (motors)	422,408	53.8
Electrochemical processes	60,323	10.1
Process heat	101,516	10.4
Facility HVAC (motors)	77,768	8.3
Facility lighting	58,013	6.5
Process cooling and refrigeration	60,389	6.0
On-site transportation (motors)	2,197	0.1
Other process	13,816	0.4
Other	45,000	
Total	8,40,787	100

Source: U.S. Department of Energy, Energy Information Agency, Manufacturing Energy Consumption Survey, 2006, Washington, DC, 2012.

another 3 years to process the results. In addition, it is only a sample survey, so the accuracy of the data is less than what would be optimal. However, it is better data than is available for most countries in the world. The most recent, detailed data on manufacturing energy end use (as of early 2013) are still the 2006 Manufacturing Energy Consumption Survey (MECS) data. Table 10.2 shows the electrical energy consumed for different end uses in 2006.

Because several of the categories mentioned earlier involve motor use, the total manufacturing energy use for motors is actually the sum of several categories:

Machine drives	50.5%
Facility HVAC	7.2%
Process cooling and refrigeration	9.3%
On-site transportation	0.1%
Total motors	67%

This gives the same results shown as the pie chart shown in Figure 10.3.

10.4.2 Electric Drives

Electric drives of one type or another use 68% of industrial electricity. Examples include electric motors, machine tools, compressors, refrigeration systems, fans, and pumps. Improvements in these applications would have a significant effect on reducing industrial electrical energy.

Motor efficiency can be improved in some cases by retrofit (modifications, better lubrication, improved cooling, heat recovery), but generally requires purchasing of more efficient units. For motors in the sizes 1–200 hp, manufacturers today supply a range of efficiency. Greater efficiency in a motor requires improved design, more costly materials, and generally greater first cost. Losses in electric drive systems may be divided into four categories:

	Typical Efficiency (%)
Prime mover (motor)	10–95
Coupling (clutches)	80–99
Transmission	70–95
Mechanical load	1–90

TABLE 10.3

Typical Electric Motor Data (1800 rpm Motors)

Size (hp)	Full Load Efficiency (%)	
	1975	1993 (EPACT)
1	76	82.5–85.5
5	84	87.5–89.5
10	85	90.2–91.7
20	85	91.7–93.0
40	85	93.0–94.1
100	91	94.5–95.4
200	90	95.4–96.2

Each category must be evaluated to determine energy management possibilities. In many applications, the prime mover will be the most efficient element of the system. Table 10.3 shows typical induction motor data, illustrating the improvement in efficiency due to the 1992 Energy Policy Act. Note that both efficiency and power factor decrease with partial load operation, which means that motors should be sized to operate at or near full load ratings.

Manufacturers have introduced new high-efficiency electric motors in recent years. Many utilities offer rebates of $5–$20/hp for customers installing these motors.

10.4.3 Electrochemical Processes

Industrial uses of electrochemical processes include electrowinning, electroplating, electrochemicals, electrochemical machining, fuel cells, welding, and batteries.

A major use of electrolytic energy is in electrowinning—the electrolytic smelting of primary metals such as aluminum and magnesium. Current methods require on the order of 13–15 kW h/kg; efforts are under way to improve electrode performance and reduce this to 10 kW h/kg. Recycling now accounts for one-third of aluminum production; this requires only about 5% of the energy required to produce aluminum from ore.

Electrowinning is also an important low-cost method of primary copper production. Another major use of electrochemical processes is in the production of chlorine and sodium hydroxide from salt brine. Electroplating and anodizing are two additional uses of electricity of great importance. Electroplating is basically the electrodeposition of an adherent coating upon a base metal. It is used with copper, brass, zinc, and other metals. Anodizing is roughly the reverse of electroplating, with the workpiece (aluminum) serving as the anode. The reaction progresses inward from the surface to form a protective film of aluminum oxide on the surface.

Fuel cells are devices for converting chemical energy to electrical energy directly through electrolytic action. Currently, they represent a small use of energy, but research is directed at developing large systems suitable for use by electric utilities for small dispersed generation plants. Batteries are another major use of electrolytic energy, ranging in size from small units with energy storage in the joule or fractional joule capacity up to units proposed for electric utility use that will store 18×10^9 J (5 MW h). Electroforming, etching, and welding are forms of electrochemical used in manufacturing and material shaping. The range of applications for these techniques stretches from microcircuits to aircraft carriers. In some applications, energy for machining is reduced, and reduction of scrap also saves energy. Welding has benefits in the repair and salvage of materials and equipment, reducing the need for energy to manufacture replacements.

10.4.4 Electric Process Heat

Electricity is widely used as a source of process heat due to ease of control, cleanliness, wide range in unit capacities (watts to megawatts), safety, and low initial cost. Typical heating applications include resistance heaters (metal sheath heaters, ovens, furnaces), electric salt bath furnaces, infrared heaters, induction and high-frequency resistance heating, dielectric heating, and direct arc electric furnaces.

Electric arc furnaces in the primary metals industry are a major use of electricity. Typical energy use in a direct arc steel furnace is about 2.0 kW h/kg. Electric arc furnaces are used primarily to refine recycled scrap steel. This method uses about 40% of the energy required to produce steel from iron ore using basic oxygen furnaces. Energy savings can be achieved by using waste heat to preheat scrap iron being charged to the furnace.

Glass making is another process that uses electric heat. An electric current flows between electrodes placed in the charge, causing it to melt. Electric motors constitute a small part of total glass production. Major opportunities for improved efficiency with electric process heat applications in general include improved heat transfer surfaces, better insulation, heat recovery, and improved controls.

10.4.5 HVAC

Heating, ventilating, and air conditioning (HVAC) is an important use of energy in the industrial sector. The environmental needs in an industrial operation can be quite different from residential or commercial operations. In some cases, strict environmental standards must be met for a specific function or process. More often, the environmental requirements for the process itself are not limiting, but space conditioning is a prerequisite for the comfort of production personnel. Chapters 5 and 6 have a more complete discussion of energy management opportunities in HVAC systems.

10.4.6 Lighting

Industrial lighting needs range from low-level requirements for assembly and welding of large structures (such as shipyards) to the high levels needed for the manufacture of precision mechanical and electronic components such as integrated circuits. Lighting uses about 20% of the U.S. electrical energy and 7% of all energy. Of all lighting energy, about 20% is industrial, with the balance being sizes of systems; energy management opportunities in industrial lighting systems are similar to those in residential/commercial systems (see Chapter 5).

10.4.7 Electric Load Analysis

The energy audit methodology is a general tool that can be used to analyze energy use in several forms and over a short or long period of time. Another useful technique, particularly for obtaining a short-term view of industrial electricity use, is an analysis based on the evaluation of the daily load curve. Normally, this analysis uses metering equipment installed by the utility and therefore available at the plant. However, special metering equipment can be installed if necessary to monitor specific process or building.

For small installations, both power and energy use can be determined from the kilowatt hour meter installed by the utility. Energy in kW h is determined by

$$E = (0.001)(K_h P_t C_t N) \text{ (kW h)} \tag{10.1}$$

where
E is the electric energy used, kW h
K_h is the meter constant, W h/revolution
P_t is the potential transformer ratio
C_t is the current transformer ratio
N is the number of revolutions of the meter disk

(The value of K_h is usually marked on the meter. P_t and C_t are usually 1.0 for small installations.) To determine energy use, the meter would be observed during an operation and the number of revolutions of the disk counted. Then, the equation can be used to determine E.

To determine the average load over some period p (h), determine E as earlier for time p and then use the relation that

$$L = \frac{E}{p} \text{ (kW)} \tag{10.2}$$

where
E is in kW h
p is in h
L is the load in kW

Larger installations will have meters with digital outputs or strip charts. Often, these will provide a direct indication of kW h and kW as a function of time. Some also indicate the reactive load (kVARs) or the power factor.

The first step is to construct the daily load curve. This is done by obtaining kW h readings each hour using the meter. The readings are then plotted on a graph to show the variation of the load over a 24 h period. Table 10.4 shows a set of readings obtained over a 24 h period in the XYZ manufacturing plant located in Sacramento, California, and operating one shift per day. These readings have been plotted in Figure 10.6.

Several interesting conclusions can be immediately drawn from this figure:

- The greatest demand for electricity occurs at 11:00.
- Through the lunch break, the third highest demand occurs.
- The ratio of the greatest demand to the least demand is approximately 3:1.
- Only approximately 50% of the energy used actually goes into making a product (54% on-shift use, 46% off-shift use).

When presented to management, these facts were of sufficient interest that a further study of electricity use was requested. Additional insight into the operation of a plant (and into the cost of purchase of electricity) can be obtained from the load analysis. Following a brief discussion of electrical load parameters, a load analysis for the XYZ company will be described.

TABLE 10.4

Kilowatt Hour Meter Readings for XYZ Manufacturing Company

Time Meter Read	Elapsed (kW h)	Notes Concerning Usage	Percentage of Total Usage (%)
1:00 (a.m.)	640		
2:00	610		
3:00	570		
4:00	570	7 h preshift use	
5:00	640		
6:00	770		
7:00	1,120		
Subtotal	4,920		17
8:00	1,470		
9:00	1,700		
10:00	1,790		
11:00	1,850		
	(Peak)	9 h on-shift use	
12:00 (noon)	1,830		
13:00 (1 p.m.)	1,790		
14:00 (2 p.m.)	1,790		
15:00 (3 p.m.)	1,760		
16:00 (4 p.m.)	1,690		
Subtotal	15,670		54
17:00 (5 p.m.)	1,470		
18:00 (6 p.m.)	1,310		
19:00 (7 p.m.)	1,210		
20:00 (8 p.m.)	1,090	8 h postshift use	
21:00 (9 p.m.)	960		
22:00 (10 p.m.)	800		
23:00 (11 p.m.)	730		
24:00 (12 a.m.)	640		
Subtotal	8,210		29
Grand totals	28,800		100

Any industrial electrical load consists of lighting, motors, chillers, compressors, and other types of equipment. The sum of the capacities of this equipment, in kW, is the *connected* load. The actual load at any point in time is normally less than the connected load since every motor is not turned on at the same time; only part of the lights may be on at any one time, and so on. Thus, the load is said to be diversified, and a measure of this can be found by calculating a *diversity factor*:

$$DV = \frac{(D_{m1} + D_{m2} + D_{m3} + \cdots)}{(D_{max})} \tag{10.3}$$

where

D_{m1}, D_{m2}, and so on are the sum of maximum demand of individual loads in kW
D_{max} is the maximum demand of plant in kW

If the individual loads do not occur simultaneously (usually they do not), the diversity factor will be greater than unity. Typical values for industrial plants are 1.3–2.5.

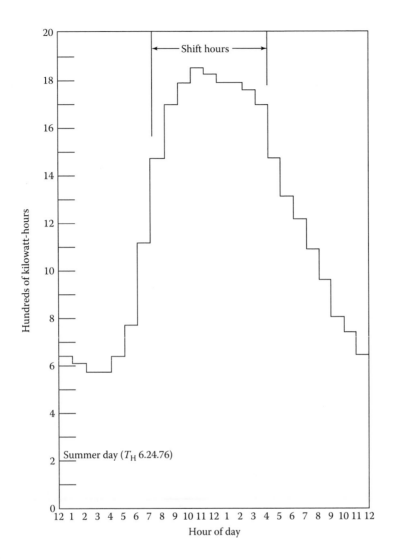

FIGURE 10.6
Daily load curve for XYZ company.

If each individual load operated to its maximum extent simultaneously, the *maximum* demand for power would be equal to the connected load, and the diversity factor would be 1.0. However, as pointed out earlier, this does not happen except for special cases. The demand for power varies over time as loads are added and removed from the system. It is usual practice for the supplying utility to specify a demand interval (usually 0.25, 0.5, or 1.0 h) over which it will calculate

$$D = \frac{E}{p} \text{ (kW)} \tag{10.4}$$

where
 D is the demand in kW
 E is the kW h used during p
 p is the demand interval in h

The demand calculated in this manner is an average value, greater than the lowest instantaneous demand during the demand interval but less than the maximum demand during the interval.

Utilities are interested in *peak demand* since this determines the capacity of the equipment they must install to meet the customer's power requirements. This is measured by a demand factor, defined as

$$DF = \frac{D_{max}}{CL} \qquad (10.5)$$

where
D_{max} is the maximum demand in kW
CL is the connected load in kW

The demand factor is normally less than unity; typical values range from 0.25 to 0.90.

Since the customer normally pays a premium for the maximum load placed on the utility system, it is of interest to determine how effectively the maximum load is used. The most effective use of the equipment would be to have the peak load occur at the start of the use period and continue unchanged throughout it. Normally, this does not occur, and a measure of the extent to which the maximum demand is sustained throughout the period (a day, month, or year) is given by the *hours use of demand*:

$$HUOD = \frac{E}{D_{max}} \text{ (h)} \qquad (10.6)$$

where
$HUOD$ is the hours use of demand in h
E is the energy used in period p, in kW h
D_{max} is the maximum demand during period p, in kW
p is the period over which $HUOD$ is determined—for example, 1 day, 1 month, or 1 year (p is always expressed in h)

The *load factor* is another parameter that measures the plant's ability to use electricity efficiently. In effect, it measures the ratio of the average load for a given period of time to the maximum load that occurs during the same period. The most effective use results when the load factor is as high as possible once E or $HUOD$ has been minimized (it is always less than 1). The load factor is defined as

$$LF = \frac{E}{(D_{max})(p)} \qquad (10.7)$$

where
LF is the load factor (dimensionless)
E is the energy used in period p in kW h
D_{max} is the maximum demand during period p in kW
p is the period over which load factor is determined (e.g., 1 day, 1 month, or 1 year) in h

Another way to determine *LF* is from the relation

$$LF = \frac{HUOD}{p} \tag{10.8}$$

Still another method is to determine the average load, *L*—kW h/*p* during *p* divided by *p*—and then use the relation

$$LF = \frac{L}{D_{max}} \tag{10.9}$$

These relations are summarized for convenience in Table 10.5.

Returning to the XYZ plant, the various load parameters can now be calculated. Table 10.6 summarizes the needed data and the results of the calculations. The most striking

TABLE 10.5

Summary of Load Analysis Parameters

Formulas	Definitions
$E = \dfrac{K_h P_t C_t N}{1000}$	E = Electric energy used in period p (kW h)
$L = \dfrac{E}{P}$	E_{max} = Maximum energy used during period p (kW h)
$DV = \dfrac{(D_{m1} + D_{m2} + D_{m3})}{D_{max}}$	K_h = Meter constant (W h/revolution)
$D = \dfrac{E}{P}, D_{max} = \dfrac{E_{max}}{P}$	P_t = Potential transformer ratio
$DF = \dfrac{D_{max}}{CL}$	C_t = Current transformer ratio
$HUOD = \dfrac{E}{D_{max}}$	N = Number of revolutions of the meter disk
$LF = \dfrac{E}{(D_{max})(p)} = \dfrac{HUOD}{p} = \dfrac{L}{D_{max}}$	L= Average load (kW) p = Period of time used to determine load, demand, electricity use, etc., normally 1 h, day, month, or year; measured in hours DV = Diversity factor (dimensionless) D_{max} = Maximum demand in period p (kW) $D_{m1}, D_{m2}, \ldots$ = Maximum demand of individual load (kW) D = Demand during period p (kW) DF = Demand factor for period p (dimensionless) CL = Connected load (kW) $HUOD$ = Hours use of demand during period p (h) LF = Load factor during period p (dimensionless)

TABLE 10.6

Data for Load Analysis of XYZ Plant

p = 24 h
E = 28,800 kW h/day
E_{max} = 1850 kW h
CL = 2792 kW
$D_{m1}, D_{m2}, \ldots$ = 53, 62, 144, 80, 700, 1420 kW

thing shown by the calculations is the hours use of demand, equal to 15.6. This is a surprise, since the plant is only operating one shift. The other significant point brought out by the calculations is the low load factor.

An energy audit of the facility was conducted, and the major loads were the evaluated number of energy management opportunities whereby both loads (kW) and energy use (kW h) could be reduced. The audit indicated that inefficient lighting (on about 12 h/day) could be replaced in the parking lot. General office lighting was found to be uniformly at 100 fc; by selective reduction and task lighting, the average level could be reduced to 75 fc or less. The air-conditioning load would also be reduced. Improved controls could be installed to automatically shut down lighting during off-shift and weekend hours (the practice had been to leave the lights on). Some walls and ceilings were selected for repainting to improve reflectance and reduced lighting energy. It was found that the air-conditioning chillers operated during weekends and off-hours; improved controls would prevent this. Also, the ventilation rates were found to be excessive and could be reduced. In the plant, compressed air system leaks, heat losses from plating tanks, and on-peak operation of the heat treat furnace represented energy and load management opportunities.

The major energy management opportunities were evaluated to have the following potential savings, with a total payback of 5.3 months, as shown in Table 10.7. The average daily savings of electricity amounted to approximately 4400 kW h/day. This led to savings of $80,000/year, with the cost of the modification being $36,000.

This can be compared to the original situation (Table 10.4). See also Figure 10.6, which shows the daily load curve after the changes have been made. The percentage of use on shift is now higher. Note that D_{max} has been improved significantly (reduced by 13%); the *HUOD* has improved slightly (about 3% lower now); and the *LF* is slightly lower. Furthermore, improvements are undoubtedly still possible in this facility; they should be directed first at reducing nonessential uses, thereby reducing *HUOD*.

So far, the discussion has dealt entirely with power and has neglected the reactive component of the load. In the most general case, the apparent power in kV A that must be supplied to the load is the sum of the active power in kW and the *reactive power* in kVAR (the reader who is unfamiliar with these terms should refer to a basic electrical engineering text):

$$|S| = \sqrt{P^2 + Q^2} \tag{10.10}$$

where
  S is the apparent power in kV A
  P is the active power in kW
  Q is the reactive power in kVAR

In this notation, the apparent power is a vector of magnitude S and angle θ where θ is commonly referred to as the phase angle and given as

$$\theta = \tan^{-1} (kVAR/kW) \tag{10.11}$$

Another useful parameter is the *power factor*, given by

$$pf = \cos \theta \tag{10.12}$$

TABLE 10.7

Sample Calculations for XYZ Plant

1. $D_{max} = \dfrac{E_{max}}{p} = \dfrac{1850 \text{ kW h}}{1 \text{ h}} = 1850 \text{ kW}$

2. $DV = \dfrac{D_{m1} + D_{m2} + D_{m3} + \cdots}{D_{max}} = \dfrac{2459}{1850} = 1.33$

3. $DF = \dfrac{D_{max}}{CL} = \dfrac{1850}{2792} = 0.66$

4. $HOUD_{(daily)} = \dfrac{E}{D_{max}} = \dfrac{28,800 \text{ kW h/day}}{1,850 \text{ kW}} = 15.6 \text{ h/day}$

5. $LF_{(daily)} = \dfrac{HUOD}{p} = \dfrac{15.6}{24.0} = 0.65$

The load parameters after the changes were made can be found:

$D_{max} = \dfrac{1850 - 235}{1 \text{ h}} = 1615 \text{ kW}$

$HOUD = \dfrac{24,400 \text{ kW h/day}}{1,615 \text{ kW}} = 15.1 \text{ h/day}$

$LF = \dfrac{15.1}{24} = 0.63$

Calculated Savings	Savings	
	kW	kW h/year
More efficient parking lot lighting	16	67,000
Reduce office lighting	111	495,000
Office lighting controls to reduce off-shift use	—	425,000
Air-conditioning controls and smaller fan motor	71	425,000
Compressed air system repairs and reduction of heat losses from plating tanks	—	200,000
Shift heat treat oven off-peak	37	—
Totals	235	1,612,000
The revised electricity use was found to be	kW h	%
Preshift	4,400	18
On shift	14,400	59
Postshift	5,600	23
Totals	24,400	100

The power factor is also given by

$$pf = \frac{|P|}{|S|} \tag{10.13}$$

The power factor is always less than or equal to unity. A high value is desirable because it implies a small reactive component to the load (Figure 10.7). A low value means the reactive component is large.

The importance of the power factor is related to the reactive component of the load. Even though the reactive component does not dissipate power (it is stored in magnetic or electric fields), the switch gear and distribution system must be sized to handle the current required by the apparent power, or the vector sum of the active and reactive components. This results in a greater capital and operating expense. The operating expense is increased due to the standby losses that occur in supplying the reactive component of the load.

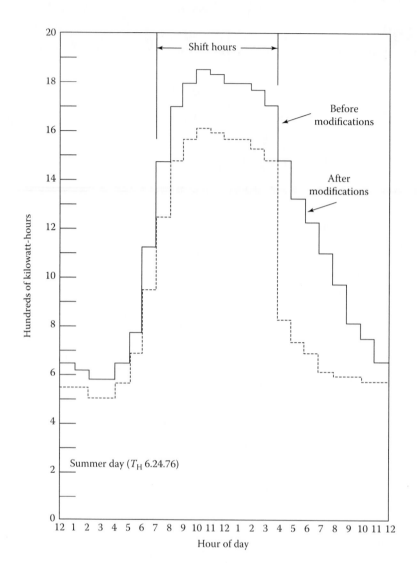

FIGURE 10.7
Daily load curve for XYZ company after modifications.

The power factor can be improved by adding capacitors to the load to compensate for part of the inductive reactance. The benefit of this approach depends on the economics of each specific case and generally requires a careful review or analysis.

These points can be clarified with an example. Consider the distribution system shown in Figure 10.8. Four loads are supplied by a 600 A bus. Load A is a distant load that has a large reactive component and a low power factor ($pf = 0.6$). To supply the active power requirement of 75 kW, an apparent power of 125 kV A must be provided, and a current of 150 A is required.

The size of the wire to supply the load is dictated by the current to be carried and voltage drop considerations. In this case, #3/0 wire that weighs 508 lb/1000 ft and has a resistance

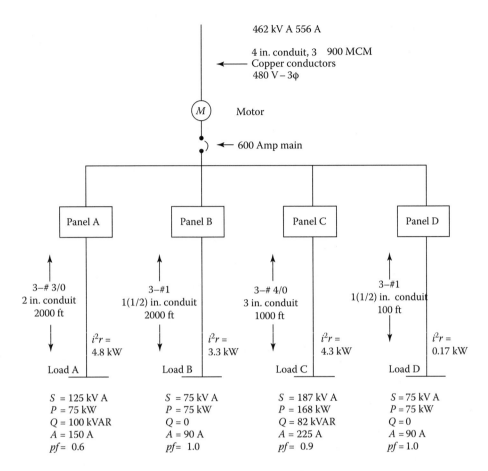

FIGURE 10.8
Electrical diagrams for building 201, XYZ company.

of 0.062 U/1000 ft is used. Since the current in this conductor is 150 A, the power dissipated in the resistance of the conductor is

$$P = \sqrt{3}i^2r = (1.732)(150^2)(0.124) \text{ W}$$

$$P = 4.8 \text{ kW}$$

Similar calculations can be made for load B, which uses #1 wire, at 253 lb/1000 ft and 0.12 Ω/1000 ft.

Now the effect of the power factor is visible. Although the active power is the same for both load A and load B, load A requires 150 A vs. 90 A for load B. The i^2r standby losses are also higher for load A as opposed to load B. The installation cost to service load B is roughly half that of load A, due to the long conduit run of load B compared to load D. For large loads that are served over long distances and operate continuously, consideration should be given to using larger wire sizes to reduce standby losses.

An estimate of the annual cost of power dissipated as heat in these conduit runs can be made if the typical operating hours of each load are known:

Load	Line Losses (kW)	Operating Hours/Year	kW h/Year
A	4.8	2000	9,600
B	3.3	2000	6,600
C	4.3	4000	17,200
D	0.17	2000	340
Total			33,740

At an average cost of 6¢/kW h (includes demand and energy costs), the losses in the distribution system alone are $2022/year. Over the life of the facility, this is a major expense for a totally unproductive use of energy.

10.4.8 Data Acquisition and Control Systems for Energy Management

Data acquisition is essential in energy management for at least three reasons: (1) baseline operational data are an absolute requirement for understanding the size and timing of energy demands for each plant, division, system, and component, and design of the energy management strategy; (2) continuing data acquisition during the course of the energy conservation effort is necessary to calculate the gains made in energy-use efficiency and to measure the success of the program; and (3) effective automatic control depends upon the accurate measurement of the controlled variables and the system operational data. More information on control systems can be found in Chapter 6.

With small, manually controlled systems, data acquisition is possible using indicating instruments and manual recording of data. With large systems, or almost any size automatically controlled system, manual data collection is impractical. The easy availability and the modest price of personal digital computers (PCs) and their present growth in speed and power as their price declines make them the preferred choice for data acquisition equipment. The only additions needed to the basic PC are the video monitor; a mass data storage device, analog-to-digital (A/D) interface cards; software for controlling the data sampling, data storage, and data presentation; and a suitable enclosure for protecting the equipment from the industrial environment.

That same computer is also suitable for use as the master controller of an automatic control system. Both functions can be carried on simultaneously, with the same equipment, with a few additions necessary for the control part of the system. These additional electronic components include multiplexers to increase the capacity of the A/D cards; direct memory access (an addition to the A/D boards), which speeds operation by allowing the data transfer to bypass the CPU and programmable logic controllers (PLCs); and digital input/output cards (I/Os), both used for controlling the equipment controllers. The most critical parts of either the data acquisition or the control system are the software. The hardware can be ordered off the shelf, but the software must be either written from scratch or purchased and modified for each particular system in order to achieve the reliable, high-performance operation that is desired. Although many proprietary program languages exist for the PCs and PLCs used for control purposes, BASIC is the most widely used. A schematic diagram of a typical data acquisition and control system is shown in Figure 10.9.

A major application has been in the control of mechanical and electrical systems in commercial and industrial buildings. These have been used to control lighting, electric demand,

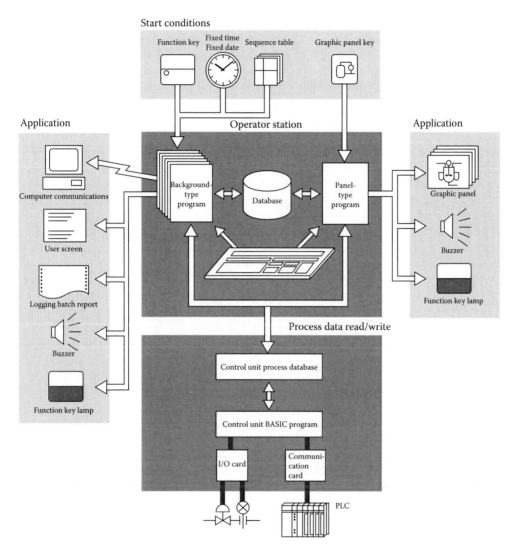

FIGURE 10.9
Typical data acquisition and control system.

ventilating fans, thermostat setbacks, air-conditioning systems, and the like. These same computer systems can also be used in industrial buildings with or without modification for process control. Several manufactures of the computer systems will not only engineer and install the system but will maintain and operate it from a remote location. Such operations are ordinarily regional. For large systems in large plants, one may be able to have the same service provided in-plant. However, there are also many small stand-alone analog control systems that can be used advantageously for the control of simple processes. Examples of these are a temperature controller using a thermocouple output to control the temperature of a liquid storage tank; a level controller, which keeps the liquid level in a tank constant by controlling a solenoid valve; and an oxygen trim system for a small boiler, which translates the measurement of the oxygen concentration in the exhaust stream into the jack shaft position, which regulates the combustion air flow to the burner.

After installing a demand limiter on an electric arc foundry cupola, the manager was able to reduce the power level from 7100 to 4900 kW with negligible effect on the production time and no effect on product quality. The savings in demand charges alone were $4400/month with an additional savings in energy costs.

10.4.9 Web-Based Facility Automation Systems

Of all recent developments affecting computerized energy management systems, the most powerful new technology to come into use in the last several years has been information technology, or IT. The combination of cheap, high-performance microcomputers, together with the emergence of high-capacity communication lines, networks, and the Internet, has produced explosive growth in IT and its application throughout the U.S. economy. Energy information and control systems have been no exception. IT- and Internet-based systems are the wave of the future. Almost every piece of equipment and almost every activity will be connected and integrated into the overall facility operation in the next several years.[22]

In particular, the future of DDC in facility automation systems (FASs) can be found on the web. Almost all FAS manufacturers see the need to move their products to the Internet. Tremendous economies of scale and synergies can be found there. Manufacturers no longer have to create the transport mechanisms for data to flow within a building or campus. They just need to make sure their equipment can utilize the network data paths already installed or designed for a facility. Likewise, with the software to display data to users, manufacturers that take advantage of presentation layer standards such as HTML and JAVA can provide the end user with a rich, graphical, and intuitive interface to their FAS using a standard web browser.

Owners will reap the benefits of Internet standards through a richer user interface, more competition among FAS providers, and the ability to use their IT infrastructure to leverage the cost of transporting data within a facility. Another area where costs will continue to fall in using Internet standards is the hardware required to transport data within a building or a campus. Off-the-shelf products such as routers, switches, hubs, and server computers make the FAS just another node of the IT infrastructure. Standard IT tools can be used to diagnose the FAS network, generate reports of FAS bandwidth on the intranet, and back up the FAS database.

The FAS of old relied heavily on a collection of separate systems that operated independently, and often with proprietary communication protocols that made expansion, modification, updating, and integration with other building or plant information and control systems very cumbersome, if not impossible. Today, the FAS is not only expected to handle all of the energy- and equipment-related tasks, but also to provide operating information and control interfaces to other facility systems, including the total facility or enterprise management system.

Measuring, monitoring, and maximizing energy savings is a fundamental task of all FAS and is the primary justification for many FAS installations. Improving facility operations in all areas, through enterprise information and control functions, is fast becoming an equally important function of the overall FAS or facility management system. The web provides the means to share information easier, quicker, and cheaper than ever before. There is no doubt that the web is having a huge impact on the FAS industry. The FAS of tomorrow will rely heavily on the web, TCP/IP, high-speed data networks, and enterprise-level connectivity. If it has not already been done, it is a good time for energy managers to get to know their IT counterparts at their facility, along with those in the accounting and maintenance departments. The future FAS will be here sooner than you think.[23]

10.4.9.1 Energy Management Strategies for Industry

Energy management strategies for industry can be grouped into three categories:

1. Operational and maintenance strategies
2. Retrofit or modification strategies
3. New design strategies

The order in which these are listed corresponds approximately to increased capital investment and increased implementation times. Immediate savings at little or no capital cost can generally be achieved by improved operations and better maintenance. Once these "easy" savings have been realized, additional improvements in efficiency will require capital investments.

10.4.9.2 Electric Drives and Electrically Driven Machinery

About 68% of industrial electricity use is for electrical motor–driven equipment. Integral horsepower (<0.75 kW) motors are more numerous. Major industrial motor loads are, in order of importance, pumps, compressors, blowers, fans, and miscellaneous integral motor applications including conveyors, DC drives, machine tools, and fractional horsepower applications.

Numerous examples of pumping in industry can be observed. These include process pumping in chemical plants, fluid movement in oil refineries, and cooling water circulation. An example of compressors is in the production of nitrogen and oxygen, two common chemicals. Large amounts of electricity are used to drive the compressors, which supply air to the process.

The typical industrial motor is a polyphase motor rated at 11.2 kW (15 hp) and having a life of about 40,000 h. The efficiencies of electric motors have increased recently as a result of higher energy prices, conservation efforts, and new government standards (the Energy Policy Act). High-efficiency motors cost roughly 20%–30% more than standard motors, but this expense is quickly repaid for motors that see continuous use.

Most efficient use of motors requires that attention be given to the following:

Optimum power: Motors operate most efficiently at rated voltage. Three-phase power supplies should be balanced; an unbalance of 3% can increase losses 25%.[1]

Good motor maintenance: Provide adequate cooling, keep heat transfer surfaces and vents clean, and provide adequate lubrication. Improved lubrication alone can increase efficiency a few percentage points.

Equipment scheduling: Turn equipment off when not in use; schedule large motor operation to minimize demand peaks.

Size equipment properly: Match the motor to the load and to the duty cycle. Motors operate most efficiently at rated load.

Evaluate continuous vs. batch processes: Sometimes a smaller motor operating continuously will be more economical.

Power factor: Correct if economics dictate savings. Motors have the best power factor at rated load.

Retrofit or new designs permit use of more efficient motors. For motors up to about 10–15 kW (15–20 hp), there are variations in efficiency. Select the most efficient motor for the job. Check to verify that the additional cost (if any) will be repaid by the savings that will accrue over the life of the installation.

In addition to reviewing the electric drive system, consider the power train and the load. Friction results in energy dissipation in the form of heat. Bearings, gears, and belt drives all have certain losses, as do clutches. Proper operation and maintenance can reduce energy wastage in these systems and improve overall efficiency.

Material shaping and forming, such as is accomplished with machine tools, requires that electrical energy be transformed into various forms of mechanical energy. The energy expenditure related to the material and to the depth and speed of the cut. By experimenting with a specific process, it is possible to establish cutting rates that are optimum for the levels of production required and are most efficient in terms of energy use. Motors are not the only part of the electric drive system that sustains losses. Other losses occur in the electric power systems that supply the motor. Electric power systems include substations, transformers, switching gear, distribution systems, feeders, power and lighting panels, and related equipment. Possibilities for energy management include the following:

Use highest voltages that are practical: For a given application, doubling the voltage cuts the required current in half and reduces the i^2r losses by a factor of 4.

Eliminate unnecessary transformers: They waste energy. Proper selection of equipment and facility voltages can reduce the number of transformers required and cut transformer losses. Remember, the customer pays for losses when the transformers are on his side of the meter. For example, it is generally better to order equipment with motors of the correct voltage, even if this costs more, than to install special transformers.

Energy losses are an inherent part of electric power distribution systems: This is primarily due to i^2r losses and transformers. The end-use conversion systems for electrical energy used in the process also contribute to energy waste. Proper design and operation of an electrical system can minimize energy losses and contribute to the reduction of electricity bills. Where long feeder runs are operated at near-maximum capacities, check to see if larger wire sizes would permit savings and be economically justifiable.

The overall power factor of electrical systems should be checked for low power factor: This could increase energy losses and the cost of electrical service, in addition to excessive voltage drops and increased penalty charges by the utility. Electrical systems studies should be made, and consideration should be given to power factor correction capacitors. In certain applications, as much as 10%–15% savings can be achieved in a poorly operating plant.

Check load factors: This is another parameter that measures the plant's ability to use electrical power efficiently. It is defined as the ratio of the actual kW h used to the maximum demand in kW times the total hours in the time period. A reduction in demand to bring this ratio closer to unity without decreasing plant output means more economical operation. For example, if the maximum demand for a given month (200 h) is 30,000 kWe and the actual kW h is 3.6×10^6 kW h, the load factor is 60%. Proper management of operations during high-demand periods, which may extend only 15–20 min, can reduce the demand during that time without curtailing production. For example, if the 30,000 kWe could be reduced to 20,000 kWe, this would increase the load factor to about 90%. Such a reduction could amount to a $20,000–$50,000 reduction in the electricity bill.

Reduce peak loads wherever possible: Many nonessential loads can be shed during the demand peak without interrupting production. These loads would include such items as air compressors, heaters, coolers, and air conditioners. Manual monitoring and control is possible but is often impractical because of the short periods of time that are normally involved and the lack of centralized control systems. Automatic power demand control systems are available.

Provide improved monitoring or metering capability, submeters, or demand recorders: While it is true that meters alone will not save energy, plant managers need feedback to determine if their energy management programs are taking effect. Often, the installation of meters on individual processes or buildings leads to immediate savings of 5%–10% by virtue of the ability to see how much energy is being used and to test the effectiveness of corrective measures.

10.4.9.3 Fans, Blowers, and Pumps

Simple control changes are the first thing to consider with these types of equipment. Switches, time clocks, or other devices can ensure that they do not operate except when needed by the process. Heat removal or process mass flow requirements will determine the size of fans and pumps. Often, there is excess capacity, either as a result of design conservatism or because of process changes subsequent to the installation of equipment. The required capacity should be checked, since excess capacity leads to unnecessary demand charges and decreased efficiency.

For fans, the volume rate of air flow Q varies in proportion to the speed of the impeller:

$$Q = c_f N \text{ (m}^3/\text{s)} \qquad (10.14)$$

where
Q is the air flow, m^3/s
c_f is a constant, m^3/r
N is the fan speed, r/s

The pressure developed by the fan varies as the square of the impeller speed. The important rule, however, is that the power needed to drive the fan varies as the cube of the speed:

$$P = P_c N^3 \text{ (W)} \qquad (10.15)$$

where
P is the input power, W
P_c is a constant, W s^3/r^3

The cubic law of pumping power indicates that if the air flow is to be doubled, eight (2^3) times as much power must be supplied. Conversely, if the air flow is to be cut in half, only one-eighth ($1/2^3$) as much power must be supplied. Air flow (and hence power) can be reduced by changing pulleys or installing smaller motors.

Pumps follow laws similar to fans, the key being the cubic relationship of power to the volume pumped through a given system. Small decreases in flow rate, such as might be obtained with a smaller pump or gotten by trimming the impeller, can save significant amounts of energy.

VSDs are another technique for reducing process energy use. VSDs permit fans, blowers, and pumps to vary speed depending on process requirements. This can lead to significant savings on noncontinuous processes. Recent improvements in solid-state electronics have caused the price of VSDs to drop substantially. This is another technology that is supported by utility rebates in many areas.

10.4.9.4 Air Compressors

Compressed air is a major energy use in many manufacturing operations. Electricity used to compress air is converted into heat and potential energy in the compressed air stream. Efficient operation of compressed air systems therefore requires the recovery of excess heat where possible, as well as the maximum recovery of the stored potential energy.

Efficient operation is achieved in these ways:

Select the appropriate type and size of equipment for the duty cycle required: Process requirements vary, depending on flow rates, pressure, and demand of the system. Energy savings can be achieved by selecting the most appropriate equipment for the job. The rotary compressor is more popular for industrial operations in the range of 20–200 kW, even though it is somewhat less efficient than the reciprocal compressor. This has been due to lower initial cost and reduced maintenance. When operated at partial load, reciprocating units can be as much as 25% more efficient than rotary units. However, newer rotary units incorporate a valve that alters displacement under partial load conditions and improves efficiency. Selection of an air-cooled vs. a water-cooled unit would be influenced by whether water or air was the preferred medium for heat recovery.

Proper operation of compressed air systems can also lead to improved energy utilization: Obviously, air leaks in lines and valves should be eliminated. The pressure of the compressed air should be reduced to a minimum. The percentage saving in power required to drive the compressor at a reduced pressure can be estimated from the fan laws described previously. For example, suppose the pressure were reduced to one half the initial value. Since pressure varies as the square of the speed, this implies the speed would be 70.7% of the initial value. Since power varies as the cube of the speed, the power would now be $0.707^3 = 35\%$ of the initial value. Of course, this is the theoretical limit; actual compressors would not do as well, and the reduction would depend on the type of compressor. Measurements indicate that actual savings would be about half the theoretical limit; reducing pressure 50% would reduce brake horsepower about 30%. To illustrate this point further, for a compressor operating at 6.89×10^5 N/m^2 (100 psi) and a reduction of the discharge pressure to 6.20×10^5 N/m^2 (90 psi), a 5% decrease in brake horsepower would result. For a 373 kW (500 hp) motor operating for 1 year, the 150,000 kW h savings per year would result in about $9000/year in electric power costs.

The intake line for the air compressor should be at the lowest temperature available: This normally means outside air. The reduced temperature of air intake results in a smaller volume of air to be compressed. The percentage horsepower saving relative to a 21°C (70°F) intake air temperature is about 2% for each 10°F drop in temperature. Conversely, input power increases by about 2% for each 10°F increase in intake air temperature.

Leakage is the greatest efficiency offender in compressed air systems: The amount of leakage should be determined and measures should be taken to reduce it. If air leakage in a plant is more than 10% of the plant demand, a poor condition exists. The amount of leakage can be determined by a simple test during off-production hours (when air-using equipment is shut down) by noting the time that the compressor operates under load compared with the total cycle. This indicates the percentage of the compressor's capacity that is used to supply the plant air leakage. Thus, if the load cycle compared with the total cycle were 60 s compared with 180 s, the efficiency would be 33%, or 33% of the compressor capacity is the amount of air leaking in m^3/min (ft^3/min).

Recover heat where feasible: There are sometimes situations where water-cooled or air-cooled compressors are a convenient source of heat for hot water, space heating, or process applications. As a rough rule of thumb, about $300\ J/m^3$ min of air compressed (~10 Btu/ft^3 min) can be recovered from an air-cooled rotary compressor.

Substitute electric motors for air motor (pneumatic) drives: Electric motors are far more efficient. Typical vaned air motors range in size from 0.15 to 6.0 kW (0.2–8 hp), cost $300–$1500, and produce 1.4–27 N m (1–20 ft lb) of torque at 620 kN/m^2 (90 psi) air pressure. These are used in manufacturing operations where electric motors would be hazardous, or where light weight and high power are essential. Inefficiency results from air system leaks and the need (compared to electric motors) to generate compressed air as an intermediate step in converting electric to mechanical energy.

Review air usage in paint spray booths: In paint spray booths and exhaust hoods, air is circulated through the hoods to control dangerous vapors. Makeup air is constantly required for dilution purposes. This represents a point of energy rejection through the exhaust air.

Examination should be made of the volumes of air required in an attempt in reduce flow and unnecessary operation. Possible mechanisms for heat recovery from the exhaust gases should be explored using recovery systems.

10.4.9.5 Electrochemical Operations

Electrochemical processes are an industrial use of electricity, particularly in the primary metals industry, where it is used in the extraction process for several important metals. Energy management opportunities include the following:

Improve design and materials for electrodes: Evaluate loss mechanisms for the purpose of improving efficiency.

Examine electrolysis and plating operations for savings: Review rectifier performance, heat loss from tanks, and the condition of conductors and connections.

Welding is another electrochemical process. Alternating current welders are generally preferable when they can be used, since they have a better power factor, better demand characteristics, and more economical operation.

Welding operations can also be made more efficient by the use of automated systems, which require 50% less energy than manual welding. Manual welders deposit a bead only 15%–30% of the time the machine is running. Automated processes, however, reduce the no-load time to 40% or less. Different welding processes should be compared in order to determine the most efficient process. Electroslag welding is suited only for metals over 1 cm (0.5 in.) thick but is more efficient than other processes.

Two other significant applications of electrolysis of concern to industry are batteries and corrosion. Batteries are used for standby power, transportation, and other applications. Proper battery maintenance and improved battery design contribute to efficient energy use.

Corrosion is responsible for a large loss of energy-intensive metals every year and thus indirectly contributes to energy wastage. Corrosion can be prevented and important economies realized, by use of protective films, cathodic protection, and electroplating or anodizing.

10.4.9.6 Electric Process Heat and Steam Systems

In as much as approximately 40% of the energy utilized in industry goes toward the production of process steam, it presents a large potential for energy misuse and fuel waste from improper maintenance and operation. Even though electrically generated steam and hot water is a small percentage of total industrial steam and hot water, the electrical fraction is likely to increase as other fuels increase in price. This makes increased efficiency even more important. Examples are as follows:

Steam leaks from lines and faulty valves result in considerable losses: These losses depend on the size of the opening and the pressure of the steam, but can be very costly. A hole 0.1 ft in diameter with steam at 200 psig can bleed $1000–$2000 worth of steam (500 GJ) in a year.

Steam traps are major contributors to energy losses when not functioning properly: A large process industry might have thousands of steam traps, which could result in large costs if they are not operating correctly. Steam traps are intended to remove condensate and noncondensable gases while trapping or preventing the loss of steam. If they stick open, orifices as large as 6 mm (0.25 in.) can allow steam to escape. Such a trap would allow 1894 GJ/year (2000 MBtu/year) of heat to be rejected to the atmosphere on a 6.89×10^5 N/m (100 psi) pressure steam line. Many steam traps are improperly sized, contributing to an inefficient operation. Routine inspection, testing, and a correction program for steam valves and traps are essential in any energy program and can contribute to cost savings.

Poor practice and design of steam distribution systems can be the source of heat waste up to 10% or more: It is not uncommon to find an efficient boiler or process plant joined to an inadequate steam distribution system. Modernization of plants results from modified steam requirements. The old distribution systems are still intact, however, and can be the source of major heat losses. Large steam lines intended to supply units no longer present in the plant are sometimes used for minor needs, such as space heating and cleaning operations, that would be better accomplished with other heat sources.

Steam distribution systems operating on an intermittent basis require a start-up warming time to bring the distribution system into proper operation. This can extend up to 2 or 3 h, which puts a demand on fuel needs. Not allowing for proper ventilating of air can also extend the start-up time. In addition, condensate return can be facilitated if it is allowed to drain by gravity into a tank or receiver and is then pumped into the boiler feed tank.

Proper management of condensate return: Proper management can lead to great savings. Lost feedwater must be made up and heated. For example, every 0.45 kg (1 lb) of steam that must be generated from 15°C feedwater instead of 70°C feedwater requires an additional 1.056×10^5 J (100 Btu) more than 1.12 MJ (1063 Btu) required or a 10% increase in fuel. A rule of thumb is that a 1% fuel saving results for every 5°C increase in feedwater temperature. Maximizing condensate recovery is an important fuel laying procedure.

Poorly insulated lines and valves due either to poor initial design or a deteriorated condition: Heat losses from a poorly insulated pipe can be costly. A poorly insulated line carrying steam at 400 psig can lose ~1000 GJ/year (10^9 Btu/year) or more per 30 m (100 ft) of pipe. At steam costs of \$2.00/GJ, this translates to a \$2000 expense/year.

Improper operation and maintenance of tracing systems: Steam tracing is used to protect piping and equipment from cold weather freezing. The proper operation and maintenance of tracing systems will not only ensure the protection of traced piping but also saves fuel. Occasionally, these systems are operating when not required. Steam is often used in tracing systems, and many of the deficiencies mentioned earlier apply (e.g., poorly operating valves, insulation, leaks).

Reduce losses in process hot water systems: Electrically heated hot water systems are used in many industrial processes for cleaning, pickling, coating, or etching components. Hot or cold water systems can dissipate energy. Leaks and poor insulation should be repaired.

10.4.9.7 Electrical Process Heat

Industrial process heat applications can be divided into four categories: direct-fired, indirect-fired, fuel, or electric. Here we shall consider electric direct-fired installations (ovens, furnaces) and indirect-fired (electric water heaters and boilers) applications. Electrical installations use metal sheath resistance heaters, resistance ovens or furnaces, electric salt bath furnaces, infrared heaters, induction and high-frequency resistance heaters, dielectric heaters, and direct arc furnaces. From the housekeeping and maintenance point of view, typical opportunities would include the following:

Repair or improve insulation: Operational and standby losses can be considerable, especially in larger units. Remember that insulation may degrade with time or may have been optimized to different economic criteria.

Provide finer controls: Excessive temperatures in process equipment waste energy. Run tests to determine the minimum temperatures that are acceptable, then test instrumentation to verify that it can provide accurate process control and regulation.

Practice heat recovery: This is an important method, applicable to many industrial processes as well as HVAC systems and so forth. It is described in more detail in the next section.

10.4.9.8 Heat Recovery

Exhaust gases from electric ovens and furnaces provide excellent opportunities for heat recovery. Depending on the exhaust gas temperature, exhaust heat can be used to raise steam or to preheat air or feedstocks. Another potential source of waste-heat recovery is the exhaust air that must be rejected from industrial operations in order to maintain health and ventilation safety standards. If the reject air has been subjected to heating and cooling processes, it represents an energy loss inasmuch as the makeup air must be modified to meet the interior conditions. One way to reduce this waste is through the use of heat wheels or similar heat exchange systems.

Energy in the form of heat is available at a variety of sources in industrial operations, many of which are not normally derived from primary heat sources. Such sources include electric motors, crushing and grinding operations, air compressors, and drying processes. These units require cooling in order to maintain proper operation. The heat from these systems can be collected and transferred to some appropriate use such as space heating or water heating.

The heat pipe is gaining wider acceptance for specialized and demanding heat transfer applications. The transfer of energy between incoming and outgoing air can be accomplished by banks of these devices. A refrigerant and a capillary wick are permanently sealed inside a metal tube, setting up a liquid-to-vapor circulation path. Thermal energy applied to either end of the pipe causes the refrigerant to vaporize. The refrigerant vapor then travels to the other end of the pipe, where thermal energy is removed. This causes the vapor to condense into liquid again, and the condensed liquid then flows back to the opposite end through the capillary wick.

Industrial operations involving fluid flow systems that transport heat such as in chemical and refinery operations offer many opportunities for heat recovery. With proper design and sequencing of heat exchangers, the incoming product can be heated with various process steams. For example, proper heat exchanger sequence in preheating the feedstock to a distillation column can reduce the energy utilized in the process.

Many process and air-conditioning systems reject heat to the atmosphere by means of wet cooling towers. Poor operation can contribute to increased power requirements.

> *Water flow and airflow should be examined to see that they are not excessive*: The cooling tower outlet temperature is fixed by atmospheric conditions if operating at design capacity. Increasing the water flow rate or the air flow will not lower the outlet temperature.
>
> *The possibility of utilizing heat that is rejected to the cooling tower for other purposes should be investigated*: This includes preheating feedwater, heating hot water systems, space heating, and other low-temperature applications. If there is a source of building exhaust air with a lower wet bulb temperature, it may be efficient to supply this to a cooling tower.

10.4.9.9 Power Recovery

Power recovery concepts are an extension of the heat recovery concept described earlier. Many industrial processes have pressurized liquid and gaseous streams at 150°C–375°C (300°F–700°F) that present excellent opportunities for power recovery. In many cases, high-pressure process stream energy is lost by throttling across a control valve.

The extraction of work from high-pressure liquid streams can be accomplished by means of hydraulic turbines (essentially diffuser-type or volute-type pumps running backward). These pumps can be either single or multistage. Power recovery ranges from 170 to 1340 kW (230–1800 hp). The lower limit of power recovery approaches the minimum economically justified for capital expenditures at present power costs.

10.4.9.10 Heating, Ventilating, and Air-Conditioning Operation

The environmental needs in an industrial operation can be quite different from those in a residential or commercial structure. In some cases, strict environmental standards must be met for a specific function or process. More often, the environmental requirements for the process itself are not severe; however, conditioning of the space is necessary for the comfort of operating personnel, and thus large volumes of air must be processed. Quite often, opportunities exist in the industrial operation where surplus energy can be utilized in environmental conditioning. A few suggestions follow:

Review HVAC controls: Building heating and cooling controls should be examined and preset.

Ventilation, air, and building exhaust requirements should be examined: A reduction of air flow will result in a savings of electrical energy delivered to motor drives and additionally reduce the energy requirements for space heating and cooling. Because pumping power varies as the cube of the air flow rate, substantial savings can be achieved by reducing air flows where possible.

Do not condition spaces needlessly: Review air-conditioning and heating operations, seal off sections of plant operations that do not require environmental conditioning, and use air-conditioning equipment only when needed. During nonworking hours, the environmental control equipment should be shut down or reduced. Automatic timers can be effective.

Provide proper equipment maintenance: Ensure that all equipment is operating efficiently. (Filters, fan belts, and bearings should be in good condition.)

Use only equipment capacity needed: When multiple units are available, examine the operating efficiency of each unit and put operations in sequence in order to maximize overall efficiency.

Recirculate conditioned (heated or cooled) air where feasible: If this cannot be done, perhaps exhaust air can be used as supply air to certain processes (e.g., a paint spray booth) to reduce the volume of air that must be conditioned.

For additional energy management opportunities in HVAC systems, see Chapter 6.

10.4.9.11 Lighting

Industrial lighting needs range from low-level requirements for assembly and welding of large structures (such as shipyards) to the high levels needed for the manufacture of precision mechanical and electronic components (e.g., integrated circuits). There are four basic housekeeping checks that should be made:

Is a more efficient lighting application possible? Remove excessive or unnecessary lamps.

Is relamping possible? Install lower-wattage lamps during routine maintenance.

Will cleaning improve light output? Fixtures, lamps, and lenses should be cleansed periodically.

Can better controls be devised? Eliminate turning on more lamps than necessary. For modification, retrofit, or new design, consideration should be given to the spectrum of high-efficiency lamps and luminaries that are available. For example, high-pressure sodium lamps are finding increasing acceptance for industrial use, with savings of nearly a factor of five compared to incandescent lamps. See Chapter 5 for additional details.

10.4.9.12 New Electrotechnologies

Electricity has certain characteristics that make it uniquely suitable for industrial processes. These characteristics include electricity's suitability for timely and precise control, its ability to interact with materials at the molecular level, the ability to apply it selectively and specifically, and the ability to vary its frequency and wavelength so as to enhance or inhibit its interaction with materials. These aspects may be said to relate to the *quality* of electricity as an energy form. It is important to recognize that different forms of energy have different qualities in the sense of their ability to perform useful work. Thus, although the Btu content of two energy forms may be the same, their ability to transform materials may be quite different.

New electrotechnologies based on the properties of electricity are now finding their way into modern manufacturing. In many cases, the introduction of electricity reduces manufacturing costs, improves quality reduces pollution, or has other beneficial results. Some examples include the following:

Microwave heating	Ion nitriding
Induction heating	Infrared drying
Plasma processing	UV drying and curing
Magnetic forming	Advanced finishes
RF drying and heating	Electron beam heating

Microwave heating is a familiar technology that exhibits the unique characteristics of electricity described earlier. First, it is useful to review how conventional heating is preformed to dry paint, anneal a part, or remove water. A source of heat is required, along with a container (oven, furnace, pot, etc.) to which the heat is applied. Heat is transferred from the container to the workpiece by conduction, radiation, convection, or a combination of these. There are certain irreversible losses associated with heat transfer in this process. Moreover, since the container must be heated, more energy is expended than is really required. Microwave heating avoids these losses due to the unique characteristics of electricity.

Timely control: There is no loss associated with the warm-up or cooldown of ovens. The heat is applied directly when needed.

Molecular interaction: By interacting at the molecular level, heat is deposited directly in the material to be heated, without having to preheat an oven, saving the extra energy required for this purpose and avoiding the losses that result from heat leakage from the oven.

Selective application: By selectively applying heat only to the material to be heated, parasitic losses are avoided. In fact, the specificity of heat applied this way can improve quality by not heating other materials.

Selective wavelength and frequency: A microwave frequency is selected that permits the microwave energy to interact with the material to be heated, and not with other materials. Typically, the frequency is greater than 2000 MHz.

Microwave heating was selected for this discussion, but similar comments could be made about infrared, ultraviolet (UV), dielectric, induction, or electron beam heating. In each case, the frequency or other characteristics of the energy form are selected to provide the unique performance required.

UV curing (now used for adhesive and finishes) is another example. The parts to be joined or coated can be prepared and the excessive adhesive removed without fear of pre-hardening. Then, the UV energy is applied, causing the adhesive to harden.

Induction heating is another example. It is similar to microwave heating except that the energy is applied at a lower frequency. Induction heating operates on the principle of inducing electric currents to flow in materials, heating them by the power dissipated in the material. The method has several other advantages. In a conventional furnace, the workpiece has to be in the furnace for a sufficient time to reach temperature. Because of this, some of the material is oxidized and lost as scale. In a typical high-temperature gas furnace, this can be 2% of the throughput. Additional product is scrapped as a result of surface defects caused by uneven heating and cooling. This can amount to another 1% of throughput. Induction heating can reduce these losses by a factor of four.

The fact that electricity can be readily controlled and carries with it a high information content through digitization or frequency modulation also offers the potential for quantum improvements in efficiency. A slightly different example is the printing industry.

Today, the old linotype technology has been replaced by electronic processes. The lead melting pots that used to operate continuously in every newspaper plant have been removed, eliminating a major energy use and an environmental hazard. Books, magazines, and newspapers can be written composed and printed entirely by electronic means. Text is processed by computer techniques. Camera-ready art is prepared by computers directly or prepared photographically and then optically scanned to create digital images. The resulting electronic files can be used in web offset printing by an electronic photochemical process. The same information can be transmitted electronically, via satellite, to a receiving station at a remote location where a high-resolution fax machine reconstitutes the image. This method is being used to simultaneously and instantaneously distribute advertising copy to multiple newspapers, using a single original. Previously, to insert an advertisement in 25 newspapers, 25 sets of photographic originals would have to be prepared and delivered, by messenger, air express, or mail, to each newspaper.

Some of the other applications of the new electrotechnologies include RF drying of plywood veneers, textiles, and other materials; electric infrared drying for automobile paint and other finishes; electric resistance melting for high-purity metals and scrap recovery; and laser cutting of wood, cloth, and other materials.

10.4.9.13 General Industrial Processes

The variety of industrial processes is so great that detailed specific recommendations are outside the scope of this chapter. Useful sources of information are found in trade journals, vendor technical bulletins, and manufacturers' association journals. These suggestions are intended to be representative, but by no means do they cover all possibilities.

> *In machining operations, eliminate unnecessary operations and reduce scrap*: This is so fundamental from a purely economic point of view that it will not be possible to find significant improvements in many situations. The point is that each additional operation and each increment of scrap also represent a needless use of energy. Machining itself is not particularly energy intensive. Even so, there are alternate technologies that can not only save energy but reduce material wastage as well. For example, powder metallurgy generates less scrap and is efficient if done in induction-type furnaces.

Use stretch forming: Forming operations are more efficient if stretch forming is used. In this process, sheet metal or extrusions are stretched 2%–3% prior to forming, which makes the material more ductile so that less energy is required to form the product.

Use alternate heat treating methods: Conventional heat treating methods such as carburizing are energy intensive. Alternate approaches are possible. For example, a hard surface can be produced by induction heating, which is a more efficient energy process. Plating, metallizing, flame spraying, or cladding can substitute for carburizing, although they do not duplicate the fatigue strengthening compressive skin of carburization or induction hardening.

Use alternative painting methods: Conventional techniques using solvent-based paints require drying and curing at elevated temperature. Powder coating is a substitute process in which no solvents are used. Powder particles are electrostatically charged and attracted to the part being painted so that only a small amount of paint leaving the spray gun misses the part and the overspray is recoverable. The parts can be cured rapidly in infrared ovens, which require less energy than standard hot air systems. Water-based paints and high-solids coatings are also being used and are less costly than solvent-based paints. They use essentially the same equipment as the conventional solvent paint spray systems so that the conversion can be made at minimum costs. New water-based emulsion paints contain only 5% organic solvent and require no afterburning. High-solids coatings are already in use commercially for shelves, household fixtures, furniture, and beverage cans, and require no afterburning. They can be as durable as conventional finishes and are cured by either conventional baking or UV exposure.

Substitute for energy-intensive processes such as hot forging: Hot forging may require a part to go through several heat treatments. Cold forging with easily wrought alloys may offer a replacement. Lowering the preheat temperatures may also be an opportunity for savings. Squeeze forging is a relatively new process in which molten metal is poured into the forging dye. The process is nearly scrap free, requires less press power, and promises to contribute to more efficient energy utilization.

Movement of materials through the plant creates opportunities for saving energy. Material transport energy can be reduced by the following methods:

Combining processes or relocate machinery to reduce transport energy: Sometimes merely relocating equipment can reduce the need to haul materials.

Turning off conveyors and other transport equipment when not needed: Look for opportunities where controls can be modified to permit shutting down of equipment not in use.

Using gravity feeds wherever possible: Avoid unnecessary lifting and lowering of products.

10.4.9.14 Demand Management

The cost of electrical energy for medium-to-large industrial and commercial customers generally consists of two components. The first component is the *energy charge*, which is based on the cost of fuel to the utility, the average cost of amortizing the utility generating plant, and on the operating and maintenance costs experienced by the utility. Energy costs for industrial users in the United States are typically in the range of $0.05–$0.10/kW h.

The second component is the *demand charge*, which reflects the investment cost the utility must make to serve the customer. Besides the installed generating capacity needed, the utility also provides distribution lines, transformers, substations, and switches whose cost depends on the size of the load being served. This cost is recovered in a demand charge, which typically is $2–$10/kW month.

Demand charges typically account for 10%–50% of the bill, although wide variations are possible depending on the type of installation. Arc welders, for example, have relatively high demand charges, since the installed capacity is great (10–30 kW for a typical industrial machine) and the energy use is low.

From the utility's point of view, it is advantageous to have its installed generating capacity operating at full load as much of the time as possible. To follow load variations conveniently, the utility operates its largest and most economical generating units continuously to meet its base load and then brings smaller (and generally more expensive) generating units online to meet peak load needs.

Today, consideration is being given to time-of-day or *peak load pricing* as a means of assigning the cost of operating peak generating capacity to those loads that require it. From the viewpoint of the utility, *load management* implies maintaining a high capacity factor and minimizing peak load demands. From the customer's viewpoint, *demand management* means minimizing electrical demands (both on- and off-peak) so as to minimize overall electricity costs.

Utilities are experimenting with several techniques for load management. Besides rate schedules that encourage the most effective use of power, some utilities have installed remotely operated switches that permit the utility to disconnect nonessential parts of the customer's load when demand is excessive. These switches are actuated by a radio signal, through the telephone lines, of over the power grid itself through a harmonic signal (ripple frequency) that is introduced into the grid.

Customers can control the demand of their loads by any of several methods:

- Manually switching off-loads ("load shedding")
- Use of timers and interlocks to prevent several large loads from operating simultaneously
- Use of controllers and computers to control loads and minimize peak demand by scheduling equipment operation
- Energy storage (e.g., producing hot or chilled water during off-peak hours and storing it for use on-peak)

Demand can be monitored manually (by reading a meter) or automatically using utility-installed equipment or customer-owned equipment installed in parallel with the utility meter. For automatic monitoring, the basic approach involves pulse counting.

The demand meter produces electronic pulses, the number of which is proportional to demand in kW. Demand is usually averaged over some interval (e.g., 15 min) for calculating cost. By monitoring the pulse rate electronically, a computer can project what the demand will be during the next demand measurement interval and can then follow a preestablished plan for shedding loads if the demand set point is likely to be exceeded.

Computer control can assist in the dispatching of power supply to the fluctuating demands of plant facilities. Large, electrically based facilities are capable of forcing large power demands during peak times that exceed the limits contracted with the utility or cause penalties in increased costs. Computer control can even out the load by shaving peaks and filling in the valleys, thus minimizing power costs. In times of emergency or fuel curtailment, operation of the plant can be programmed to provide optimum production and operating performance under prevailing conditions. Furthermore, computer monitoring and control provide accurate and continuous records of plant performance.

It should be stressed here that many of these same functions can be carried out by manual controls, time clocks, microprocessors, or other inexpensive devices. Selection of a computer system must be justified economically on the basis of the number of parameters to be controlled and the level of sophistication required. Many of the benefits described here can be obtained in some types of operations without the expense of a computer.

10.5 Thermal Energy Management in Industry

10.5.1 Importance of Fuel Use and Heat in Industry

The U.S. manufacturing sector depends heavily on fuels for the conversion of raw materials into usable products. Industry uses a wide range of fuels, including natural gas, petroleum, coal, and renewables. The petroleum forms used include distillate fuel oil, residual fuel oil, gasoline, LPG, and others. How efficiently energy is used, its cost, and its availability consequently have a substantial impact on the competitiveness and economic health of the U.S. manufacturers. More efficient use of fuels lowers production costs, conserves limited energy resources, and increases productivity. Efficient use of energy also has positive impacts on the environment—reductions in fuel use translate directly into decreased emissions of pollutants such as sulfur oxides, nitrogen oxides, particulates, and greenhouse gases (e.g., carbon dioxide).

From Figure 10.10, it can be seen that fuel use in manufacturing is just over 70% of the total energy used in manufacturing on an end-use basis. Figure 10.11 shows the percentage of each fuel used for boiler fuel and process heat combined. Of this total, 53% is used for boiler fuel, and 47% for direct process heating. Thus, there is a huge potential for energy management and energy efficiency improvement related to the use fuels and thermal energy in industry. Section 10.5 discusses the use of improved and new equipment and technology to accomplish some of these reductions in energy use and cost.

Energy efficiency can be defined as the effectiveness with which energy resources are converted into usable work. Thermal efficiency is commonly used to measure the efficiency of energy conversion systems such as process heaters, steam systems, engines, and power generators. Thermal efficiency is essentially the measure of the efficiency and completeness of fuel combustion, or, in more technical terms, the ratio of the net work supplied to the heat supplied by the combusted fuel. In a gas-fired heater, for example, thermal efficiency is equal to the total heat absorbed divided by the total heat supplied; in an automotive engine, thermal efficiency is the work done by the gases in the cylinder divided by the heat energy of the fuel supplied.[1]

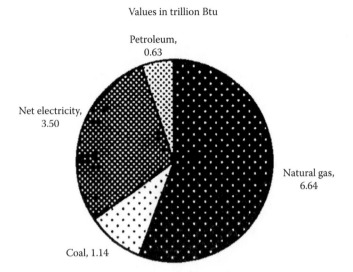

FIGURE 10.10
Manufacturing energy consumption by fuel (end-use data). (From U.S. Department of Energy, Energy Information Agency, Manufacturing Energy Consumption Survey, 2006, Washington, DC, 2012.)

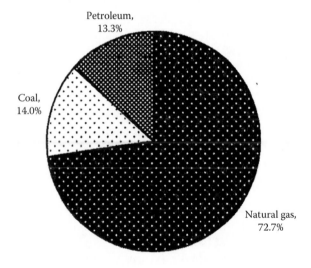

FIGURE 10.11
Manufacturing energy use for boiler fuel and process heating end-use basis. (From U.S. Department of Energy, Energy Information Agency, Manufacturing Energy Consumption Survey, 1998, Washington, DC, 2002.)

Energy efficiency varies dramatically across industries and manufacturing processes, and even between plants manufacturing the same products. Efficiency can be limited by mechanical, chemical, or other physical parameters, or by the age and design of equipment. In some cases, operating and maintenance practices contribute to lower-than-optimum efficiency. Regardless of the reason, less-than-optimum energy efficiency implies that not all of the energy input is being converted to useful work—some is released as lost energy. In the manufacturing sector, these energy losses amount to several quadrillion Btus (quadrillion British thermal units, or quads) and billions of dollars in lost revenues every year.

Typical thermal efficiencies of selected energy systems and industrial equipment[23] are provided in the following table:

Power generation	25%–44%
Steam boilers (natural gas)	80%
Steam boilers (coal and oil)	84%–85%
Waste-heat boilers	60%–70%
Thermal cracking (refineries)	58%–61%
EAF steelmaking	56%
Paper drying	48%
Kraft pulping	60%–69%
Distillation column	25%–40%
Cement calciner	30%–70%
Compressors	10%–20%
Pumps and fans	55%–65%
Motors	90%–95%

Boiler losses represent energy lost due to boiler inefficiency. In practice, boiler efficiency can be as low as 55%–60% or as high as 90%. The age of the boiler, maintenance practices, and fuel type are contributing factors to boiler efficiency. It is assumed that the greater losses are in steam pipes (20%), with small losses incurred in other fuel transmission lines (3%) and electricity transmission lines (3%). Losses in steam pipes and traps have been reported to be as high as from 20% to 40%. A conservative value of 20% was used for steam distribution losses in this study.[1]

10.5.2 Boiler Combustion Efficiency Improvement

Boilers and other fuel-fired equipment, such as ovens and kilns, combust fuel with air for the purpose of releasing chemical energy as heat. For an industrial boiler, the purpose is to generate high-temperature and high-pressure steam to use directly in a manufacturing process or to operate other equipment such as steam turbines to produce shaft power. As shown earlier in Figure 10.11, the predominant boiler fuel is natural gas. The efficiency of any combustion process is dependent on the amount of air that is used in relation to the amount of fuel and how they are mixed. Air is about 20% oxygen, so approximately 5 units of air must be brought into the boiler for every 1 unit of oxygen that is needed. Controlling this air–fuel mixture, and minimizing the amount of excess air while still obtaining safe mixing of the air and fuel, is key to ensuring a high combustion efficiency in the boiler.[24]

10.5.2.1 Combustion Control

The stoichiometric equation for the combustion of methane, the principal constituent of natural gas, with air is

$$CH_4 + 2O_2 + 7.52N_2 \rightarrow CO_2 + 2H_2O + 7.52N_2 \qquad (10.16)$$

The stoichiometric equation is the one representing the exact amount of air necessary to oxidize the carbon and hydrogen in the fuel to carbon dioxide and water vapor. However, it is necessary to provide more than the stoichiometric amount of air since the mixing of fuel and air is imperfect in the real combustion chamber. Thus, the combustion equation for hydrocarbon fuels becomes

$$C_xH_y + \phi\left(\frac{x+y}{4}\right)O_2 + 3.76\phi\left(\frac{x+y}{4}\right)N_2 \rightarrow xCO_2 + y/2H_2O$$

$$+ (\phi - 1)\left(\frac{x+y}{4}\right)O_2 + 3.76\phi\left(\frac{x+y}{4}\right)N_2 \tag{10.17}$$

Note that for a given fuel, nothing in the equation changes except the parameter f, the equivalence ratio, as the fuel–air ratio changes.

As ϕ is increased beyond the optimal value for good combustion, the stack losses increase, and the heat available for the process decreases. As the equivalence ratio increases for a given flue temperature and a given fuel, more fuel must be consumed to supply a given amount of heat to the process.

The control problem for the furnace or boiler is to provide the minimum amount of air for good combustion over a wide range of firing conditions and a wide range of ambient temperatures. The most common combustion controller uses the ratio of the pressure drops across orifices, nozzles, or venturis in the air and fuel lines. Since these meters measure volume flow, a change in the temperature of combustion air with respect to fuel, or vice versa, will affect the equivalence ratio of the burner. Furthermore, since the pressure drops across the flowmeters are exponentially related to the volume flow rates, control dampers must have very complicated actuator motions. All the problems of ratio controllers are eliminated if the air is controlled from an oxygen meter. These are now coming into more general use as reasonably priced, high-temperature oxygen sensors become available. It is possible to control to any set value of percentage oxygen in the products; that is,

$$\%O_2 = \frac{\phi - 1}{\left((x+y/2)/(x+y/4)\right) + 4.76\phi - 1} \tag{10.18}$$

Figure 10.12 is a nomograph from the Bailey Meter Company[32] that gives estimates of the annual dollar savings resulting from the reduction of excess air to 15% for gas-, oil-, or coal-fired boilers with stack temperatures from 300°F to 700°F. The fuel savings are predicted on the basis that as excess air is reduced, the resulting reduction in mass flow of combustion gases results in reduced gas velocity and thus a longer gas residence time in the boiler. The increased residence time increases the heat transfer from the gases to the water. The combined effect of lower exhaust gas flows and increased heat exchange effectiveness is estimated to be 1.5 times greater than that due to the reduced mass flow alone.

As an example, assume the following data pertaining to an oil-fired boiler. Entering the graph at the top abscissa with 6.2% O_2, we drop to the oil fuel line and then horizontally to the 327°C (620°F) flue gas temperature line. Continuing to the left ordinate, we can see that 6.2% O_2 corresponds to 37.5% excess air. Dropping vertically from the intersection of

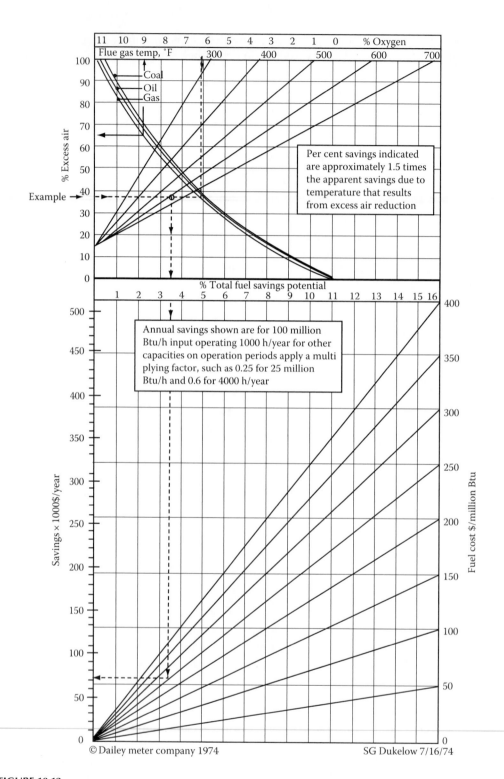

FIGURE 10.12
Nomograph for estimating savings from the adjustment of burners. (From Dukelow, S.G., Bailey Meter Company, Wickliffe, OH, 1974. With permission.)

the flue gas temperature line and the excess air line, we note a 3.4% total fuel savings. Fuel costs are as follows:

Burner capacity	63 GJ/h (60 × 10⁶ Btu/h)
Annual operating hours	6,200
Fuel cost	$0.38/L ($1.44/gal)
Heating value fuel	42.36 MJ/L (152,000 Btu/gal)
Percent O_2 in exhaust gases	6.2%
Stack temperature	327°C (620°F)

$$10^9 \text{ J/GJ } (10^6 \text{ Btu/million Btu}) \times \frac{\$0.38/\text{L}}{74,236,500 \text{ J/L}} \left(\frac{1.44/\text{gal}}{152,000 \text{ Btu/gal}} \right)$$

$$= \$8.96/\text{GJ } (9.48/\text{million Btu})$$

Continuing the vertical line to intersect the $2.50/million Btu and then moving to left ordinate show a savings of $140,000/year for 8000 h of operation, 100 × 10⁶ Btu/h input, and $5.00/million Btu fuel cost. Adjusting that result for the assumed operating data,

$$\text{Annual savings} = \frac{6,200}{8,000} \times \frac{60 \times 10^6}{100 \times 10^6} \times \frac{9.48}{5.00} \times \$140,000 = \$123,430/\text{year}$$

This savings could be obtained by installing a modern oxygen controller, an investment with approximately a 1-year payoff, or from heightened operator attention with frequent flue gas testing and manual adjustments. Valuable sources of information concerning fuel conservation in boilers and furnaces are given by the DOE/OIT website.

10.5.2.2 Waste-Heat Management

Waste heat as generally understood in industry is the energy rejected from any process to the atmospheric air or to a body of water. It may be transmitted by radiation, conduction, or convection, but often it is contained in gaseous or liquid streams emitted to the environment. Almost 50% of all fuel energy used in the United States is transferred as waste heat to the environment, causing thermal pollution as well as chemical pollution of one sort or another. It has been estimated that half of that total may be economically recoverable for useful heating functions.

What must be known about waste-heat streams in order to decide whether they can become useful? Here is a list along with a parallel list of characteristics of the heat load that should be matched by the waste-heat supply.

Waste-heat supply
Quantity
Quality
Temporal availability of supply
Characteristics of fluid
Heat load
Quantity required
Quality required
Temporal availability of load
Special fluid requirements

Let us examine the particular case of a plant producing ice-cream cones. All energy quantities are given in terms of 15.5°C reference temperature. Sources of waste heat include the following:

- Products of combustion from 120 natural gas jets used to heat the molds in the carousel-type baking machines. The stack gases are collected under an insulated hood and released to the atmosphere through a short stack. Each of six machines emits 236.2 m³/h of stack gas at 160°C. Total source rate is 161,400 kJ/h or 3874 MJ/day for a three-shift day.

- Cooling water from the jackets, intercoolers, and aftercoolers of two air compressors used to supply air to the pneumatic actuators of the cone machines; 11.36 L/min of water at 48.9°C is available. This represents a source rate of 96 MJ/h. The compressors run an average of 21 h/production day. Thus, this source rate is 2015 MJ/day.

- The water chillers used to refrigerate the cone batter make available—at 130°F—264 MJ/h of water heat. This source is available to heat water to 48.9°C using desuperheaters following the water chiller compressors. The source rate is 6330 MJ/day.

- At 21.2°C, 226 m³/min of ventilating air is discharged to the atmosphere. This is a source of rate of less than 22.2 MJ/h or 525 MJ/day.

Uses for waste heat include the following:

- For cleanup operations during 3 h of every shift or during 9 of every 24 h, 681 L/h of hot water at 82.2°C is needed. Total daily heat load is 4518 MJ.

- Heating degree-days total in excess of 3333 annually. Thus, any heat available at temperatures above 21.1°C can be utilized with the aid of runaround systems during the 5(1/2)-month heating season. Estimated heating load per year is 4010 GJ.

- Total daily waste heat available—12.74 GJ/day.

- Total annual waste heat available—3.19 TJ/year.

- Total annual worth of waste heat (at $5.00/GJ for gas)—$15,893.

- Total daily heat load—this varies from a maximum of 59.45 GJ/day at the height of the heating season to the hot water load of 4.52 GJ/day in the summer months.

Although the amount of waste heat from the water chillers is 40% greater than the load needed for hot water heating, the quality is insufficient to allow its full use, since the hot water must be heated to 82°C and the compressor discharge is at a temperature of 54°C.

However, the chiller waste heat can be used to preheat the hot water. Assuming 13°C supply water and a 10°C heat exchange temperature approach, the load that can be supplied by the chiller is

$$\frac{49-13}{82-13} \times 4.52 = 2.36 \text{ GJ/day}$$

Since the cone machines have an exhaust gas discharge of 3.87 GJ/day at 160°C, the remainder of the hot water heating load of 2.17 GJ/day is available. Thus, a total saving of 1129 GJ/year in fuel is possible with a cost saving of $5645 annually based on $5.00/GJ gas. The investment costs will involve the construction of a common exhaust heater for the cone machines, a desuperheater for each of the three water chiller compressors, a gas-to-liquid

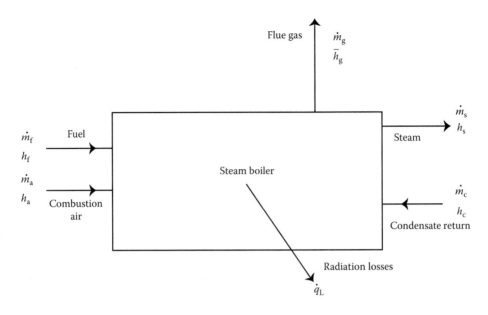

FIGURE 10.13
Heat balance on steam boiler.

heat exchanger following the cone-machine exhaust heater, and possibly an induced draft fan following the heat exchanger, since the drop in exhaust gas temperature will decrease the natural draft due to hot-gas buoyancy.

It is necessary to almost match four of four characteristics. Not exactly, of course, but the closer means thermodynamic availability of the waste heat. Unless the energy of the waste stream is sufficiently hot, it will be impossible to even transfer it to the heat load, since spontaneous heat transfer occurs only from higher to lower temperature.

The quantity and quality of energy available from a waste-heat source or for a heat load are studied with the aid of a heat balance. Figure 10.13 shows the heat balance for a steam boiler. The rates of enthalpy entering or leaving the system fluid streams must balance with the radiation loss rate from the boiler's external surfaces. Writing the first law equation for a steady-flow-steady-state process

$$\dot{q}_L = \dot{m}_f h_f + \dot{m}_a h_a + \dot{m}_c h_c - \dot{m}_s h_s - \dot{m}_g h_g \tag{10.19}$$

and referring to the heat-balance diagram, one sees that the enthalpy flux $\dot{m}_g h_g$, leaving the boiler in the exhaust gas stream, is a possible source of waste heat. A fraction of that energy can be transferred in a heat exchanger to the combustion air, thus increasing the enthalpy flux $\dot{m}_g h_a$ and reducing the amount of fuel required. The fraction of fuel that can be saved is given in the equation

$$\frac{\dot{m}_f \dot{m}_{f'}}{\dot{m}_f} = 1 - \left[\frac{K_1 - (1+\phi)\bar{C}_p T_g}{K_1 - (1+\phi)\bar{C}'_p T'_g} \right] \tag{10.20}$$

where the primed values are those obtained with waste-heat recovery. K_1 represents the specific enthalpy of the fuel–air mixture, $h_f + \phi h_a$, which is presumed to be the same with or without waste-heat recovery, ϕ is the molar ratio of air to fuel, and $\bar{C}_p$ is the specific heat averaged over the exhaust gas components. Figure 10.14, which is derived from

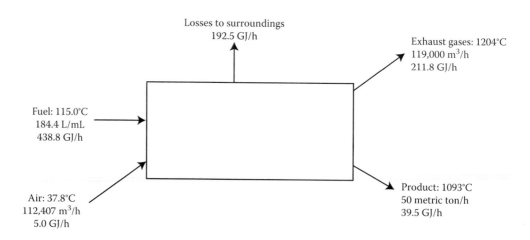

FIGURE 10.14
Heat balance for a simple continuous steel tube furnace.

Equation 10.20, gives possible fuel savings from using high-temperature flue gas to heat the combustion air in industrial furnaces.

It should be pointed out that the use of recovered waste heat to preheat combustion air, boiler feedwater, and product to be heat treated confers special benefits not necessarily accruing when the heat is recovered to be used in another system. The preheating operation results in less fuel being consumed in the furnace, and the corresponding smaller air consumption means even smaller waste heat being ejected from the stacks.

Table 10.8 shows heat balances for a boiler with no flue gas–heat recovery, with a feedwater preheater (economizer) installed and with an air preheater, respectively.

It is seen that air preheater alone saves 6% of the fuel and the economizer saves 9.2%. Since the economizer is cheaper to install than the air preheater, the choice is easy to make for an industrial boiler. For a utility boiler, both units are invariably used in series in the exit gas stream.

Table 10.9 is an economic study using 2005 prices for fuel, labor, and equipment. At that time, it was estimated that a radiation recuperator fitted to a fiberglass furnace would cost $820,200 and effect a savings of $90,435/year, making for a payoff period of approximately 2.8 years. This assumes of course that the original burners, combustion control system, and so on could be used without modification.

At this point, we can take some time to relate waste-heat recovery to the combustion process itself. We can first state categorically that the use of preheated combustion air improves combustion conditions and efficiency at all loads and in a newly designed installation permits a reduction in size of the boiler or furnace. It is true that the increased mixture temperature that accompanies air preheat results in some narrowing of the mixture stability limits, but in all practical furnaces, this is of small importance.

In many cases, low-temperature burners may be used for preheated air, particularly if the air preheat temperature is not excessive and if the fuel is gaseous. However, the higher volume flow of preheated air in the air piping may cause large enough increases in pressure drop to require a larger combustion air fan. Of course, larger diameter air piping can prevent the increased pressure drop, since air preheating results in reduced quantities of fuel and combustion air. For high preheat temperatures, alloy piping, high-temperature insulation, and water-cooled burners may be required. Since many automatic combustion-control systems sense volume flows of air and/or fuel, the correct control settings will

TABLE 10.8

Heat Balances for a Steam Generator

Case	Name	Input Streams Temperature (°C)	Flow Rate	Energy	Name	Output Streams Temperature (°C)	Flow Rate	Energy
Without economizer or air preheater	Natural gas	26.7	3,611 m³/h	134.78 GJ/h	Steam	185.6	45,349 kg/h	126.23
	Air	26.7	35,574 m³/h	560.32 MJ/h	Flue gas	372.2	41,185 m³/h	19.49
	Makeup water	10.0	7,076 kg/h	297.25 MJ/h	Surface losses	—	—	1.21
	Condensate return	82.2	41,277 kg/h	14.21 GJ/h	Blowdown	185.6	2,994 kg/h	2.36
With air preheater	Natural gas	26.7	3,395 m³/h	126.72 GJ/h	Steam	185.6	45,359 kg/h	126.23
	Air	232.2	32,114 m³/h	478.90 GJ/h	Flue gas	260.0	35,509 m³/h	11.42
	Makeup water	10	70,767 kg/h	297.25 MJ/h	Surface losses	—	—	1.21
	Condensate return	82.2	41,277 kg/h	14.21 GJ/h	Blowdown	185.6	2,994 kg/h	2.36
With economizer	Natural gas	26.7	3278 m³/h	122.37 GJ/h	Steam	185.6	45,359 kg/h	126.23
	Air	26.7	31,013 m³/h	462.48 MJ/h	Flue gas	176.7	34,281 m³/h	7.08
	Makeup water	101	7,076 kg/h	297.25 MJ/h	Surface loss	—	—	1.21
	Condensate water	82.2	41,277 kg/h	12.21 GJ/h	Blowdown	185.6	2,994 kg/h	2.36

TABLE 10.9

Cost–Fuel Savings Analysis of a Fiberglass Furnace Recuperator

Operation	Continuous
Fuel input	19.42 GJ/h
Fuel	No. 3 fuel oil
Furnace temperature	1482°C
Flue gas temperature entering recuperator	1204°C
Air preheat (at burner)	552°C
Fuel savings = 37.4%	
Q = 7.26 GJ/h or 173.6 L/h of oil	
Fuel cost savings estimation	
Per GJ = $5.00	
Per hour = $36.30	
Per year (8,000 h) = $290,435	
Cost of recuperator	$421,400
Cost of installation, related to recuperator	$398,800
Total cost of recuperator installation	$820,200
Approximate payback time	2.82 years

change when preheated air is used. Furthermore, if air preheat temperature varies with furnace load, then the control system must compensate for this variation with an auxiliary temperature-sensing control. On the other hand, if control is based on the oxygen content of the flue gases, the control complications arising from gas volume variation with temperature is obviated. This is the preferred control system for all furnaces, and only cost prevents its wide use in small installations. Burner operation and maintenance for gas burners is not affected by preheating, but oil burners may experience accelerated fuel coking and resulting plugging from the additional heat being introduced into the liquid fuel from the preheated air. Careful burner design, which may call for water cooling or for shielding the fuel tip from furnace radiation, will always solve the problems. Coal-fired furnaces may use preheated air up to temperatures that endanger the grates or burners. Again, any problems can be solved by special designs involving water cooling if higher air temperatures can be obtained and/or desired.

The economics of waste-heat recovery today range from poor to excellent depending upon the technical aspects of the application as detailed earlier, but the general statement can be made that, at least for most small industrial boilers and furnaces, standard designs and/or off-the-shelf heat exchangers prove to be the most economic. For large systems, one can often afford to pay for special designs, construction, and installations. Furthermore, the applications are often technically constrained by material properties and space limitations and, as shall be seen later, always by economic considerations. For further information on heat exchangers, see Chapter 10.

10.5.2.3 Heating, Ventilating, and Air Conditioning

HVAC, while not usually important in the energy-intensive industries, may be responsible for the major share of energy consumption in the light manufacturing field, particularly in high-technology companies and those engaged primarily in assembly.

Because of air pollution from industrial processes, many HVAC systems require 100% outside ventilating air. Furthermore, ventilating air requirements are often much in excess of those in residential and commercial practice. An approximate method for calculating the total heat required for ventilating air in kJ per heating season is given by

$$E_v \text{ (kJ)} = 60 \times 24 \text{ (min/day)} \times (1.2 \times 0.519) \text{ (kJ/m}^3\text{K)} \times \text{SCMM} \times \text{DD}$$
$$= 896.8 \times \text{SCMM} \times \text{DD} \tag{10.21}$$

where
    SCMM is the standard cubic meter per minute of total air entering plant including unwanted infiltration
    DD is the heating degree-days (°C)

This underestimates the energy requirement, because degree-days are based on 18.33°C reference temperature, and indoor temperatures are ordinarily held 1.6°–3.9° higher. For a location with 3333 degree-days, each year the heating energy given by Equation 10.13 is about 17% low.

Savings can be effected by reducing the ventilating air rate to the actual rate necessary for health and safety and by ducting outside air into direct-fired heating equipment such as furnaces, boilers, ovens, and dryers. Air infiltration should be prevented through a program of building maintenance to replace broken windows, doors, roofs, and siding, and by campaigns to prevent unnecessary opening of windows and doors.

Additional roof insulation is often economic, particularly because thermal stratification makes roof temperatures much higher than average wall temperatures. Properly installed vertical air circulators can prevent the vertical stratification and save heat loss through the roof. Windows can be double glazed, storm windows can be installed, or windows can be covered with insulation. Although the benefits of natural lighting are eliminated by this measure, it can be very effective in reducing infiltration and heat transfer losses.

Waste heat from ventilating air itself, from boiler and furnace exhaust stacks, and from air-conditioning refrigeration compressors can be recovered and used to preheat make-up air. Consideration should also be given to providing spot ventilation in hazardous locations instead of increasing general ventilation air requirements.

As an example of the savings possible in ventilation air control, a plant requiring 424.5 CMM outside air flow is selected. A gas-fired boiler with an energy input of 0.0165 GJ is used for heating and is supplied with room air.

$$\text{Combustion air} = \frac{16.5 \times 10^6}{37,281} \times \frac{12}{60} = 88.52 \text{ CMM}$$

for a fuel with 37,281 kJ/m³ heating value and an air fuel ratio of 12 m³ air/m³ fuel. The number of annual degree-days was 3175.

A study showed that the actual air supplied through infiltration and air handlers was 809 m³/min. An outside air duct was installed to supply combustion air for the boiler, and the actual ventilating air supply was reduced to the required 424 m³/min. The fuel saving that resulted using Equation 10.21 was

$$896.8(809 - 424)3175 = 1096.2 \text{ GJ}$$

worth $5482 in fuel at $5.00/GJ for natural gas.

10.5.2.4 Modifications of Unit Processes

The particular process used for the production of any item affects not only the cost of production but also the quality of the product. Since the quality of the product is critical in customer acceptance and therefore in sales, the unit process itself cannot be considered a prime target for the energy conservation program. That does not say that one should ignore the serendipitous discovery of a better and cheaper way of producing something. Indeed, one should take instant advantage of such a situation, but that clearly is the kind of decision that management could make without considering energy conservation at all.

10.5.2.5 Optimizing Process Scheduling

Industrial thermal processing equipment tends to be quite massive compared to the product treated. Therefore, the heat required to bring the equipment to steady-state production conditions may be large enough to make start-ups fuel intensive. This calls for scheduling this equipment so that it is in use for as long periods as can be practically scheduled. It also may call for idling the equipment (feeding fuel to keep the temperature close to production temperature) when it is temporarily out of use. The fuel rate for idling may be between 10% and 40% of the full production rate for direct-fired equipment. Furthermore, the stack losses tend to increase as a percentage of fuel energy released. It is clear that overfrequent start-ups and long idling times are wasteful of energy and add to production costs. The hazards of eliminating some of that waste through rescheduling must not be taken lightly. For instance, a holdup in an intermediate heating process can slow up all subsequent operations and make for inefficiency down the line. The breakdown of a unit that has a very large production backlog is much more serious than that of one having a smaller backlog. Scheduling processes in a complex product line is a very difficult exercise and perhaps better suited to a computerized PERT program than to an energy conversation engineer. That does not mean that the possibilities for saving through better process scheduling should be ignored. It is only a warning to move slowly and take pains to find the difficulties that can arise thereby.

A manufacturer of precision instruments changed the specifications for the finishes of over half of his products, thereby eliminating the baking required for the enamel that had been used. He also rescheduled the baking process for the remaining products so that the oven was lighted only twice a week instead of every production day. A study is now proceeding to determine if electric infrared baking will not be more economic than using the gas-fired oven.

10.5.2.6 Cogeneration of Process Steam and Electricity

In-plant (or on-site) electrical energy cogeneration is nothing new. It has been used in industries with large process steam loads for many years, both in the United States and in Europe. It consists of producing steam at a higher pressure than required for process use, expanding the high-pressure steam through a backpressure turbine to generate electrical energy and then using the exhaust steam as process steam. Alternatively, the power system may be a diesel engine that drives an electrical generator. The diesel engine exhaust is then stripped of its heat content as it flows through a waste-heat boiler where steam is generated for plant processes. A third possibility is operation of a gas turbine generator to supply electric power and hot exhaust gases, which produce process steam in a waste-heat boiler. As will be seen later, the ratio of electric power to steam-heat rate varies markedly from one of these types of systems to the next. In medium-to-large industrial plants, the cogeneration of electric power and process steam is economically feasible provided certain plant energy characteristics are present. In small plants or in larger plants with small process steam loads, cogeneration is not economic because of the large capital expenditure involved. Under few circumstances is the in-plant generation of electric power economic without a large process steam requirement. A small industrial electric plant cannot compete with an electric utility unless the generation required in-plant exceeds the capacity of the utility. In remote areas where no electric utility exists, or where its reliability is inferior to that of the on-site plant, the exception can be made.

Cogeneration if applied correctly is not only cost effective, but also fuel conserving. That is, the fuel for the on-site plant is less than that used jointly by the utility to supply the plant's electric energy and that used on-site to supply process steam. Figures 10.15 and 10.16 illustrate the reasons for and the magnitude of the savings possible. However, several conditions must be met in order that an effective application is possible. First, the ratio of process steam heat rate to electric power must fall close to these given in the following table:

Heat Engine Type	E_{stream}/E_{elect}
Steam turbine	2.3
Gas turbine	4.0
Diesel engine	1.5

The table is based upon overall electric plant efficiencies to 30%, 20%, and 40% respectively, for steam turbine, gas turbine, and diesel engine. Second, it is required that the availability

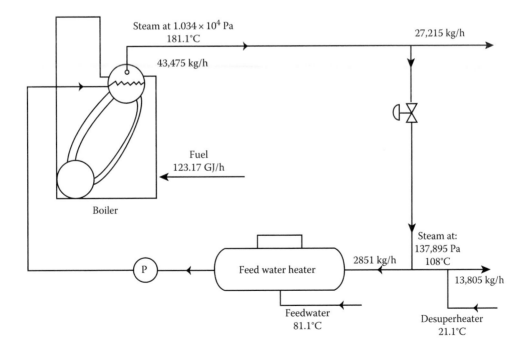

FIGURE 10.15
Steam plant schematic before adding electrical generation.

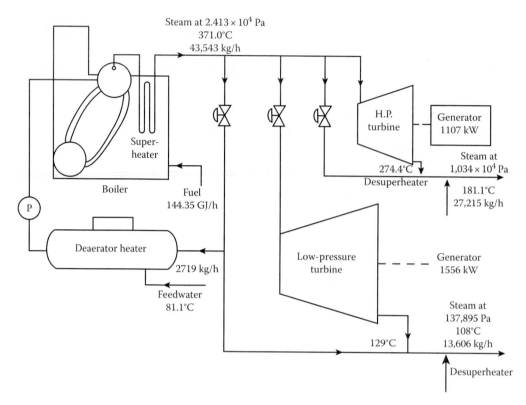

FIGURE 10.16
Steam plant schematic after installing electrical generation.

of the steam load coincides closely with the availability of the electric load. If these temporal availabilities are out of phase, heat storage systems will be necessary, and the economy of the system altered. Third, it is necessary to have local electric utility support. Unless backup service is available from your utility, the cost of building in redundancy is too great. This may be the crucial factor in some cases.

10.5.2.7 Commercial Options in Waste-Heat Recovery Equipment

The equipment that is useful in recovering waste heat can be categorized as heat exchangers, heat storage systems, combination heat storage–heat exchanger systems, and heat pumps.

Heat exchangers certainly constitute the largest sales volume in this group. They consist of two enclosed flow paths and a separating surface that prevents mixing, supports any pressure difference between the fluids of the two fluids, and provides the means through which heat is transferred from the hotter to the cooler fluid. These are ordinarily operated at steady-state steady-flow condition. The fluids may be gases, liquids, condensing vapors, or evaporating liquids, and occasionally fluidized solids. For more information, see Chapter 10.

Radiation recuperators are high-temperature combustion-air preheaters used for transferring heat from furnace exhaust gases to combustion air. As seen in Figure 10.17, they consist of two concentric cylinders, the inner one as a stack for the furnace and the

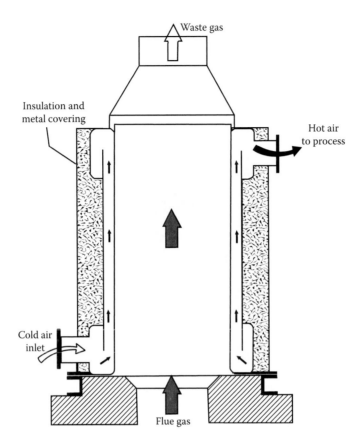

FIGURE 10.17
Metallic radiation recuperator.

concentric space between the inner and outer cylinders as the path for the combustion air, which ordinarily moves upward and therefore parallel to the flow of the exhaust gases. With special construction materials, these can handle 1355°C furnace gases and save as much as 30% of the fuel otherwise required. The main problem in their use is damage due to overheating for reduced air flow or temperature excursions in the exhaust gas flow.

Convective air preheaters are corrugated metal or tubular devices that are used to preheat combustion air in the moderate temperature range (121°C–649°C) for ovens, furnaces, boilers, and gas turbines, or to preheat ventilating air from sources as low in temperature as 21°C. Figures 10.18 and 10.19 illustrate typical construction. These are often available in modular design so that almost any capacity and any degree of effectiveness can be obtained by multiple arrangements. The biggest problem is keeping them clean.

Economizer is the name traditionally used to describe the gas-to-liquid heat exchanger used to preheat the feedwater in boilers from waste heat in the exhaust gas stream. These often take the form of loops, spiral, or parallel arrays of finned tubing through which the feedwater flows and over which the exhaust gases pass. They are available in modular form to be introduced into the exhaust stack or into the breeching. They can also be used in reverse to heat air or other gases with waste heat from liquid streams.

A more recent development is the use of condensing economizers that are placed in the exhaust stream following high-temperature economizers. They are capable of extracting

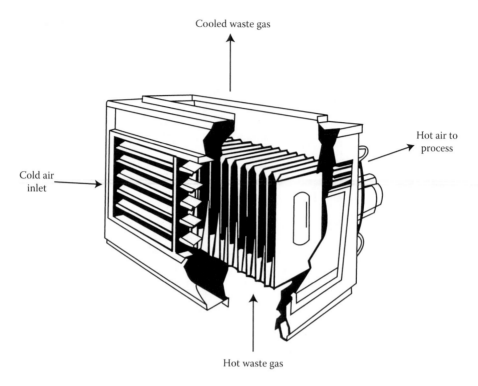

FIGURE 10.18
Air preheater.

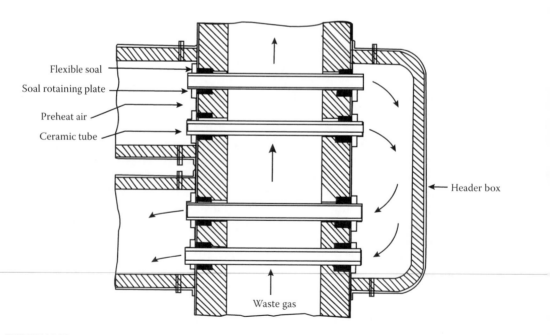

FIGURE 10.19
Ceramic tube recuperator.

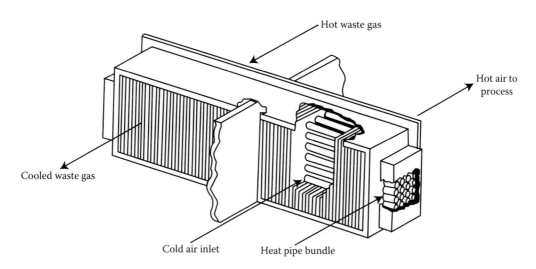

Hot waste gas

Hot air to process

Cooled waste gas

Cold air inlet

Heat pipe bundle

FIGURE 10.20
Heat pipe recuperator.

an additional 6%–8% of the fuel input energy from the boiler exhaust gases. However, they are used only under certain restricted conditions. Obviously, the cooling fluid must be at a temperature below the dew point of the exhaust stream. This condition is often satisfied when boilers are operated with 100% make-up water. A second, less restrictive condition is that the flue gases be free of sulfur oxides. This is normally the case for natural gas-fired boilers. Otherwise, the economizer tubes will be attacked by sulfurous and/or sulfuric acid. Acid corrosion can be slowed down markedly by the use of all-stainless steel construction, but the cost of the equipment is increased significantly.

Heat-pipe arrays are often used for air-to-air heat exchangers because of their compact size. Heat-transfer rates per unit area are quite high. A disadvantage is that a given heat pipe (i.e., a given internal working substance) has a limited temperature range for efficient operation. The heat pipe transfers heat from the hot end by evaporative heating and at the cold end by condensing the vapor. Figure 10.20 is a sketch of an air preheater using an array of heat pipes.

Waste-heat boilers are water-tube boilers, usually prefabricated in all but the largest sizes, used to produce saturated steam from high-temperature waste heat in gas streams. The boiler tubes are often finned to keep the dimensions of the boiler smaller. They are often used to strip waste heat from diesel engine exhausts, gas turbine exhausts, and pollution control incinerators or afterburners. Figure 10.21 is a diagram of the internals of a typical waste-heat boiler. Figure 10.22 is a schematic diagram showing a waste-heat boiler for which the evaporator is in the form of a finned-tube economizer. Forced water circulation is used giving some flexibility in placing the steam drum and allowing the use of smaller tubes. It also allows the orientation of the evaporator to be either vertical or horizontal. Other advantages of this design are the attainment of high boiler efficiencies, a more compact boiler, less cost to repair or retube, the ability to make superheated steam using the first one or more rows of downstream tubes as the superheater, and the elimination of thermal shock, since the evaporator is not directly connected to the steam drum.

Heat storage systems, or regenerators, once very popular for high-temperature applications, have been largely replaced by radiation recuperators because of the relative simplicity of the latter. Regenerators consist of twin flues filled with open ceramic checkerwork.

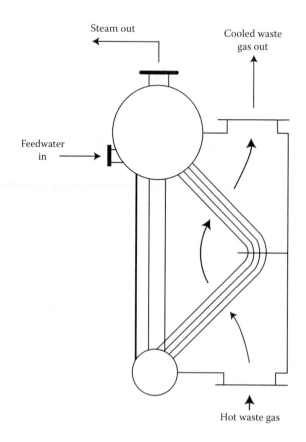

FIGURE 10.21
Waste-heat boiler.

The high-temperature exhaust of a furnace flowing through one leg of a swing valve to one of the flues heated the checkerwork, while the combustion air for the furnace flowed through the second flue in order to preheat it. When the temperatures of the two masses of checkerwork were at proper levels, the swing valve was thrown and the procedure was continued, but with reversed flow in both flues. Regenerators are still used in some glass- and metal-melt furnaces, where they are able to operate in the temperature range 1093°C–1649°C. It should be noted that the original application of the regenerators was to achieve the high-melt temperatures required with low-heating-value fuel.

A number of ceramic materials in a range of sizes and geometric forms are available for incorporation into heat storage units. These can be used to store waste heat in order to remedy time discrepancies between source and load. A good example is the storage of solar energy in a rock pile so that it becomes available for use at night and on cloudy days. Heat storage, other than for regenerators in high-temperature applications, has not yet been used a great deal for waste-heat recovery but will probably become more popular as more experience with it accumulates.

Combination heat storage unit–heat exchangers called heat wheels are available for waste-heat recovery in the temperature range 0°C–982°C. The heat wheel is a porous flat cylinder that rotates within a pair of parallel ducts, as can be observed in Figure 10.23. As the hot gases flow through the matrix of the wheel, they heat one side of it, which then gives up that heat to the cold gases as it passes through the second duct. Heat-recovery

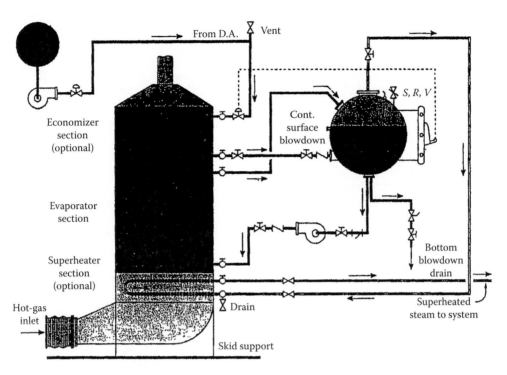

FIGURE 10.22
Schematic diagram of a finned tube waste-heat boiler. (Courtesy of Canon Technology.)

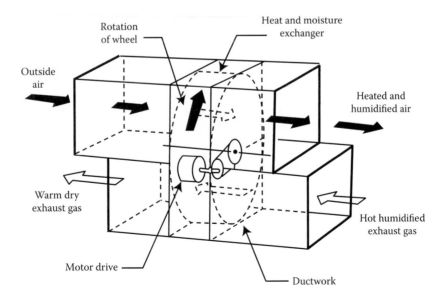

FIGURE 10.23
Heat wheel.

efficiencies range to 80% in low- and moderate-wire matrix of the same material. In the high-temperature range, the material is ceramic. In order to prevent cross-contamination of the fluid streams, a purge section, cleared by fresh air, can be provided. If the matrix of the wheel is covered with a hygroscopic material, latent heat as well as sensible heat can be recovered. Problems encountered with heat wheels include freeze damage in winter, seal wear, and bearing maintenance in high-temperature applications.

The heat pump is a device operating on a refrigeration cycle that is used to transfer energy from a low-temperature source to a higher-temperature load. It has been highly developed as a domestic heating plant using energy from the air or from well but has not been used a great deal for industrial applications. The COP (or the ratio of heat delivered to work input) for an ideal Carnot refrigeration cycle equals $T_H/(T_H - T_L)$, where T_H is the load temperature and T_L is the source temperature. It is obvious that when the temperature difference $T_H - T_L$ becomes of the order of T_H, the heat could be derived almost as cheaply from electric resistance heating. However, for efficient refrigeration machines and a small temperature potential to overcome, the actual COP is favorable to moderate-cost waste energy. The heat pump can be used to transfer waste heat from and to any combination of liquid, gas, and vapor.

10.6 Role of New Equipment and Technology in Industrial Energy Efficiency

10.6.1 Industrial Energy Savings Potential

The beginning of this chapter mentioned several studies of industrial energy efficiency that showed that a 20% reduction could be accomplished in a relatively easy and cost-effective manner. The purpose of this section is to provide the equipment, technology, and operational changes that could lead to industrial energy savings on the order of 20% or more.

10.6.2 The U.S. DOE Energy-Loss Study and the NAM Efficiency and Innovation Study

The most recent major study on the potential for improving energy efficiency in industry was conducted for the U.S. Department of Energy, OIT, for their ITP in December 2004. This study was called the "Energy Use, Loss and Opportunities Analysis: U.S. Manufacturing and Mining."[1] This study was then used by NAM together with the Alliance to Save Energy to produce a report on the "Efficiency and Innovation in U.S. Manufacturing Energy Use."[3] Both the U.S. DOE/OIT and NAM conclude that there is a significant opportunity for reducing industrial energy use by 20%.

As stated in the NAM report

> [i]ndustry's best R&D options for reducing energy costs were summarized in a study sponsored by the U.S. DOE. This study identifies energy efficiency opportunities that yield energy, economic and environmental benefits, primarily for large volume, commodity/process industries. Opportunities were prioritized to reflect the magnitude of potential savings, broadness of suitability across industries, and feasibility to implement. In total, these energy-saving opportunities represent 5.2 quadrillion Btu—21% of primary energy consumed by the manufacturing sector. These savings equate to almost $19 billion for manufacturers, based on 2004 energy prices and consumption volumes.

TABLE 10.10

Best Opportunities for Future Industrial Energy Savings

	Total Energy Savings		Total Cost Savings	
Type of Opportunity	**(trillion Btu)**	**% of Total**	**($mill.)**	**% of Total**
Top R&D opportunities for energy savings in commodity/process manufacturing; initiatives that provide the largest energy and dollar savings				
Waste-heat and energy recovery	1831	35	6,408	34
Improvements to boilers, fired systems, process heaters, and cooling opportunities	907	17	3,077	16
Energy system integration and best practices opportunities	1438	28	5,655	30
Energy source flexibility and combined heat and power	828	16	3,100	16
Improved sensors, controls, automation and robotics for energy systems	191	4	630	3
Totals	5195		18,870	

Source: U.S. DOE, *Annual Report: Technology, Delivery, Industry of the Future*, U.S. Department of Energy, Office of Industrial Technologies, Industrial Technology Program, Washington, DC, 2004, see references section (DOE-TP); NAM, *Efficiency and Innovation in U.S. Manufacturing Energy Use*, National Association of Manufacturers, Washington, DC, 2005.

Table 10.10 summarizes these leading opportunities. An expanded version of this information appears in Table 10.11.

10.6.3 The ACEEE Fan and Pump Study

In April 2003, the ACEEE released a report on their study "Realizing Energy Efficiency Opportunities in Industrial Fan and Pump Systems." They concluded that fans and pumps account for more than a quarter of industrial electricity consumption and that optimization of the operation of these fans and pumps could achieve electricity savings ranging from 20% to well over 50% of this category of use.[10] This report says that most optimization projects involve greater engineering costs than equipment costs, but the average payback for a good optimization project is about 1.2 years, with the cost of saved energy on the order of $0.012/kW h. In addition, these estimates do not account for productivity gains known to exist at many of the plant sites, which are sometimes as much as two to five times the energy savings.

This ACEEE report contains some excellent data on motor systems end use that is very difficult to find in general. Their data show that 40% of industrial motor use is for fans and pumps. Because the MECS data from EIA show that about 68% of electric use in industry is motors, this leads to the fraction of industrial electric use from fans and pumps to be $(0.68) \times (0.4) = 0.32$, or 32%. The end-use data from ACEEE are reproduced in Table 10.12.

To see the final impact of this estimate of industrial energy savings, now apply the ACEEE estimates of 20%–50% savings on fan and pump energy to the 32% fraction of fan and pump contribution to the total industrial electricity use. Thus, the overall savings are in the range $(0.2) \times (0.32) = 6.4\%$ to $(0.5) \times (0.32) = 16\%$. Just this opportunity alone could result in achieving over half of the 20% savings contained in the U.S. DOE/OIT study.

Table 10.13 presents the ACEEE estimates of the relative magnitude of electricity consumed by fans and pump systems for the important industries on a national basis.

TABLE 10.11

Top R&D Opportunities for Industrial Energy Savings

Top R&D Opportunities for Energy Initiatives That Provide the Largest Energy: Type of Opportunity	Savings in Commodity/Process Manufacturing and Dollar Savings: Leading Industry Recipients	Total Energy Savings (trillion Btu)	Total Energy Savings (% of Total)	Total Cost Savings ($mil.)	Total Cost Savings (% of Total)[a]
Waste-heat and energy recovery		1,831	35	6,408	34
• From gases and liquids, including hot gas cleanup and dehydration of liquid waste streams	Chemicals, petroleum, forest products	851	16	2,271	12
• From drying processes	Chemicals, forest products, food processing	377	7	1,240	7
• From gases in metals and nonmetallic minerals manufacture (excluding calcining), including hot gas cleanup	Iron and steel, cement	235	5	1,133	6
• From by-product gases	Petroleum, iron, and steel	132	3	750	4
• Using energy export and colocation (fuels from pulp mills, forest bio-refineries, colocation of energy sources/sinks)	Forest products	105	2	580	3
• From calcining (not flue gases)	Cement, forest products	74	1	159	1
• From metal quenching/cooling processes	Iron and steel, cement	57	1	275	1
Improvements to boilers, fired systems, process heaters, and cooling opportunities		907	17	3,077	16
Advanced industrial boilers	Chemicals, forest products, petroleum, steel, food processing	400	8	1,090	6
Improved heating/heat transfer systems (heat exchangers, new materials, improved heat transport)	Petroleum, chemicals	260	5	860	5
Improved heating/heat transfer for metals, melting, heating, annealing (cascade heating, batch to continuous process, improved heat channeling, modular systems)	Iron and steel, metal casting, aluminum	190	4	915	5
Advanced process cooling and refrigeration	Food processing, chemicals, petroleum, and forest products	57[b]	1	212	1
Energy system integration and best practices opportunities		1,438	28	5,655	30
* Steam best practices (improved generation, distribution, and recovery), not including advanced boilers	All manufacturing	310	6	850	5
* Pump system optimization	All manufacturing	302[b]	6	1,370	7

(Continued)

TABLE 10.11 (*Continued*)

Top R&D Opportunities for Industrial Energy Savings

Top R&D Opportunities for Energy Initiatives That Provide the Largest Energy: Type of Opportunity	Savings in Commodity/ Process Manufacturing and Dollar Savings: Leading Industry Recipients	Total Energy Savings (trillion Btu)	(% of Total)	Total Cost Savings ($mil.)	(% of Total)[a]
* Energy system integration	Chemicals, petroleum, forest products, iron and steel, food, aluminum	260	5	860	5
* Energy-efficient motors and rewind practices	All manufacturing	258[b]	5	1,175	6
* Compressed air system optimization	All manufacturing	163[b]	3	740	4
* Optimized materials processing	All manufacturing	145[b]	3	660	3
Energy source flexibility and combined heat and power		828	16	3,100	16
* Combined heat and power on-site in manufacturers' central plants, producing both thermal and electricity needs	Forest products, chemicals, food processing, metals, machinery	634	12	2,000	11
Energy source flexibility (heat-activated power generation, waste steam for mechanical drives, indirect vs. direct heat vs. steam)	Chemicals, petroleum, forest products, iron and steel	194	4	1,100	6
Improved sensors, controls, automation and robotics for energy systems	Chemicals, petroleum, forest products, iron and steel, food, cement, aluminum	191	4	630	3
Totals		5,195		18,870	

Source: U.S. DOE, *Annual Report: Technology, Delivery, Industry of the Future*, U.S. Department of Energy, Office of Industrial Technologies, Industrial Technology Program, Washington, DC, 2004.

Note: All are R&D opportunities except for items denoted by an asterisk (*), which are near-term best practices, applicable to current assets.

[a] Totals may not add up due to rounding.

[b] Energy savings figures include the corresponding recapture of losses inherent in electricity generation, transmission, and distribution.

TABLE 10.12

National Industrial Motor Systems Energy End Use

Pumps	25%
Materials processing	22%
Compressed air	16%
Fans	14%
Material handling	12%
Refrigeration	7%
Other	4%

Source: Elliott, R.N. and Nadel, S., *Realizing Energy Efficiency Opportunities in Industrial Fan and Pump Systems*, Report A034, American Council for an Energy-Efficient Economy, Washington, DC, 2003.

TABLE 10.13

Characterization of Industrial Fan and Pump Load in the United States

NAICS	Industry	Electricity Demand 1997	Pumps (%)	Fans and Blowers (%)	Total Motors (%)	Motor Electricity	Fans/Pumps Share of Electricity (%)	Fans/Pumps Electricity Use
11	Agriculture	16,325	25	20	75	12,244	45	7,346
22	Mining	85,394	7	21	90	76,854	29	24,363
311	Food mfg.	66,166	11	5	81	53,756	16	10,809
314	Textile product mills	5,135	14	15	82	4,221	30	1,523
321	Wood product mfg.	21,884	4	10	80	17,464	14	3,064
322	Paper mfg.	119,627	28	16	84	101,078	44	52,636
324	Petroleum and coal products mfg.	69,601	51	13	85	59,369	63	44,061
325	Chemical mfg.	212,709	18	8	73	154,693	26	54,797
326	Plastics and rubber mfg.	52,556	9	4	66	34,847	13	6,729
327	Nonmetallic minerals product mfg.	37,416	4	4	65	24,328	8	3,037
331	Primary metal mfg.	172,518	2	4	26	44,855	6	10,351
332	Fabricated metal product mfg.	49,590	7	5	65	32,462	12	6,149
333	Machinery mfg.	27,295	8	4	67	18,391	12	3,330
334	Computer and electronic product mfg.	40,099	2	3	54	21,783	4	1,801
336	Transportation equipment mfg.	54,282	4	6	64	34,629	11	5,753
	Total	1,030,598				690,974		235,750
		Fraction of total elec.				67%		23%

10.6.4 The LBL/ACEEE Study of Emerging Energy-Efficient Industrial Technologies

In October 2000, ACEEE released a report of a study it did in conjunction with staff from Lawrence Berkeley Laboratories (LBL), where they identified 175 emerging energy-efficient technologies, and honed this list down to 32 technologies that had a high likelihood of success and a high energy savings.[25] An interesting aspect of this study is that it shows that the United States is not running out of technologies to improve energy efficiency and economic and environmental performance, and will not run out in the future. The study shows that many of the technologies have important nonenergy benefits, ranging from reduced environmental impact to improved productivity. Several technologies have reduced capital costs compared to the current technology used by those industries. Nonenergy benefits such as these are frequently a motivating factor in bringing this kind of technology to market.

The LBL/ACEEE list of 32 most beneficial technologies is shown in Table 10.14.

TABLE 10.14

Technologies with High Energy Savings and a High Likelihood of Success

Technology	Code	Total Energy Savings	Likelihood of Success	Recommended Next Steps
Efficient cell retrofit designs	Alum-2	High	High	Demo
Advanced lighting technologies	Lighting-1	High	High	Dissemination, demo
Advance ASD designs	Motorsys-1	High	High	R&D
Membrane technology wastewater	Other-3	High	High	Dissemination, R&D
Sensors and controls	Other-5	High	High	R&D, demo, dissemination
Black liquor gasification	Paper-1	High	High	Demo
Near-net-shape casting/strip casting	Steel-2	High	High	R&D
New EAF furnace processes	Steel-3	High	High	Field test
Oxy-fuel combustion in reheat furnace	Steel-4	High	High	Field test
Advanced CHP turbine systems	Utilities-1	High	High	Policies
Autothermal reforming—ammonia	Chem-7	High	Medium	Dissemination
Membrane technology—food	Food-3	High	Medium	Dissemination, R&D
Advanced lighting design	Lighting-2	High	Medium	Dissemination, demo
Compressed air system management	Motorsys-3	High	Medium	Dissemination
Motor system optimization	Motorsys-5	High	Medium	Dissemination, training
Pump efficiency improvement	Motorsys-6	High	Medium	Dissemination, training
High-efficiency/low-NO_x burners	Other-2	High	Medium	Dissemination, demo
Process integration (pinch analysis)	Other-4	High	Medium	Dissemination
Heat recovery—paper	Paper-5	High	Medium	Demo
Impulse drying	Paper-7	High	Medium	Demo
Smelting reduction processes	Steel-5	High	Medium	Demo
Advanced reciprocating engines	Utilities-2	High	Medium	R&D, demo
Fuel cells	Utilities-3	High	Medium	Demo
Microturbines	Utilities-4	High	Medium	R&D, demo
Inert anodes/wetted cathodes	Alum-4	High	Medium	R&D
Advanced forming	Alum-1	Medium	High	R&D
Plastics recovery	Chem-8	Medium	High	Demo
Continuous melt silicon crystal growth	Electron-1	Medium	High	R&D
100% Recycled glass cullet	Glass-1	Medium	High	Demo
Anaerobic waste water treatment	Other-1	Medium	High	Dissemination, demo
Dry sheet forming	Paper-4	Medium	High	R&D, demo
Biodesulfurization	Refin-1	Medium	High	R&D, demo

Note: Technologies in this table are listed in alphabetical order based on industry sector.

10.7 Conclusion

Energy is the lifeblood of industry; it is used to convert fuels to thermal, electric, or motive energy to manufacture all the products of daily life. Using this energy efficiently is a necessity to keep industries competitive, clean, and at their peak of productivity. Energy management programs that improve the operational efficiency and the technological efficiency of industry are critical to the long-term success of industry and manufacturing in the United States. One important result in this area has been a recognition that the United States is not running out of technologies to improve industrial energy efficiency, productivity, and environmental performance, and it is not going to run out in the foreseeable future. A substantial opportunity to the country's industrial energy use by over 20% is currently available using better operational procedures and using improved equipment in industrial plants. These savings to industry are worth almost $19 billion at 2004 energy prices. With crude oil prices edging toward $70 in late summer 2005, this dollar savings amount should be substantially higher. It is time to capture the benefits of this opportunity.

References

1. U.S. Department of Energy (U.S. DOE), September 2012. EIA annual energy review for 2011, Washington, DC.
2. U.S. DOE Energy Information Administration, 1995. Manufacturing energy consumption survey: Changes in energy efficiency 1985–1991. U.S. Department of Energy, Office of Industrial Technologies, Industrial Technology Program, Washington, DC.
3. NAM, 2005. Efficiency and innovation in U.S. manufacturing energy use. National Association of Manufacturers, Washington, DC.
4. U.S. DOE, 2004. Annual report: Technology, delivery, industry of the future. U.S. Department of Energy, Office of Industrial Technologies, Industrial Technology Program, Washington, DC.
5. U.S. DOE, 1990. The National energy strategy, Chapter on Industrial energy use. U.S. Department of Energy, Washington, DC.
6. Oak Ridge National Laboratory, 2000. Scenarios for a clean energy future, Interlaboratory Working Group. Oak Ridge National Laboratory, Oak Ridge, TN.
7. Price, L. and Worrell, E. 2004. Improving industrial energy efficiency in the U.S.: Technologies and policies for 2010–2050. *Proceedings of the 10–50 Solution: Technologies and Policies for a Low Carbon Future*. Lawrence Berkeley National Laboratory, Berkeley, CA.
8. Elliott, R. N. and Nadel, S. 2003. Realizing energy efficiency opportunities in industrial fan and pump systems, Report A034. American Council for an Energy-Efficient Economy, Washington, DC.
9. Aspenson, R. L. 1989. Testimony to the U.S. Department of Energy on the National Energy Strategy, hearings on energy and productivity, Providence, RI.
10. 3M Company. April 2013. Improving energy efficiency. http://www.MMM.com.
11. Capehart, L. C. and Capehart, B. L. 1994. Writing user-friendly energy audit reports. *Strategic Planning for Energy and the Environment*, 14(2), 17–26 (Published by the Association of Energy Engineers, Atlanta, GA).
12. Capehart, B. L. and Capehart, L. C. 1994. Improving industrial energy audit analyses. *Proceedings of the ACEEE Summer Study of Industrial Energy Use*. ACEEE, Washington, DC.
13. Pawlik, K.-D., Capehart, L. C., and Capehart, B. L. 2001. Analyzing facility energy use: A balancing act. *Strategic Planning for Energy and the Environment*, 21(2), 8–23.
14. Air-Conditioning & Refrigeration Institute. June 2005. ARI Unitary Directory. http://www.ari.org.
15. Nadel, S., Elliott, R. N., Shepard, M., Greenberg, S., and Katz, G. 2002. *Energy-Efficient Motor Systems: A Handbook on Technology, Program and Policy Opportunities*, 2nd edn. American Council for an Energy-Efficient Economy, Washington, DC.
16. Washington State Energy Office. June 2005. MotorMaster electric motor selection software and database. http://www.oit.doe.gov/bestpractices/software_tools.shtml.
17. Hoshide, R. K. 1994. Electric motor do's and don't's. *Energy Engineering*, 91(1), 6–24 (Published by the Association of Energy Engineers, Atlanta, GA).

18. Stebbins, W. L. 1994. Are you certain you understand the economics for applying ASD systems to centrifugal loads? *Energy Engineering*, 91(1), 25–44 (Published by the Association of Energy Engineers, Atlanta, GA).
19. Vaillencourt, R. R. 1994. Simple solutions to VSD pumping measures. *Energy Engineering*, 91(1), 45–59 (Published by the Association of Energy Engineers, Atlanta, GA).
20. Kempers, G. 1995. DSM pitfalls for centrifugal pumps and fans. *Energy Engineering*, 92(2), 15–23 (Published by the Association of Energy Engineers, Atlanta, GA).
21. Webb, M. (Senior Engineer). January 1995. Personal communication. Virginia Power Company, Roanoke, VA.
22. Capehart, B. L., ed. 2004. *Information Technology for Energy Managers: Understanding Web Based Energy Information and Control Systems.* Fairmont Press, Atlanta, GA.
23. Capehart, B. L. and Capehart, L. C., eds. 2005. *Web Based Energy Information and Control Systems: Case Studies and Applications.* Fairmont Press, Atlanta, GA.
24. Turner, W. C., ed. 2012. *Energy Management Handbook.* Fairmont Press, Atlanta, GA.
25. Martin, N. and Elliott, R. N. 2000. *Emerging Energy Efficient Technologies.* American Council for an Energy-Efficient Economy, Washington, DC.
26. U.S. Department of Energy, Energy Information Agency. 2012. Manufacturing energy consumption survey, 2006, Washington, DC.
27. U.S. DOE, Energy Information Agency. September 2014. Monthly energy review, Washington, DC.
28. U.S. Department of Energy. 2005. Office of Industrial Technologies, Industrial Technology Program, Washington, DC.
29. U.S. Department of Energy, Energy Information Agency. 2002. Manufacturing Energy Consumption Survey, 2006, Washington, DC.
30. U.S. Department of Energy. 2005. Office of Industrial Technologies, Washington, DC, http://www.oit.doe.gov/iac/schools.shtml.
31. U.S. Department of Energy, Energy Information Agency. 2002. Manufacturing Energy Consumption Survey, 1998, Washington, DC.
32. Dukelow, S.G. 1974. Bailey Meter Company, Wickliffe, OH.

11

Electric Motor Systems Efficiency

Aníbal T. de Almeida, Steve F. Greenberg, and Prakash Rao

CONTENTS

11.1 Introduction

Motor systems are by far the most important type of electrical load, ranging from small fractional horsepower motors incorporated in home appliances to multi-megawatt motors driving pumps and fans in power plants. Motors consume over half of total U.S. electricity, and in industry they are responsible for about two-thirds of the electricity consumption. In the commercial and residential sectors, motors consume slightly less than half of the electricity. The cost of powering motors is immense, roughly $100 billion a year in the United States alone. There is a vast potential for saving energy and money by increasing the efficiency of motors and motor systems.

11.1.1 Motor Types

Motors produce useful work by causing the shaft to rotate. Motors have a rotating part, the rotor, and a stationary part, the stator. Both parts produce magnetic fields, either through windings excited by electric currents or through the use of permanent magnets (PMs). It is the interaction between these two magnetic fields that is responsible for the torque generation, as shown in Figure 11.1.

There are a wide variety of electric motors, based on the type of power supply (AC or DC) that feeds the windings, as well as on different methods and technologies to generate the magnetic fields in the rotor and in the stator. Figure 11.2 presents the most important types of motors.

Because of their low cost, high reliability, and fairly high efficiency, most of the motors used in large home appliances, industry, and commercial buildings are induction motors. Figure 11.3 shows the operating principle of a three-phase induction motor.

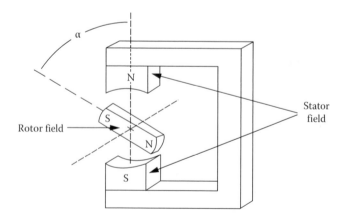

FIGURE 11.1

Torque generation in a motor. The generated torque is proportional to the strength of each magnetic field and depends on the angle between the two fields. Mathematically, torque equals $|B_{rotor}| \times |B_{stator}| \times \sin \alpha$, where B refers to a magnetic field. (Reprinted from Nadel, S. et al., *Energy-Efficient Motor Systems: A Handbook on Technologies, Programs, and Policy Opportunities*, 1st edn., American Council for an Energy-Efficient Economy, Washington, DC, 1992. With permission.)

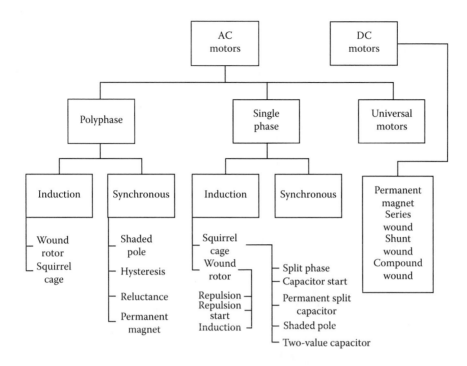

FIGURE 11.2
Motor types. (From EPRI, Electric motors: Markets, trends and application, EPRI Report TR-100423, Electric Power Research Institute, Palo Alto, CA, 1992.)

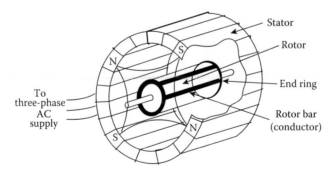

FIGURE 11.3
Operation of a four-pole squirrel-cage induction motor. Rotating magnetic field is created in the stator by AC currents carried in the stator winding. Three-phase voltage source results in the creation of north and south magnetic poles that revolve or *move around* the stator. The changing magnetic field from the stator induces current in the rotor conductors, in turn creating the rotor magnetic field. Magnetic forces in the rotor tend to follow the stator magnetic fields, producing rotary motor action.

Synchronous motors, where the rotor spins at the same speed as the stator magnetic field, are used in applications requiring constant speed, high operating efficiency, and controllable power factor. Efficiency and power factor are particularly important above 1000 hp. Although DC motors are easy to control, both in terms of speed and torque, they are expensive to produce and have modest reliability. DC motors are used for some industrial and electric traction applications, but their importance is dwindling.

11.2 Motor Systems Efficiency

The efficiency of a motor-driven process depends upon several factors, which may include

- Motor efficiency
- Motor speed controls
- Proper sizing
- Power supply quality
- Distribution losses
- Transmission
- Maintenance
- Driven equipment (pump, fan, etc.) mechanical efficiency

It must be emphasized that the design of the process itself influences the overall efficiency (units produced/kWh or service produced/kWh) to a large extent. In fact, in many systems the largest opportunity for increased efficiency is in improved use of the mechanical energy (usually in the form of fluids or solid materials in motion) in the process. Comprehensive programs to address motor-system energy use start with the process and work back toward the power line, optimizing each element in turn, as well as the overall system. Outlining such a program is beyond the scope of this discussion; see, for example, Baldwin (1989) for the benefits that propagate all the way back to the power plant. Additionally, the U.S. Department of Energy (U.S. DOE) has published several *Energy Tips* sheets on motor system efficiency (USDOE, 2012). Topics include estimating motor efficiency, motor drives, and electrical system considerations. Additional publications detailing motor and motor system efficiency characteristics, how to identify motor system losses, and system efficiency improvement opportunities are available from the U.S. DOE (USDOE, 2014a–c).

11.2.1 Motor Efficiency

Figure 11.4 shows the distribution of the losses of an induction motor as a function of the load. At low loads, the core magnetic losses (hysteresis and eddy currents) are dominant, whereas at higher loads, the copper resistive ("Joule" or I^2R) losses are the most important. Mechanical losses are also present in the form of friction in the bearings and windage.

11.2.1.1 Motor Efficiency: Energy Efficient, Premium Efficient, and Beyond

After World War II and until the early 1970s, there was a trend to design inefficient motors, which minimized the use of raw materials (copper, aluminum, and silicon steel). These induction motors had lower initial costs and were more compact than the previous generations of motors, but their running costs were higher. When electricity prices started escalating rapidly in the mid-1970s, most of the large motor manufacturers added a line of higher-efficiency motors to their selection. Such motors feature optimized design, more generous electrical and magnetic circuits, and higher quality materials (Baldwin, 1989). Incremental efficiency improvements are still possible with the use of superior materials (e.g., amorphous silicon steel) and optimized computer-aided design techniques.

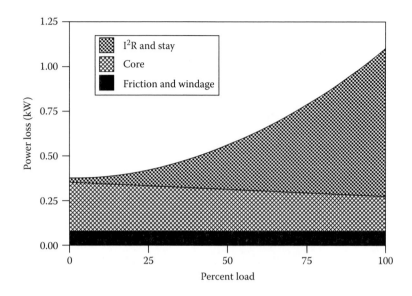

FIGURE 11.4
Variation of losses with load for a 10 hp motor. (Reprinted from Nadel, S. et al., *Energy-Efficient Motor Systems: A Handbook on Technologies, Programs, and Policy Opportunities*, 1st edn., American Council for an Energy-Efficient Economy, Washington, DC, 1992. With permission.)

In 1997, the U.S. Energy Policy Act (EPAct) put in place mandatory efficiency standards for many general-purpose motors. Also in the 1990s, the Consortium for Energy Efficiency (CEE) developed a voluntary premium-efficiency standard, which evolved into the NEMA Premium designation (Nadel et al., 2002; NEMA, 2000). Premium efficiency motors offer an efficiency improvement over EPAct, which typically ranges from 4% for a 1 hp motor to 2% for a 150 hp motor (Figure 11.5). Due to long motor lives, many motors in use are less efficient than EPAct, and thus there is an even larger difference between them and premium efficiency motors. Premium efficiency motors normally cost around 15%–25% more than standard motors, which translates into a price premium of $8–$40/hp. In new applications, and for motors with a large number of operating hours, the paybacks are normally under 4 years for Premium vs. EPAct motors, and under 2 years for Premium vs. older, standard-efficiency motors.

In 2007, the U.S. Energy Independence and Security Act (EISA) raised the minimum efficiency standards set by EPAct to meet efficiency requirements equivalent to NEMA Premium effective December 2010. Additionally, EISA also expanded the number of motors covered by EPAct. In 2014, U.S. DOE passed a final rule expanding the number of motors required to meet the energy efficiency guidelines outlined in NEMA Table 12-12, which is the table used to designate a premium efficiency motor. For example, where EISA requires 1–200 hp general purpose Design B motors to meet NEMA Premium efficiency levels, the 2014 rule expands the covered size range to 500 hp (USDOE 2014c).

An important measure for the wide market acceptance of high efficiency motors is the availability of harmonized standards, dealing with motor performance testing, efficiency classification, and display of ratings (Brunner et al., 2013, De Almeida 2013). This also applies to adjustable-speed drives (ASDs; see later section). In PR China and in the EU, high-efficiency/IE2 motors are mandatory since 2011, and Premium/IE3 motors will be mandatory in 2015 in the EU (De Almeida et al., 2014). The IEC60034-30 efficiency

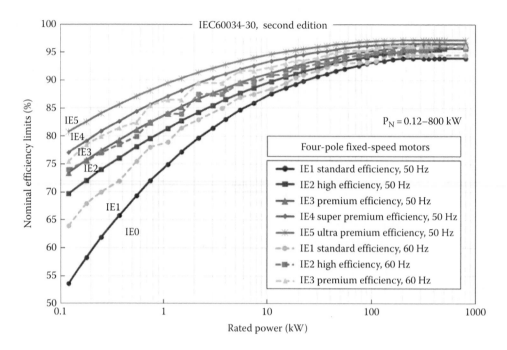

FIGURE 11.5
Planned revision (second edition) of IEC60034-30 nominal efficiency classes limits, for four-pole motors (0.12–800 kW power range). (From De Almeida, A. et al., *IEEE Indust. Appl. Mag.*, 17(1), 12, January/February 2011; De Almeida, A. et al., *IEEE Trans. Indust. Appl.*, 50(2), March/April, 1274, 2014; IEC60034-2-1, 2nd edn., 2/1687/CDV: Rotating electrical machines—Part 2-1: Standard method for determining losses and efficiency from tests [excluding machines for traction vehicles], 2013.)

classification standard (IEC60034-30, 2008) is being revised, and while the IE4* Super-Premium efficiency class is now well established, a new IE5 Ultra-Premium efficiency class is being considered in the Second Edition[†] (IEC/TS60034-31, 2010). However, there is a question with regard to its achievability (De Almeida et al., 2013).

In Figure 11.5, the IE1, IE2, IE3, IE4, and IE5 classes of the revised IEC60034-30 standard are shown for four-pole, 50/60 Hz motors (IEC60034-30, 2008). The motor nominal efficiency should be determined according to IEC60034-2-1 (IEC60034-2-1, 2007; IEC6003-2-1, 2013).

11.2.1.2 *Efficiency of Rewound Motors*

When a motor fails, the user has the options of having the motor rebuilt, buying a new standard motor, or buying a NEMA Premium motor. Except for large motors with low annual operating hours, it is typically very cost-effective to replace the failed motor with a Premium motor. Although motor rebuilding is a low-cost alternative, the efficiency of a rebuilt motor can be substantially decreased by the use of improper methods for stripping the old winding. On average, the efficiency of a motor decreases by about 1% each time the motor is rewound.

---

[*] The designation of the energy-efficiency class in the IEC60034-30 standard consists of the letters "IE" (International Energy-Efficiency), directly followed by a numeral representing the classification.

[†] The Second Edition will be denoted as IEC60034-30-1, for line-start motors and IEC60034-30-2 for ASD-fed motors.

The use of high temperatures (above 350°C) during the rewinding process can damage the interlaminar insulation and distort the magnetic circuit with particular impact on the air gap shape, leading to substantially higher core and stray losses. Before any motor is rewound, it should be checked for mechanical damage and the condition of the magnetic circuit should be tested with an electronic iron core loss meter. There are techniques available to remove the old windings, even the ones coated with epoxy varnish, which do not exceed 350°C (Dreisilker, 1987).

11.2.2 Recent Motor Developments

In the low horsepower range, the induction motor is being challenged by new developments in motor technology such as PM and reluctance motors, which are as durable as induction motors and are more efficient. These advanced motors do not have losses in the rotor and feature higher torque and power/weight ratio. In fractional-horsepower motors, such as the ones used in home appliances, the efficiency improvements can reach 10%–15%, compared with single-phase induction motors. Compared to the shaded-pole motors commonly used in small fans, improved motor types can more than double motor efficiency.

11.2.2.1 Permanent-Magnet Motors

Over the last few decades, there has been substantial progress in the area of PM materials. Figure 11.6 shows the relative performance of several families of magnetic materials. High-performance PM materials such as neodymium–iron–boron alloys, with a large energy

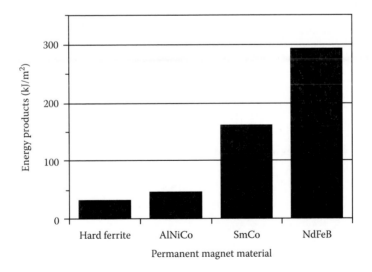

FIGURE 11.6
The evolution of permanent-magnet materials, showing the increasing magnetic energy density (*energy product*). Ferrites were developed in the 1940s and AlNiCos (aluminum, nickel, and cobalt) in the 1930s. The rare-earth magnets were developed beginning in the 1960s (samarium–cobalt) and in the 1980s (neodymium–iron–boron). The higher the energy density, the more compact the motor design can be for a given power rating. (Reprinted from Nadel, S. et al., *Energy-Efficient Motor Systems: A Handbook on Technologies, Programs, and Policy Opportunities*, 1st edn., American Council for an Energy-Efficient Economy, Washington, DC, 1992. With permission.)

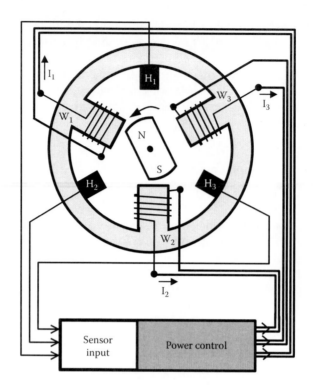

FIGURE 11.7
Control system scheme of a brushless DC motor. This motor is also known as an electronically commutated permanent-magnet (PM) motor. The motor is composed of three sets of stator windings arranged around the PM rotor. AC power is first converted to DC, and then switched to the windings by the power control unit, which responds to both an external speed command and rotor position feedback from H_1, H_2, and H_3, which are magnetic position sensors. If a DC power supply is available, it can be used directly by the power control unit in place of the AC supply and converter. The function of the commutator and brushes in the conventional DC motor is replaced by the control unit and power switches. The PM rotor follows the rotating magnetic field created by the stator windings. The speed of the motor is easily changed by varying the frequency of switching.

density and moderate cost, offer the possibility of achieving high efficiency and compact lightweight motors.

In modern designs, the PMs are used in the rotor. The currents in the stator windings are switched by semiconductor power devices based on the position of the rotor, normally detected by Hall sensors, as shown in Figure 11.7. The rotor rotates in synchronism with the rotating magnetic field created by the stator coils, leading to the possibility of accurate speed control. Because these motors have no brushes, and with suitable control circuits they can be fed from a DC supply, they are sometimes called brushless DC motors.

11.2.2.2 Switched Reluctance Motors

Switched reluctance motors are also synchronous motors whose stator windings are commutated by semiconductor power switches to create a rotating field. The rotor has no windings, being made of iron with salient poles. The rotor poles are magnetized by the influence of the stator rotating field. The attraction between the magnetized poles and the rotating field creates a torque that keeps the rotor moving at synchronous speed. Figure 11.8 shows the structure of a switched reluctance motor.

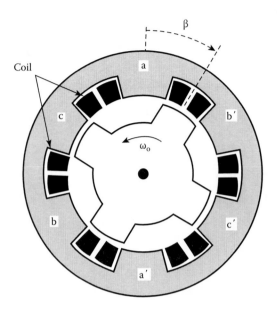

FIGURE 11.8
Schematic view of a switched reluctance motor. The configuration shown is a 6/4 pole. A rotating magnetic field is produced by switching power on and off to the stator coils in sequence, thus magnetizing poles a–a′, b–b′, and c–c′ in sequence. Switching times are controlled by microprocessors with custom programming.

Switched reluctance motors are more efficient than induction motors, are simple to build, and are robust, and if mass produced, their price can compete with induction motors. Switched reluctance motors can also be used in high-speed applications (above the 3600 rpm possible with induction or synchronous motors operating on a 60 Hz AC supply) without the need for gears.

11.2.2.3 Synchronous Reluctance Motors

In synchronous reluctance motors (SynRMs), the stator has windings (similar to those used in the squirrel-cage induction motors [SCIMs]) and the rotor is cylindrical (as distinguished from Figure 11.8) but with anisotropic magnetic structure (magnetic reluctance varies with the flux direction) (ABB, 2012, 2013; Boglietti et al., 2005). In these motors, when fed by an ASD, speed and torque control is possible and large energy savings may be realized in variable-flow pump and fan applications.

Similar to permanent-magnet synchronous motors (PMSMs), it is possible to design SynRMs for (as illustrated in Figure 11.9 from left to right) the same frame size and better efficiency, an in-between solution with slight higher efficiency and slightly smaller frame size, or smaller frame size and equal efficiency.

Figure 11.10 shows the expected loss reduction when comparing the same and smaller frame sizes to a reference SCIM frame size (De Almeida et al., 2013). The new rotor has neither magnets nor windings, and thus suffers virtually no power losses—which makes it uniquely cool. For the same frame size, there is typically a 25% reduction in the overall losses with respect to IE3-Class/NEMA Premium SCIMs. Therefore, there is the possibility of a significant efficiency gain if the standard frame size is used.

It should be mentioned that the commercially available models of SynRM and drive packages are optimized for variable-speed pump and fan applications to achieve the

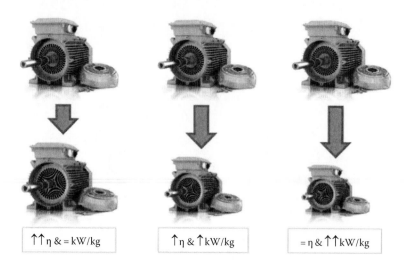

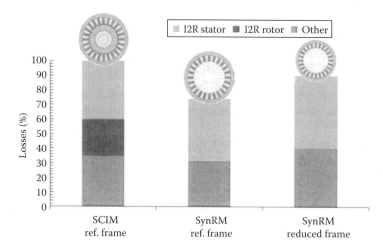

FIGURE 11.9
Frame size vs. efficiency and power density in SynRMs.

FIGURE 11.10
Same stator size, different rotor type: loss reduction (note: stator copper losses slightly increase and rotor losses are reduced to zero).

highest efficiency. Since SynRM have no rotor cage losses, their overall losses are lower and, consequently, the efficiency is higher. For the same efficiency, SynRM can be up to two frame sizes smaller than conventional SCIM. The output and efficiency performance of SynRM are comparable to a PMSM drive, but with construction techniques similar to those used in SCIMs, bringing the best of both worlds to users.

11.2.2.4 Promising Industrial Motor Technologies

The Super-Premium efficiency class (NEMA Premium or IE3 mentioned earlier) motor technologies can be divided into fixed- and variable-speed applications. For fixed-speed applications, the existing SCIM and line-start (motors that can start and run directly from grid power without an ASD) permanent magnet (LSPM) technologies are already within the IE4 class. Although not commercially available, SynRM technology is likely to evolve to a line-start solution by means of embedding an auxiliary cage in the rotor (e.g., embedded in the air spaces of the rotor), as was done for the PM technology. This type of motor would be useful in constant speed applications and would not require an ASD.

Presently for variable-speed applications, the existing ASD-fed PMSMs (with rare-earth PMs and silicon steel core or with ferrite PMs and amorphous metal core) and SynRMs are the best options. In such applications, it is important to note that the efficiency at partial speed and torque is very important since the operation cycle typically includes long periods with speeds and torques lower than nominal. Therefore (as with any variable load motor system), when analyzing the energy performance of the drive system, the efficiency maps are much more relevant than a single efficiency point for the rated speed and torque.

11.2.3 Motor Speed Controls

AC induction and synchronous motors are essentially constant-speed motors. Most motor applications would benefit if the speed could be adjusted to the process requirements. This is especially true for new applications where the processes can be designed to take advantage of the variable speed. The potential benefits of speed variation include increased productivity and product quality, less wear in the mechanical components, and substantial energy savings.

In many pump, fan, and compressor applications, the mechanical power grows roughly with the cube of the fluid flow; to move 80% of the nominal flow only half of the power is required. Fluid-flow applications are therefore excellent candidates for motor speed control.

Conventional methods of flow control have used inefficient throttling devices such as valves, dampers, and vanes. These devices have a low initial cost but introduce high running costs due to their inefficiency. Figure 11.11 shows the relative performance of different techniques to control flow produced by a fan. In Figure 11.11, it should be noted that in VAV systems static pressure typically acts as a feedback mechanism for controlling fan speed. Consequently, there will be some power consumption at 0 CFM with the result that overall savings is less than the theoretical condition.

Motor system operation can be improved through the use of several speed-control technologies, such as those covered in the following three sections.

11.2.3.1 Mechanical and Eddy-Current Drives

Mechanical speed control technologies include hydraulic transmissions, adjustable sheaves, and gearboxes. Eddy-current drives work as induction clutches with controlled slip (Magnusson, 1984).

Both mechanical drives and eddy drives have relatively low importance. They suffer from low efficiency, bulkiness, limited flexibility, or limited reliability when compared with other alternatives; in the case of mechanical drives, they may require regular maintenance.

Mechanical and eddy drives are not normally used as a retrofit due to their space requirements. Their use is more and more restricted to the low horsepower range where their use may be acceptable due to the possible higher cost of ASDs.

11.2.3.2 Multispeed Motors

In applications where only a few operating speeds are required, multispeed motors may provide the most cost-effective solution. These motors are available with a variety of

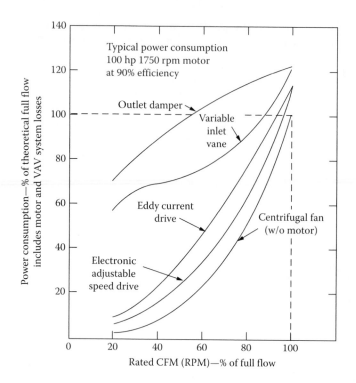

FIGURE 11.11

Comparison of several techniques for varying air flow in a variable air volume (VAV) ventilation system. The curve on the lower right represents the power required by the fan itself, not including motor losses. Electronic adjustable-speed drives are the most efficient VAV control option, offering large savings compared to outlet dampers or inlet vanes, except at high fractions of the rated fan speed. (From Greenberg, S. et al., Technology assessment: Adjustable speed motors and motor drives, Lawrence Berkeley Laboratory Report LBL-25080, Lawrence Berkeley Laboratory, University of California, Berkeley, CA, 1988.)

torque-speed characteristics (variable torque, constant torque, and constant horsepower) (Andreas, 1992), to match different types of loads. Double-winding motors can provide up to four speeds but they are normally bulkier (one frame size larger) than single-speed motors for the same horsepower rating. Pole-amplitude modulated (PAM) motors are single-winding, two-speed SCIMs that provide a wide range of speed ratios (Pastor, 1986). Because they use a single winding they have the same frame size of single-speed motors for the same horsepower and are thus easy to install as a retrofit. PAM motors are available with a broad choice of speed combinations (even ratios close to unity), being especially suited and cost-effective for those fan and pump applications that can be met by a two-speed duty cycle.

11.2.3.3 Electronic Adjustable-Speed Drives

Induction motors operate with a torque-speed relation as shown in Figure 11.12. The speed of the motor is very nearly proportional to the frequency of the AC power supplied to it; thus, the speed can be varied by applying a variable-frequency input to the motor. Electronic ASDs (Bose, 1986) achieve this motor input by converting the fixed frequency power supply (50 or 60 Hz), normally first to a DC supply and then to a continuously variable frequency/variable voltage (Figure 11.13). ASDs are thus able to continuously change the speed of AC motors. Electronic ASDs have no moving parts (sometimes with the

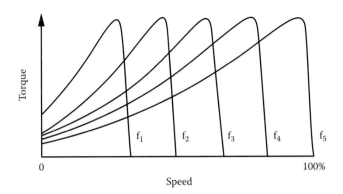

FIGURE 11.12
Speed–torque curves for an induction motor ($f_1 < f_2 < f_3 < f_4 < f_5$ and f_5 = normal line frequency). Normal operation of the motor is in the nearly vertical part of the curves to the right of the *knee* (known as the *breakdown* or *pullout* torque).

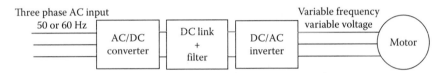

FIGURE 11.13
General inverter-based adjustable-speed drive power circuit with motor load.

exception of a cooling fan), presenting high reliability and efficiency and low maintenance requirements. Because ASDs are not bulky and have flexible positioning requirements, they are generally easy to retrofit.

Electronic ASDs are the dominant motor speed control technology at the present and for the foreseeable future. Developments in the past two decades in the areas of microelectronics and power electronics make possible the design of efficient, compact, and increasingly cost-competitive electronic ASDs. As ASDs control the currents/voltages fed to the motor through power semiconductor switches, it is possible to incorporate motor protection features, soft-start and remote control, at a modest cost. By adding additional power switches and controlling circuitry (Figure 11.14), ASDs can provide regenerative braking,

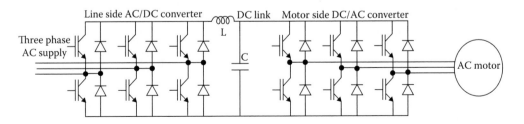

FIGURE 11.14
Power circuitry of a pulse-width modulation variable speed drive with regenerative capacity and power factor control. Whereas conventional adjustable-speed drives use diode rectifiers in the input stage, regenerative units use insulated gate bipolar transistors (IGBTs) at both the input and output stages to enable bidirectional power flow.

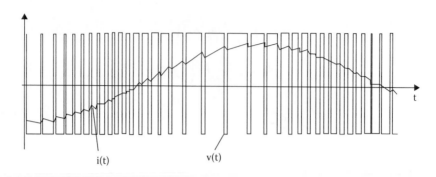

FIGURE 11.15

Pulse-width modulation for synthesizing a sinusoidal output. Output voltage, the average of which resembles the current waveform i(t), is varied by changing the width of the voltage pulses v(t). Output frequency is varied by changing the length of the cycle.

slowing the driven load and feeding power back into the AC supply. Such regenerative braking capability can increase the energy efficiency of applications such as elevators, downhill conveyors, and electric transportation.

Across the range of motor applications, no single ASD technology emerges as a clear winner when compared with other ASD types. Pulse-width modulation (PWM) voltage-source inverters ASDs dominate in the low to medium horsepower range (up to several hundred horsepower) due to their lower cost and good overall performance. Figure 11.15 shows how the variable-frequency/variable-voltage waveform is synthesized by a PWM ASD.

In the range above several hundred horsepower, the choice of ASD technology depends on several factors including the type of motor, horsepower, speed range, and control requirements (Greenberg et al., 1988). Table 11.1 presents a general classification of the most widely used adjustable-speed motor drive technologies.

It should be noted that electronic ASDs have internal power losses, and they also adversely affect the efficiency of the motors they control. The losses within the drive are due to the AC to DC conversion losses, internal resistance in the DC bus filter, and switching losses in the DC to AC inverter section. As a function of rated output power, they are typically 94%–97% efficient at full load, 91%–95% at 50% load, and 86%–94% at 25% load, over the power range of 3–100 hp, respectively, though there is significant variation in reported losses (Krukowski and Wray, 2013). There is no standard for how the losses are determined, so there is likely even more uncertainty than the reports suggest. Since ASDs provide a nonsinusoidal waveform to the motor, the motor efficiency is adversely affected, with typical degradation (relative to across-the-line operation) of 1% at full load, 3% at 50% load, and 1% at 25% load, with large variations depending on the specific motor used (Burt et al., 2008). The industry has recognized a need to address these effects, and there is now a standard for how losses are determined (ANSI/AHRI, 2013). This standard only applies to ASD and motor combinations and is the overall ASD and motor efficiency (i.e., the efficiency is the shaft power output from the motor divided by the electrical power input to the ASD); the standard does not attempt to segregate the losses within the system.

11.2.4 Motor Oversizing

Motors are often oversized as a result of the compounding of successive safety factors in the design of a system (Smeaton, 1988). The magnetic losses, friction, and windage losses are practically constant as a function of the load. Therefore, motors that are oversized

TABLE 11.1

Adjustable-Speed Motor Drive Technologies

Technology	Applicability (R = Retrofit; N = New)	Cost[b]	Comments
Motors			
Multispeed (incl. PAM[a]) motors	Fractional—500 hp PAM; fractional—2,000 + hp R, N	1.5–2 times the price of single-speed motors	Larger and less efficient than one-speed motors. PAM is more promising than multiwinding. Limited number of available speeds.
Direct-current motors	Fractional—10,000 hp N Shaft-applied drives (on motor output)	Higher than AC induction motors	Easy speed control. More maintenance required.
Mechanical			
Hydraulic drive	5–10,000 hp N	Large variation	5:1 speed range. Low efficiency below 50% speed.
Eddy-current drive	Fractional ~2,000 + hp N Wiring-applied drives (on motor input)	$900–$60/hp (for 1–150 hp)	Reliable in clean areas, Relatively long life. Low efficiency below 50% speed.
Electronic ASDs			
Voltage-source inverter	Fractional—1500 hp R, N	$300–$80/hp (for 1–300 hp)	Multimotor capability. Can generally use existing motor. PWM[c] appears most promising.
Current-source inverter	100–100,000 hp R, N	$120–$50/hp (for 100–20,000 hp)	Larger and heavier than VSI Industrial applications, including large synchronous motors.
Others	Fractional—100,000 hp R, N	Large variation	Includes cycloconverters, wound rotor, and variable voltage. Generally for special industrial applications.

[a] PAM means pole amplitude modulated.

[b] The prices are listed from high to low to correspond with the power rating, which is listed from low to high. Thus, the lower the power rating, the higher the cost per horsepower.

[c] PWM means pulse width modulation.

(working all the time below 50% of capacity) present not only lower efficiency (as shown in Figure 11.16) but also poor power factor (NEMA, 1999). The efficiency drops significantly when a motor operates lightly loaded (below 40% for a standard motor). The power factor drops continuously from full load. The decrease in performance is especially noticeable in small motors and standard-efficiency motors.

It is therefore essential to size new motors correctly and to identify motors that run grossly underloaded all the time. In the last case, the economics of replacement by a correctly sized motor should be considered. Considerations include the load at which the motor is most efficient, difference in efficiency between current and correctly sized motor, and practicality of mounting smaller sized motor. In medium or large industrial plants, where a stock of motors is normally available, oversized motors may be exchanged for the correct size versions.

11.2.5 Power Quality

Electric motors, and in particular induction motors, are designed to operate with optimal performance when fed by symmetrical three-phase sinusoidal waveforms with the

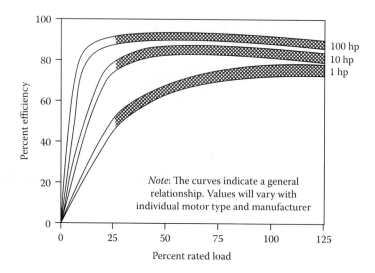

FIGURE 11.16
Typical efficiency vs. load curves for 1800 rpm, three-phase 60 Hz Design B squirrel-cage induction motors. (Reprinted from Nadel, S. et al., *Energy-Efficient Motor Systems: A Handbook on Technologies, Programs, and Policy Opportunities*, 2nd edn., American Council for an Energy-Efficient Economy, Washington, DC, 2002. With permission.)

nominal voltage value. Deviations from these ideal conditions may cause significant deterioration of the motor efficiency and lifetime. Possible power quality problems include voltage unbalance, undervoltage or overvoltage, and harmonics and interference. Harmonics and interference can be caused by, as well as affect, motor systems.

11.2.5.1 Voltage Unbalance

Induction motors are designed to operate at their best with three-phase balanced sinusoidal voltages. When the three-phase voltages are not equal, the losses increase substantially. Phase unbalance is normally caused by an unequal distribution of the single-phase loads (such as lighting) on the three phases or by faulty conditions. An unbalanced supply can be mathematically represented by two balanced systems rotating in opposite directions. The system rotating in the opposite direction to the motor induces currents in the rotor that heat the motor and decrease the torque. Even a modest phase unbalance of 2% can increase the losses by 25% (Cummings et al., 1985).

When a phase unbalance is present, the motor must be derated according to Figure 11.17.

11.2.5.2 Voltage Level

When an induction motor is operated above or below its rated voltage, its efficiency and power factor change. If the motor is underloaded, a voltage reduction may be beneficial, but for a properly sized motor, the best overall performance is achieved at the rated voltage. The voltage fluctuations are normally associated with ohmic (IR) voltage drops or with reactive power (poor power factor) flow in the distribution network (see Section 11.2.6).

11.2.5.3 Harmonics and Electromagnetic Interference

When harmonics are present in the motor supply, they heat the motor and do not produce useful torque. This in turn affects the motor lifetime and causes a derating of the motor

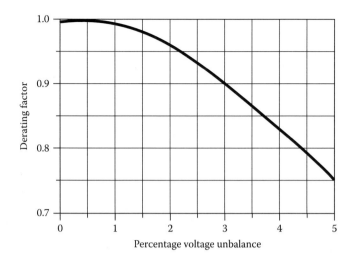

FIGURE 11.17
Derating factor due to unbalanced voltage for integral-horsepower motors. (Reprinted from Nadel, S. et al., *Energy-Efficient Motor Systems: A Handbook on Technologies, Programs, and Policy Opportunities*, 1st edn., American Council for an Energy-Efficient Economy, Washington, DC, 1992. With permission.)

capacity. This is also true when motors are supplied by ASDs that generate the harmonics themselves. The use of premium-efficiency motors can alleviate these problems due to their higher efficiency and thermal capacity; there are also motors specially designed for use with ASDs known as inverter-duty motors.

Reduction of harmonics is also important for the benefit of other consumer and utility equipment. Harmonics, caused by nonlinear loads such as the semiconductor switches in ASDs, should be reduced to an acceptable level as close as possible to the source. The most common technique uses inductive/capacitive filters at the ASD input circuit to provide a shunt path for the harmonics and to perform power factor compensation.

IEEE Standard No. 519 (IEEE, 1992) contains guidelines for harmonic control and reactive power compensation of power converters. The cost of the harmonic filter to meet this standard is typically around 5% of the cost of the ASD.

ASD power semiconductor switches operate with fast switching speeds to decrease energy losses. The fast transitions in the waveforms contain high-frequency harmonics, including those in the radio-frequency range. These high-frequency components can produce electromagnetic interference (EMI) through both conduction and radiation. The best way to deal with EMI is to suppress it at the source. Radiated EMI is suppressed through shielding and grounding of the ASD enclosure. Proper ASD design, the use of a dedicated feeder, and the use of a low-pass input filter (an inductor; often called a *line reactor*), will normally suppress conducted EMI.

11.2.6 Distribution Losses

11.2.6.1 Cable Sizing

The currents supplied to the motors in any given installation will produce Joule (I^2R) losses in the distribution cables and transformers of the consumer. Correct sizing of the cables will not only allow a cost-effective minimization of those losses but also help decrease the voltage drop between the transformer and the motor. The use of the National Electrical Code for sizing conductors leads to cable sizes that prevent overheating and allow adequate starting current to the motors, but can be far from an energy-efficient design. For example, when feeding a 100 hp motor located at 150 m from the transformer with a cable sized using NEC, about 4% of the power will be lost in heating the cable (Howe et al., 1999). Considering a 2-year payback, it is normally economical to use a cable one wire size larger than the one required by the NEC.

11.2.6.2 Reactive Power Compensation

In most industrial consumers, the main reason for a poor power factor is the widespread application of oversized induction motors. Correcting oversizing can thus contribute in many cases to a significant improvement of the power factor.

Reactive power compensation, through the application of correction capacitors, not only reduces the losses in the network but also allows full use of the power capacity of the power system components (cables, transformers, circuit breakers, etc.). In addition, voltage fluctuations are reduced, thus helping the motor to operate closer to its design voltage.

11.2.7 Mechanical Transmissions

The transmission subsystem transfers the mechanical power from the motor to the motor-driven equipment. To achieve overall high efficiency, it is necessary to use simple, properly maintained transmissions with low losses. The choice of transmission is dependent upon many factors including speed ratio desired, horsepower, layout of the shafts, and type of mechanical load.

Transmission types available include direct shaft couplings, gearboxes, chains, and belts. Belt transmissions offer significant potential for savings. About one-third of motor transmissions use belts (Howe et al., 1999). Several types of belts can be used such as V-belts, cogged V-belts, and synchronous belts.

V-belts have efficiencies in the 90%–96% range. V-belt losses are associated with flexing, slippage, and a small percentage due to windage. With wear, the V-belt stretches and needs retensioning, otherwise the slippage increases and the efficiency drops. Cogged V-belts have lower flexing losses and have better gripping on the pulleys, leading to 2%–3% efficiency improvement when compared with standard V-belts.

Synchronous belts can be 98%–99% efficient as they have no slippage and have low flexing losses; they typically last over twice as long as V-belts, leading to savings in avoided replacements that more than offset their extra cost. Figure 11.18 shows the relative performance of V-belts and synchronous belts. The efficiency gains increase with light loads.

11.2.8 Maintenance

Regular maintenance (such as inspection, adjustment, cleaning, filter replacement, lubrication, and tool sharpening) is essential to maintain peak performance of the mechanical parts and to extend their operating lifetime. Both under- and overlubrication can cause higher friction losses in the bearings and shorten the bearing lifetime. Additionally, overgreasing can cause the accumulation of grease and dirt on the motor windings, leading to overheating and premature failure.

The mechanical efficiency of the driven equipment (pump, fan, cutter, etc.) directly affects the overall system efficiency. Monitoring wear and erosion in this equipment is

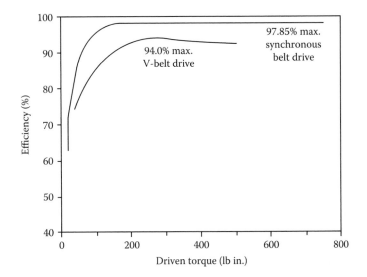

FIGURE 11.18
Efficiency vs. torque for V-belts and synchronous belts in a typical application. (Reprinted from Nadel, S. et al., *Energy-Efficient Motor Systems: A Handbook on Technologies, Programs, and Policy Opportunities*, 1st edn., American Council for an Energy-Efficient Economy, Washington, DC, 1992. With permission.)

especially important as its efficiency can be dramatically affected. For example, in chemical process industries, the erosion of the pump impeller will cause the pump efficiency to drop sharply; a dull cutter will do the same to a machine tool.

Cleaning the motor casing is also relevant because its operating temperature increases as dust and dirt accumulates on the case. The same can be said about providing a cool environment for the motor. A temperature increase leads to an increase of the windings' resistivity and therefore to larger losses. An increase of 25°C (45°F) in the motor temperature increases the Joule losses by 10%.

11.3 Energy-Saving Applications of ASDs

Typical loads that may benefit from the use of ASDs include those covered in the following four sections.

11.3.1 Pumps and Fans

In many pumps and fans where there are significant variable-flow requirements (i.e., where a significant portion of the system operating hours occurs at various partial loads), substantial savings can be achieved, as the power is roughly proportional to the cube of the flow (and thus speed of the motor). The use of ASDs instead of throttling valves with pumps shows similar behavior to that for fans in Figure 11.11. In applications where the majority of the system operation occurs at a single partial load, other techniques, such as pump impeller trimming and changing fan speed, may lead to greater savings than installation of an ASD. The use of ASDs can in many cases allow fans to be driven directly from the motor, eliminating the belt drive with its losses and maintenance requirements.

11.3.2 Centrifugal Compressors and Chillers

Most industrial compressed air systems have significant savings potential. Well-engineered efficiency improvements yield verified savings in the range of 15%–30% of system energy consumption (Fraunhofer Institute, 2000). Furthermore, in the United States, it is estimated that 21% of the overall compressed air system energy use from 15 industrial sectors could be saved through the implementation of cost-effective energy-saving measures (McKane and Hasanbeigi, 2010).

Centrifugal compressors and chillers can take advantage of motor controls in the same way as other centrifugal loads (pumps and fans). The use of wasteful throttling devices or the on–off cycling of the equipment can be largely avoided, resulting in both energy savings and extended equipment lifetime. Since motors driven by ASDS can operate at higher speeds than those operating at line frequency, speed-up gearboxes can be eliminated, saving energy, first cost, and maintenance cost.

Savings from compressed air measures are mostly coincident with electric system peak periods. Plant air systems normally have a large duty factor, typically operating 5000–8000 h/year. Thus, energy and demand reductions are very likely to occur at system peaks and contribute to system reliability. Compressed air system efficiency improvements are highly cost-effective and additionally lead to reduced plant downtime. Many projects have been identified with significant energy and demand reduction potential and paybacks less than 2 years (XENERGY, 2001).

11.3.3 Conveyors

The use of speed controls, in both horizontal and inclined conveyors, allows the matching of speed to material flow. As the conveyor friction torque is constant, energy savings are obtained when the conveyor is operated at reduced speed. In long conveyors, such as those found in power plants and in the mining industry, the benefits of soft-start without the need for complex auxiliary equipment are also significant (De Almeida et al., 2005).

For horizontal conveyors (Figure 11.19), the torque is approximately independent of the transported load (it is only friction dependent). Typically, the materials handling output of a conveyor is controlled through regulating the material input, and the torque and speed are roughly constant. But if the materials input to the conveyor is changed, it is possible to

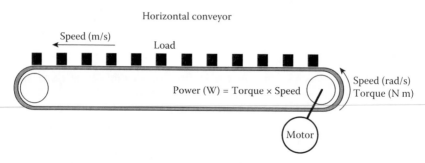

FIGURE 11.19
Power required by a conveyor.

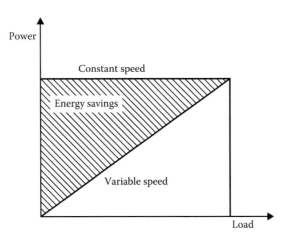

FIGURE 11.20
Energy savings in a conveyor using speed control, in relation to the typical constant speed.

reduce the speed (the torque is the same), and, as can be seen in Figure 11.20, significant energy savings can be achieved, proportional to the speed reduction.

11.3.4 High-Performance Applications

AC motors have received much attention in recent years as a proposed replacement for DC motors in high-performance speed control applications, where torque and speed must be independently controlled. Induction motors are much more reliable, more compact, more efficient, and less expensive to buy and maintain than DC motors. As induction motors have no carbon brush commutation, they are especially suitable for corrosive and explosive environments. In the past, induction motors have been difficult to control as they behave as complex nonlinear systems. However, the appearance on the market of powerful and inexpensive microprocessors has made it possible to implement in real time the complex algorithms required for induction motor control.

Field-oriented control, also called vector control, allows accurate control of the speed and torque of induction motors, in a way similar to DC motor control (Leonhard, 1984). The motor current and voltage waveforms, together with motor position feedback, are processed in real time, allowing the motor current to be decomposed into speed producing component and into a torque producing component. Vector control operation principle is represented in Figure 11.21 and is being applied to a wide variety of high-performance applications described next.

Rolling mills were one of the strongholds of DC motors, due to the accurate speed and torque requirements. With present ASD technology, AC drives can outperform DC drives in all technical aspects (reliability, torque/speed performance, maximum power, efficiency) and are capable of accurate control down to zero speed.

The availability of large diameter, high torque, and low speed AC drives makes them suitable for use in applications like ball mills and rotary kilns without the need for gearboxes. This area was also a stronghold of DC drives. Again, AC drives have the capability to offer superior performance in terms of reliability, power density, overload capability, efficiency, and dynamic characteristics.

AC traction drives can also feature regenerative braking. AC traction drives are already being used in trains, rapid transit systems, and ship propulsion and are the proper choice for the electric automobile.

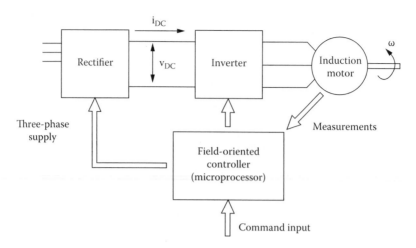

FIGURE 11.21
Schematic of a vector-control drive (also known as a field-oriented control). (Reprinted from Nadel, S. et al., *Energy-Efficient Motor Systems: A Handbook on Technologies, Programs, and Policy Opportunities*, 1st edn., American Council for an Energy-Efficient Economy, Washington, DC, 1992. With permission.)

DC drives have traditionally been used with winders in the paper and steel industry in order to achieve the constant tension requirements as the winding is performed. Sometimes the constant tension is obtained by imposing friction, which wastes energy. AC drives can also be used to replace DC drives, saving energy, and achieving better overall performance.

The use of field-oriented AC drives in machine tools and robotics allows the stringent requirements regarding dynamic performance to be met. Positioning drives can produce a peak torque up to 10 times the rated torque and make it possible to adjust the speed to very low values.

In machine tools, AC drives can provide accurate higher spindle speeds than DC drives without the need for gearboxes and their associated losses. The ASDs can also adjust the voltage level when the spindle drive is lightly loaded, providing further savings. In robotics, the higher power density and superior dynamics of AC drives are important advantages.

11.4 Energy and Power Savings Potential; Cost-Effectiveness

The energy and peak power savings potential for any motor-related technology depends on a number of factors, including the characteristics of the motor, the motor drive, the driven equipment, and the load (Nadel et al., 2002). Since all of this information is seldom available, it is difficult to determine the effect of even a single application; it is far more difficult to determine the savings potential for diverse applications, across all sectors, for an entire nation.

A 2010 UNIDO study made an initial effort to categorize the energy saving potential—both technical and cost effective—of measures frequently undertaken to improve the energy efficiency of industrial compressed air, pumping, and fan systems in several countries/regions (the United States, Canada, the European Union, Thailand, Vietnam, and Brazil) through combining available data and expert opinion. For example, the study finds that 25% of 2008 motor system energy used in the United States could be saved through implementing cost-effective measures. However, the study highlighted the lack of current installed motor system efficiency data as a substantial barrier to the analysis. The full report is available online (McKane and Hasanbeigi, 2010).

A 2013 study by the U.S. DOE estimated that the technical annual energy savings potential from upgrading motors and implementing ASDs in the residential sector is approximately 0.54 quads of the 4.73 quads of motor driven energy consumption (includes electricity generation, transmission, and distribution losses) in the sector. Similarly, the report estimated that the technical annual energy savings from upgrading motors in the commercial sector is 0.46 quads of the total 4.87 quads (includes electricity generation, transmission, and distribution losses) of motor driven energy consumption in the sector and 0.53 quads could be saved through implementing ASDs. The report states that total savings potential in the commercial sector from upgrading motors and implementing ASDs would be less than the sum of the individual saving estimates.

This section estimates the energy and power savings that could be realized nationwide in the United States through the application of certain efficiency technologies in the residential, commercial, industrial, and utility sectors. Table 11.2 lists the major motor-driven end-uses, estimates of their energy use as a percentage of total national electricity use (based on Howe et al., 1999; Nadel et al., 2002), and the potential energy savings from ASDs expressed as a fraction of existing use.

TABLE 11.2

Electric Motor Usage and Potential Adjustable-Speed Drive Savings by Sector and End Use

Sector	End Use	Usage (% of Total U.S. Usage)	Potential ASD Savings (% of Usage for Each End Use)
Residential	Refrigeration	3.0	10
	Space heating	2.1	20
	Air-conditioning	3.4	15
	HVAC dist. fan	1.8	25
	Others	1.6	5
	Residential total	11.9	15
Commercial	Refrigeration	2.2	10
	HVAC compressors	3.3	15
	HVAC distribution	4.2	25
	Others	0.7	20
	Commercial total	10.4	18
Industrial	Refrigeration	1.5	10
	Pumps	5.7	20
	Fans	3.2	20
	Compressed air	3.6	15
	Material handling	2.8	15
	Material process	5.2	15
	Others	1.0	0
	Industrial total	23.1	16
Utilities	Pumps and fans	4.9	15
	Material handling and processing	2.2	15
	Utilities total	7.1	19

11.4.1 Potential Savings in the Residential Sector

About 38% of residential electricity is used in motor systems. The primary motor technology for realizing energy and power savings in the residential sector is the electronic ASD. Heat pumps, air conditioners, forced-air furnaces, and washing machines with ASDs have already been introduced into the market. Other appliances, such as refrigerators, freezers, heat-pump water heaters, and evaporative coolers, are also potential candidates for adjustable-speed controls. Most of the energy-saving potential of ASDs in the home is associated with the use of refrigerant compressors for cooling or heating (as in heat pumps, air conditioners, refrigerators, and freezers). In all of these applications, ASDs can reduce energy consumption by matching the speed of the compressor to the instantaneous thermal load. Given the assumed savings potential for each end use, the overall savings is about 15% of the sector's motor electricity.

Several improvements in home appliance technology that are likely to become common over the next few years will complement the use of ASDs:

- *High-efficiency compressor and fan motors*: The use of PM and reluctance motors can increase the efficiency of the motor by 5%–15%, when compared with conventional squirrel-cage single-phase induction motors; as noted earlier, even larger savings are possible with many small fan motors. PM AC motors are used in the latest ASD-equipped furnace and heat pump.
- *Rotary and scroll compressors*: The use of rotary (in small applications) or scroll compressors in place of reciprocating compressors can take full advantage of the speed variation potential of ASDs.
- *Larger heat exchangers with improved design*: Improved heat exchangers increase the efficiency by decreasing the temperature difference of the compressor thermal cycle.

11.4.2 Potential Savings in the Commercial Sector

An estimated 37% of commercial electricity use is for motor-driven end uses (Howe et al., 1999); the percentage for peak power is higher. The savings potential of ASDs in air-conditioning and ventilation applications was estimated by running DOE-2 computer simulations on two representative building types in five U.S. cities (Eto and de Almeida, 1988). Comparisons were made between the cooling and ventilation systems with and without ASDs. The results indicate ventilation savings of approximately 25% for energy and 6% for peak power, and cooling savings of about 15% and 0%, respectively. In Table 11.2, it is assumed that these energy results can be applied nationwide.

The estimated 10% energy savings for refrigeration are shown in Table 11.2; an estimated 5% savings in peak power should also be attainable. Other motor efficiency measures (discussed in Section 11.2) combined can capture approximately 10% more energy and demand savings, with an overall potential savings of about 18% of the sector's motor system electricity.

11.4.3 Potential Savings in the Industrial and Utility Sectors

About 70% of industrial and 89% of the utility sector electricity use is for motor systems. In Table 11.2, the fluid moving end-use savings are estimated at 20%, except for utilities, where the system requirements of municipal water works limit the savings. Compressed air and materials applications are assumed to have 15% potential.

As most industries are nonseasonal, with flat load profiles during operating hours, the peak savings are similar to energy savings. When other motor efficiency measures (see Section 11.2) are combined, approximately 10% more energy and demand savings can be obtained, resulting in a total of 18% combined savings for the motor systems in these sectors.

11.4.4 Cost-Effectiveness of ASDs

The price of ASD equipment, in terms of cost/horsepower, is a function of the horsepower range, the type of AC motor used, and the additional control and protection facilities offered by the electronic ASD. ASD installation costs vary tremendously depending on whether the application is new or retrofit, available space, weather protection considerations, labor rates, etc. Thus, there is a huge range of installed costs possible for any given ASD size, and the costs listed in Table 11.3 necessarily have large uncertainties. These numbers are presented in their original units of Euros and kW.

To determine whether an ASD is cost-effective for any given application, the following need to be taken into account:

- First cost (acquisition and installation)
- System operating load profile (number of hours per year at each level of load)
- Motor and drive performance curves as a function of load
- Cost of electricity
- Maintenance requirements
- Reliability
- Secondary benefits (less wear on equipment, less operating noise, regeneration capability, improved control, soft-start, and automatic protection features)
- Secondary problems (power factor, harmonics, and interference)

A careful analysis should weigh the value of the benefits offered by each option against the secondary problems, such as power quality, that may impose extra costs for filters and power factor correction capacitors.

Comparing the cost of conserved energy to the cost of electricity is a crude way to assess the cost-effectiveness of energy efficiency measures. More accurate calculations would account for the time at which conservation measures save energy relative to the utility system peak demand, and relate these *load shape characteristics* to baseload, intermediate, and peaking supply resources. See Koomey et al. (1990a,b) for more details.

TABLE 11.3

Typical ASD Unit and Installation Costs for Various Sizes

Power (kW)	Unit Price (€)	Installation Costs (€)
0.12–0.75	200	50%–300%
0.75–2.1	280	50%–250%
7.5–45	1,130	50%–200%
76–110	5,320	50%–175%
375–1000	41,790	50%–150%

Source: CEMEP, Personal communication with Bernhard Sattler, Secretary CEMEP Working Group Low Voltage AC Motors, 2013.

References

ABB Review, 2012. Motoring ahead, *The Corporate Journal*, 1, 11, 56–61.

ABB, 2013. High output synchronous reluctance motor and drive package—Optimized cost of ownership for pump and fan applications.

Andreas, J., 1992. *Energy-Efficient Electric Motors: Selection and Application*. New York: Marcel Dekker.

ANSI/AHRI, October 2013. ANSI/AHRI standard 1210 (I-P) with addendum 1: 2011 standard for performance rating of variable frequency drives. Arlington, VA: Air-Conditioning, Heating & Refrigeration Institute.

Baldwin, S. F. 1989. Energy-efficient electric motor drive systems. In: Johansson, T. B. et al., eds., *Electricity: Efficient End-Use and New Generation Technologies, and Their Planning Implications*. Lund, Sweden: Lund University Press.

Boglietti, A., Cavagnino, A., Pastorelli, M., and Vagati, A., 2005. Experimental comparison of induction and synchronous reluctance motors performance. In: *Industry Applications Conference, 2005. Fortieth IAS Annual Meeting*, Hong Kong, Vol. 1, pp. 474–479.

Bose, B., 1986. *Power Electronics and AC Drives*. Englewood Cliffs, NJ: Prentice Hall.

Brunner, C., Evans, C., and Werle, R., 2013. Standard format for IEC standards—Learning from motor standards for the other electric equipment. In: *Eighth Energy Efficiency in Motor Driven Systems (EEMODS'13)*, Rio de Janeiro, Brazil.

Brunner, C., Waide, P., and Jakob, M., 2011. Harmonized standards for motors and systems—Global progress report and outlook. In: *Seventh Energy Efficiency in Motor Driven Systems (EEMODS'11)*, Alexandria, VA.

Burt, C. M., Piao, X., Gaudi, F., Busch, B., and Taufik, N. F. N., 2008. Electric motor efficiency under variable frequencies and loads. *Journal of Irrigation and Drainage Engineering*, 134(2), 129–136.

CEMEP, 2013. Personal communication with Bernhard Sattler, Secretary CEMEP Working Group Low Voltage AC Motors.

Cummings, P., Dunki-Jacobs, J., and Kerr, R., 1985. Protection of induction motors against unbalanced voltage operation. *IEEE Transactions on Industry Applications*, IA-21, 4.

De Almeida, A., Ferreira, F., and Baoming, G., 2013. Beyond induction motors—Technology trends to move up efficiency. In: *IEEE Industrial & Commercial Power Systems Conference*, Stone Mountain, GA, April 30–May 3, 2013 (*IEEE Transactions on Industry Applications*, 2014, accepted for publication).

De Almeida, A., Ferreira, F., and Both, D., January–February 2005. Technical and economical considerations in the application of variable speed drives with electric motor systems. *IEEE Industrial Applications Transactions*, 41(1), 188–199.

De Almeida, A., Ferreira, F., and Duarte, A., 2014. Technical and economical considerations on super high-efficiency three-phase motors. *IEEE Transactions on Industry Applications*, 50(2), March/April, 1274–1285

De Almeida, A., Ferreira, F. J. T. E., and Fong, J., January/February 2011. Standards for efficiency of electric motors. *IEEE Industry Applications Magazine*, 17(1), 12–19.

Dreisilker, H., August 1987. Modern rewind methods assure better rebuilt motors. *Electrical Construction and Maintenance*, 84, 31, 36.

EPRI, 1992. Electric motors: Markets, trends and application. EPRI Report TR-100423. Palo Alto, CA: Electric Power Research Institute.

Eto, J. and de Almeida, A. 1988. Saving electricity in commercial buildings with adjustable speed drives. *IEEE Transactions on Industrial Applications*, 24(3), 439–443.

Fraunhofer Institute, 2000. Compressed air systems market transformation study XVII/4.1031/Z/98-266. Prepared for the European Commission, Brussels, Belgium.

Greenberg, S., Harris, J. H., Akbari, H., and de Almeida, A. 1988. Technology assessment: Adjustable speed motors and motor drives. Lawrence Berkeley Laboratory Report LBL-25080. Berkeley, CA: Lawrence Berkeley Laboratory, University of California.

Howe, B., Lovins, A., Houghton, D., Shepard, M., and Stickney, B. 1999. *Drivepower Technology Atlas.* Boulder, CO: E-Source.

IEC60034-2-1, 2007. 1st edn.: Rotating electrical machines—Part 2-1: Standard method for determining losses and efficiency from tests (excluding machines for traction vehicles).

IEC60034-2-1, 2013. 2nd edn., 2/1687/CDV: Rotating electrical machines—Part 2-1: Standard method for determining losses and efficiency from tests (excluding machines for traction vehicles).

IEC60034-30, 2008. 1st edn.: Rotating electrical machines—Part 30: Efficiency classes of single-speed, three-phase, cage-induction motors (IE-code).

IEC60034-30, November 2011. 2nd edn., Draft, WG31/2CD: Rotating electrical machines—Part 30: Efficiency classes of single-speed, three-phase, cage-induction motors (IE-code).

IEC/TS60034-31, 2010. 1st edn.: Rotating electrical machines—Part 31: Selection of energy-efficient motors including variable speed applications—Application guide.

IEEE, 1992. IEEE guide for harmonic control and reactive compensation of static power converters. IEEE Standard 519. New York: Institute of Electrical and Electronics Engineers.

Koomey, J., Rosenfeld, A., and Gadgil, A., 1990a. Conservation screening curves to compare efficiency investments to power plants: Applications to commercial sector conservation programs. In: *Proceedings of the 1990 ACEEE Summer Study on Energy Efficiency in Buildings.* Asilomar, CA: American Council for an Energy Efficient Economy.

Koomey, J., Rosenfeld, A. H., and Gadgil, A. K., 1990b. Conservation screening curves to compare efficiency investments to power plants. *Energy Policy*, 18(8), 774.

Kurkowski, A. and Wray, C., June 2013. Standardizing data for VFD efficiency. *ASHRAE Journal*, 55(6), 16.

Leonhard, W., 1984. *Control of Electrical Drives.* New York: Springer-Verlag.

Magnusson, D., 1984. Energy economics for equipment replacement. *IEEE Transactions on Industry Applications*, IA-20, 2.

McKane, A. and Hasanbeigi, A., 2010. *Motor Systems Efficiency Supply Curves.* Vienna, Austria: UNIDO.

Nadel, S., Elliot, R.N., Shepard, M., Greenberg, S., Katz, G., and de Almeida, A., 1992. *Energy-Efficient Motor Systems: A Handbook on Technologies, Programs, and Policy Opportunities*, 1st edn. Washington, DC: American Council for an Energy-Efficient Economy.

Nadel, S., Elliot, R.N., Shepard, M., Greenberg, S., Katz, G., and de Almeida, A., 2002. *Energy-Efficient Motor Systems: A Handbook on Technologies, Programs, and Policy Opportunities*, 2nd edn. Washington, DC: American Council for an Energy-Efficient Economy.

NEMA, 1999. NEMA Standards Publication MG10-1994 (R1999), Energy management guide for selection and use of polyphase motors. Washington, DC: National Electrical Manufacturers' Association.

NEMA, 2000. NEMA Standards Publication MG1-1998, Revision 1. Motors and generators. Washington, DC: National Electrical Manufacturers' Association.

Pastor, C. E., April 30–May 2, 1986. *Motor Application Considerations: Single Speed, Multi-Speed, Variable Speed.* Oakland, CA: Pacific Gas and Electric Company Conference.

Smeaton, R., 1988. *Motor Application and Maintenance Handbook.* New York: McGraw-Hill.

USDOE, 2012. Motor systems tip sheets. http://www.energy.gov/eere/amo/motor-systems (accessed on March 6, 2015).

USDOE, 2013. *Energy Savings Potential and Opportunities for High Efficiency Electric Motors in Residential and Commercial Equipment.* Washington, DC.: United States Department of Energy.

USDOE, 2014a. *Improving Motor and Drive System Performance—A Sourcebook for Industry.* Washington, D.C.: United States Department of Energy.

USDOE, 2014b. *Continuous Energy Improvement in Motor Driven Systems—A Guidebook for Industry.* Washington, D.C.: United States Department of Energy.

USDOE, 2014c. *Premium Efficiency Motor Selection and Application Guide—A Handbook for Industry.* Washington, D.C.: United States Department of Energy.

XENERGY Inc., 2001. National market assessment: Compressed air system efficiency services. Washington, DC: U.S. Department of Energy.

Bibliography

Arthur D. Little, Inc., 1999. Opportunities for energy savings in the residential and commercial sectors with high-efficiency electric motors, final report. Prepared for the U.S. Department of Energy, Washington, D.C.

BPA/EPRI, 1993. Electric motor systems sourcebook. Olympia, WA: Bonneville Power Administration, Information Clearinghouse.

CEE, 2013. *Motor Efficiency, Selection, and Management: A Guidebook for Industrial Efficiency Programs.* Boston, MA: Consortium for Energy Efficiency.

De Almeida, A., Ferreira, F., Fong, J., and Conrad, B., 2008. Electric motor ecodesign and global market transformation. In: *Proceedings of the IEEE Industrial & Commercial Power Systems Conference*, Clearwater Beach, FL, May 4–8, 2008.

De Almeida, A., Ferreira, F., Fonseca, P., Falkner, H., Reichert, J., West, M., Nielsen, S., and Both, D., 2001. *VSDs for Electric Motor Systems.* Institute of Systems and Robotics, University of Coimbra, Coimbra, Portugal (European Commission, Directorate-General for Transport and Energy, SAVE II Programme).

EN50347, August 2001. General purpose three-phase induction motors having standard dimensions and outputs—Frame numbers 56 to 315 and flange numbers 65 to 740. EN Standards.

EPRI, 1992. *Adjustable Speed Drive Directory*, 3rd edn. Palo Alto, CA: Electric Power Research Institute.

EPRI, 1993. *Applications of AC Adjustable Speed Drives.* Palo Alto, CA: Electric Power Research Institute.

IEC60072-1, 1991. 6th edn.: Dimensions and output series for rotating electrical machines—Part 1: Frame numbers 56 to 400 and flange numbers 55 to 1080.

Jarc, D. and Schieman, R., 1985. Power line considerations for variable frequency drives. *IEEE Transactions on Industry Applications*, IA-21, 5.

Lovins, A. and Howe, B., 1992. *Switched Reluctance Motor Systems Poised for Rapid Growth.* Boulder, CO: E-Source.

Moore, T., March 1988. The advanced heat pump: All the comforts of home… and then some. *EPRI Journal.* Palo Alto, CA: Electric Power Research Institute, Vol. 13, pp. 4–13.

Waide, P. and Brunner, C., 2011. Energy efficiency policy opportunities for electric motor-driven systems. Paris, France: International Energy Agency.

XENERGY Inc., 1998. United States industrial electric motor market opportunities assessment. Washington, DC: U.S. Department of Energy.

12

Energy Storage Technologies[*]

Jeffrey P. Chamberlain, Roel Hammerschlag, and Christopher P. Schaber

CONTENTS

---

[*] Updated by Jeffrey P. Chamberlain from Chapter 18.1 of *Handbook of Energy Efficiency and Renewable Energy* (F. Kreith and D.Y. Goswami, eds.), Boca Raton, FL: CRC Press, 2007.

12.1 Overview of Storage Technologies

The availability of affordable and reliable electrical energy storage technologies is crucial to the worldwide effort to transform our electricity and transportation systems and break society's century-long dependence on fossil fuels.

Concerns about energy security and climate change are driving demand for renewable energy generation and storage systems as an alternative to current technologies. While grid energy generation using wind and solar technologies is becoming more common, the lack of cost-effective, high-capacity storage systems has severely limited the use of these technologies. Similarly, shortcomings in storage systems have limited the shift in transportation fuel from petroleum to electricity.[1] Because the current state of knowledge does not allow us to overcome these limitations, revolutionary advances in science and engineering are needed.[2]

The electric grid is undergoing a transformation that requires electricity storage in order to realize its objectives of low-carbon, reliable operation. Tomorrow's grid uses a highly diverse generation mix with inflexible (i.e., unpredictable) renewable energy sources in both centralized and distributed deployments. While battery-based storage would avoid reliability issues, the cost of batteries (~$500/kWh for a Na–S battery) is five times that of other storage technologies.[3]

In parallel, widespread market penetration of electric vehicles (EVs) will require a five-fold cost reduction, from $500–$600/kWh to $100–$125/kWh. To achieve a 350-mile driving range, EV batteries also must operate at pack-level energy densities of 400 Wh/kg or more, rather than today's 80–100 Wh/kg.

Energy storage will play a more and more critical role in an efficient and renewable energy future, much more so than it does in today's fossil-based energy economy. There are two principal reasons that energy storage will grow in importance with increased development of renewable energy:

1. Many important renewable energy sources are intermittent, and generate when weather dictates, rather than when energy demand dictates.
2. Many transportation systems require energy to be carried with the vehicle.*

---

* This is almost always true for private transportation systems, and usually untrue for public transportation systems, which can rely on rails or overhead wires to transmit electric energy. However, some public transportation systems such as buses do not have fixed routes and also require portable energy storage.

Energy can be stored in many forms: as mechanical energy in rotating, compressed, or elevated substances, as thermal or electrical energy waiting to be released from chemical bonds, or as electrical charge ready to travel from positive to negative poles on demand.

Storage media that can take and release energy in the form of electricity have the most universal value, because electricity can efficiently be converted either to mechanical or heat energy, while other energy conversion processes are less efficient. Electricity is also the output of three of the most promising renewable energy technologies: wind turbines, solar thermal, and photovoltaics. Storing this electricity in a medium that naturally accepts electricity is favored, because converting the energy to another type usually has a substantial efficiency penalty.

Still, some applications can benefit from mechanical or thermal technologies. Examples are when the application already includes mechanical devices or heat engines that can take advantage of the compatible energy form, lower environmental impacts that are associated with mechanical and thermal technologies, or low cost resulting from simpler technologies or efficiencies of scale.

In this chapter, we group the technologies into five categories: direct electric, electrochemical, mechanical, direct thermal, and thermochemical. Table 12.1 is a summary of all of the technologies covered. Each is listed with indicators of appropriate applications, which are further explained in Section 12.3.

12.2 Principal Forms of Stored Energy

The storage media discussed in this chapter can accept and deliver energy in three fundamental forms: electrical, mechanical, and thermal. Electrical and mechanical energies are both considered high-quality energy, because they can be converted to either of the other two forms with fairly little energy loss (e.g., electricity can drive a motor with only about 5% energy loss, or a resistive heater with no energy loss).

The quality of thermal energy storage depends on its temperature. Usually, thermal energy is considered low quality, because it cannot be easily converted to the other two forms. The theoretical maximum quantity of useful work W_{max} (mechanical energy) extractable from a given quantity of heat Q is

$$W_{max} = \frac{(T_1 - T_2)}{T_1 \times Q},$$

where
  T_1 is the absolute temperature of the heat
  T_2 is the surrounding, ambient absolute temperature

Any energy storage facility must be carefully chosen to accept and produce a form of energy consistent with either the energy source or the final application. Storage technologies that accept and/or produce heat should as a rule only be used with heat energy sources or with heat applications. Mechanical and electric technologies are more versatile, but in most cases, electric technologies are favored over mechanical, because electricity is more easily transmitted, because there is a larger array of useful applications, and because the construction cost is typically lower.

TABLE 12.1

Overview of Energy Storage Technologies and Their Applications

	Utility Shaping	Power Quality	Distributed Grid	Automotive
Direct electric				
Ultracapacitors		✓		✓
SMES		✓		
Electrochemical				
Batteries				
Lead–acid	✓	✓	✓	
Lithium-ion	✓	✓	✓	✓
Nickel–cadmium	✓	✓		
Nickel metal hydride				✓
Zebra				✓
Sodium–sulfur	✓	✓		
Flow batteries				
Vanadium redox	✓			
Polysulfide bromide	✓			
Zinc bromide	✓			
Electrolytic hydrogen				✓
Mechanical				
Pumped hydro	✓			
Compressed air	✓			
Flywheels		✓		✓
Direct thermal				
Sensible heat				
Liquids			✓	
Solids			✓	
Latent heat				
Phase change	✓		✓	
Hydration–dehydration	✓			
Chemical reaction	✓		✓	
Thermochemical				
Biomass solids	✓		✓	
Ethanol	✓			✓
Biodiesel				✓
Syngas	✓			✓

12.3 Applications of Energy Storage

In Table 12.1, each technology is classified by its relevance in one to four different, principal applications.

Utility shaping is the use of very large capacity storage devices in order to answer electric demand, when a renewable resource is not producing sufficient generation. An example would be nighttime delivery of energy generated by a solar thermal plant during the prior day.

Power quality is the use of very responsive storage devices (capable of large changes in output over very short timescales) to smooth power delivery during switching events, short outages, or plant run-up. Power quality applications can be implemented at central generators, at switchgear locations, and at commercial and industrial customers' facilities. Uninterruptible power supplies (UPS) are an example of this category.

Distributed grid technologies enable energy generation and storage at customer locations, rather than at a central (utility) facility. The distributed grid is an important, enabling concept for photovoltaic technologies, which are effective at a small scale and can be installed on private homes and commercial buildings. When considered in the context of photovoltaics, the energy storage for the distributed grid is similar to the utility shaping application insofar that both are solutions to an intermittent, renewable resource, but distributed photovoltaic generation requires small capacities in the neighborhood of a few tens of MJ, while utility shaping requires capacities in the TJ range.* Renewable thermal resources (solar and geothermal) can also be implemented on a distributed scale, and require household-scale thermal storage tanks. For the purposes of this chapter, district heating systems are also considered a distributed technology.

Automotive applications include battery-electric vehicles (EVs), hybrid gasoline-electric vehicles (HEVs), plug-in hybrid electric vehicles (PHEVs), and other applications that require mobile batteries larger than those used in today's internal combustion engine cars. A deep penetration of automotive batteries also could become important in a distributed grid. Large fleets of EVs or PHEVs that are grid-connected when parked would help enable renewable technologies, fulfilling utility shaping and distributed grid functions as well as their basic automotive function.

Additional energy storage applications exist, most notably portable electronics and industrial applications. However, the four applications described here make up the principal components that will interact in a significant way with the global energy grid.

12.4 Specifying Energy Storage Devices

Every energy storage technology, regardless of category, can be roughly characterized by a fairly small number of parameters. Self-discharge time, unit size, and efficiency serve to differentiate the various categories. Within a category, finer selections of storage technology can be made by paying attention to cycle life, specific energy, specific power, energy density, and power density.

Self-discharge time is the time required for a fully charged, noninterconnected storage device to reach a certain depth of discharge (DOD). DOD is typically described as a percentage of the storage device's useful capacity, so that, for instance, 90% DOD means 10% of the device's energy capacity remains. The relationship between self-discharge time and DOD is rarely linear, so self-discharge times must be measured and compared at a uniform DOD. Acceptable self-discharge times vary greatly: from a few minutes for some power quality applications to years for devices designed to shape annual power production.

Unit size describes the intrinsic scale of the technology, and is the least well defined of the parameters listed here. If the unit size is small compared to the total required capacity

---

* Storage capacities in this chapter are given in units of MJ, GJ and TJ. 1 MJ = 0.28 kWh, 1 GJ = 280 kWh, and 1 TJ = 280 MWh.

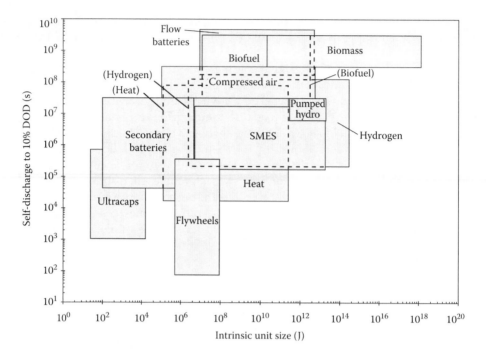

FIGURE 12.1

All storage technologies, mapped by self-discharge time and unit size. Not all hidden lines are shown. Larger self-discharge times are always more desirable, but more or less important depending on the application. Intrinsic unit size does not have a desirability proportional to its value, but rather must be matched to the application.

of a project, complexity and supply shortages can increase the cost relative to technologies with a larger unit size. Some technologies have a fairly large unit size that prohibits small-scale energy storage.

Figure 12.1 maps all of the technologies discussed in this chapter, according to their unit size and 10% self-discharge time. The gamut of technologies available covers many orders of magnitude on each axis, illustrating the broad choice available. Utility shaping applications require a moderate self-discharge time and a large unit size; power quality applications are much less sensitive to self-discharge time but require a moderate unit size. Distributed grid and automotive applications both require a moderate self-discharge time and a moderate unit size.

Efficiency is the ratio of energy output from the device to the energy input. Like energy density and specific energy, the system boundary must be carefully considered when measuring efficiency. It is particularly important to pay attention to the form of energy required at the input and output interconnections, and to include the entire system necessary to attach to those interconnections. For instance, if the system is to be used for shaping a constant-velocity, utility wind farm, then presumably, both the input and output will be AC electricity. When comparing a battery with a fuel cell in this scenario, it is necessary to include the efficiencies of an AC-to-DC rectifier for the battery, an AC-powered hydrogen generation system for the fuel cell system, and DC-to-AC converters associated with both systems.

Efficiency is related to self-discharge time. Technologies with a short self-discharge time will require constant charging in order to maintain a full charge; if discharge occurs much later than charge in a certain application, the apparent efficiency will be lower, because a significant amount of energy is lost in maintaining the initial, full charge.

Cycle life is the number of consecutive charge–discharge cycles a storage installation can undergo while maintaining the installation's other specifications within certain, limited ranges. Cycle life specifications are made against a chosen DOD depending on the application of the storage device. In some cases, for example, pressurized hydrogen storage in automobiles, each cycle will significantly discharge the hydrogen canister and the appropriate DOD reference might be 80% or 90%. In other cases, for example, a battery used in a HEV, most discharge cycles may consume only 10% or 20% of the energy stored in the battery. For most storage technologies, cycle life is significantly larger for shallow discharges than deep discharges, and it is critical that cycle life data be compared across a uniform DOD assumption.

Specific energy is a measure of how heavy the technology is. It is measured in units of energy per mass, and in this chapter, we will always report this quantity in MJ/kg. The higher the specific energy, the lighter the device. Automotive applications require high specific energies; for utility applications, specific energy is relatively unimportant, except where it impacts construction costs.

Energy density is a measure of how much space the technology occupies. It is measured in units of energy per volume, and in this chapter, we will always report this quantity in megajoule per liter (MJ/L). The higher the energy density, the smaller is the device. Again, this is most important for automotive applications, and rarely important in utility applications. Typical values for energy density associated with a few automotive-scale energy technologies are listed in Table 12.2, together with cycle life and efficiency data.

Energy density and specific energy estimates are dependent on the system definition. For example, it might be tempting to calculate the specific energy of a flow battery technology by dividing its capacity by the mass of the two electrolytes. But it is important to also include the mass of the electrolyte storage containers, and of the battery cell for a fair and comparable estimate of its specific energy. Thus, the energy density and specific energy are dependent on the size of the specific device; large devices benefit from efficiency of scale with a higher energy density and specific energy.

Specific power and *power density* are the power correlates to specific energy and energy density.

TABLE 12.2

Nominal Energy Density, Cycle Life, and Efficiency of Automotive Storage Technologies

	Energy Density (MJ/L)	**Cycle Life at 80% DOD**[a]	**Electric Efficiency (%)**
Ultracapacitors	0.2	50,000	95
Li-ion batteries	1.8	2,000	85
NiMH batteries	0.6	1,000	80
H_2 at 350 bar	3.0	n/a[b]	47
H_2 at 700 bar	5.0	n/a	45
Air at 300 bar	<0.1	n/a	37
Flywheels	<0.1	20,000	80
Ethanol	23.4	n/a	n/a

Note: Electric efficiencies are calculated for electric-to-electric conversion and momentary storage.

[a] Depth of discharge.

[b] Not applicable.

12.5 Specifying Fuels

A *fuel* is any (relatively) homogenous substance that can be combusted to produce heat. Though the energy contained in a fuel can always be extracted through combustion, other processes may be used to extract the energy (e.g., reaction in a fuel cell). A fuel may be gaseous, liquid, or solid. All energy storage technologies in the thermochemical category store energy in a fuel. In the electrochemical category, electrolytic hydrogen is a fuel.

A fuel's *lower heating value* (*LHV*) is the total quantity of sensible heat released during combustion of a designated quantify of fuel. For example, in the simplest combustion process, that of hydrogen,

$$2H_2 + O_2 \rightarrow 2H_2O(vapor) + LHV$$

or for the slightly more complex combustion of methane,

$$CH_4 + 2O_2 \rightarrow CO_2 + 2H_2O(vapor) + LHV.$$

In this chapter, the quantity of fuel is always expressed as a mass, so that LHV is a special case of specific energy. Like specific energy, LHV is expressed in units of MJ/kg in this chapter.

Higher heating value (*HHV*) is the LHV, plus the latent heat contained in the water vapor resulting from combustion.* For the examples of hydrogen and methane, this means

$$2H_2 + O_2 \rightarrow 2H_2O(liquid) + HHV$$

and

$$CH_4 + 2O_2 \rightarrow CO_2 + 2H_2O(liquid) + HHV$$

The latent heat in the water vapor can be substantial, especially for the hydrogen-rich fuels typical in renewable energy applications. Table 12.3 lists LHVs and HHVs of fuels

TABLE 12.3

Properties of Fuels

	Chemical Formula	Density (g/L)	LHV (MJ/kg)	HHV (MJ/kg)
Methanol	CH_3OH	794	19.9	22.7
Ethanol	C_2H_5OH	792	26.7	29.7
Methane	CH_4	0.68	49.5	54.8
Hydrogen	H_2	0.085	120	142
Dry syngas, airless process[a]	$40H_2 + 21CO + 10CH_4 + 29CO_2$	0.89	11.2	12.6
Dry syngas, air process[a]	$25H_2 + 16CO + 5CH_4 + 15CO_2 + 39N_2$	0.99	6.23	7.01

Source: All except syngas from U.S. Department of Energy, Properties of fuels, Alternative Fuels Data Center 2004.

[a] Chemical formulae and associated properties of syngas are representative; actual composition of syngas will vary widely according to manufacturing process.

---

* The concepts of sensible and latent heat are explained further in Section 12.9.

discussed in this chapter; in the most extreme case of molecular hydrogen, the HHV is some 18% higher than the LHV. Recovery of the latent heat requires controlled condensation of the water vapor.

In this chapter, all heating values are reported as HHV rather than LHV. HHV is favored for two reasons. One, its values allow easier checking of energy calculations with the principle of energy conservation. Two, when examining technologies for future implementation, it is wise to keep an intention of developing methods for extracting as much of each energy source's value as possible.

12.6 Electrochemical Energy Storage

12.6.1 Secondary Batteries

A secondary battery allows electrical energy to be converted into chemical energy, stored, and converted back to electrical energy. Batteries are made up of three basic parts: a negative electrode, positive electrode, and an electrolyte (Figure 12.2). The negative electrode gives up electrons to an external load, and the positive electrode accepts electrons from the load. The electrolyte provides the pathway for charge to transfer between the two electrodes. Chemical reactions between each electrode and the electrolyte remove electrons from the positive electrode and deposit them on the negative electrode. This can be written as an overall chemical reaction that represents the states of charging and discharging of a battery. The speed at which this chemical reaction takes place is related to the *internal resistance* that dictates the maximum power at which the batteries can be charged and discharged.

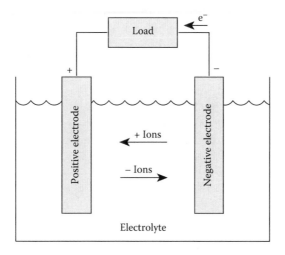

FIGURE 12.2
Schematic of a generalized secondary battery. Directions of electron and ion migration shown are for discharge, so that the positive electrode is the cathode and the negative electrode is the anode. During charge, electrons and ions move in the opposite directions and the positive electrode becomes the anode, while the negative electrode becomes the cathode.

Some batteries suffer from the *memory effect* in which a battery exhibits a lower discharge voltage under a given load than is expected. This gives the appearance of lowered capacity but is actually a voltage depression. Such a voltage depression occurs when a battery is repeatedly discharged to a partial depth and recharged again. This builds an increased internal resistance at this partial DOD, and the battery appears as a result to only be dischargeable to the partial depth. The problem, if and when it occurs, can be remedied by deep discharging the cell a few times. Most batteries considered for modern renewable applications are free from this effect, however.

12.6.1.1 Lead-Acid

Lead-acid is one of the oldest and most mature battery technologies. In its basic form, the lead-acid battery consists of a lead (Pb) negative electrode, a lead dioxide (PbO_2) positive electrode, and a separator to electrically isolate them. The electrolyte is dilute sulfuric acid (H_2SO_4), which provides the sulfate ions for the discharge reactions. The chemistry is represented by

$$PbO_2 + Pb + 2H_2SO_4 \leftrightarrow 2PbSO_4 + 2H_2O.$$

(In all battery chemistries listed in this chapter, left-to-right indicates battery discharge and right-to-left indicates charging.)

There are three main types of lead-acid batteries: the flooded cell, the sealed gel cell, and the sealed absorbed glass mat (AGM) lead-acid battery. The wet cell has a liquid electrolyte, which must be replaced occasionally to replenish the hydrogen and oxygen that escape during the charge cycle. The sealed gel cell has a silica component added to the electrolyte to stiffen it. The AGM design uses a fiberglass-like separator to hold electrolyte in close proximity to the electrodes, thereby increasing efficiency. For both the gel and AGM configurations, there is a greatly reduced risk of hydrogen explosion and corrosion from disuse. These two types do require a lower charging rate, however. Both the gel cells and the AGM batteries are sealed and pressurized so that oxygen and hydrogen, produced during the charge cycle, are recombined into water.

The lead-acid battery is a low-cost and popular storage choice for power quality applications. Its application for utility shaping, however, has been very limited due to its short cycle life. A typical installation survives 1500 deep cycles at a maximum.[4] Yet, lead-acid batteries have been used in a few commercial and large-scale energy management applications. The largest one is a 140 GJ system in Chino, California, built in 1988. Lead-acid batteries have a specific energy of only 0.18 MJ/kg, so they would not be a viable automobile option apart from providing the small amount of energy needed to start an engine. It also has a poor energy density at around 0.25 MJ/L. The advantages of the lead-acid battery technology are low cost and high power density.

12.6.1.2 Lithium Ion

Lithium ion and lithium polymer batteries, while primarily used in the portable electronics market, are likely to have future use in many other applications. The cathode in these batteries is a lithiated metal oxide ($LiCoO_2$, $LiMO_2$, etc.), and the anode is made of graphitic carbon with a layer structure. The electrolyte consists of lithium salts (such as $LiPF_6$) dissolved in organic carbonates; an example of Li-ion battery chemistry is

$$Li_xC + Li_{1-x}CoO_2 \leftrightarrow LiCoO_2 + C.$$

When the battery is charged, lithium atoms in the cathode become ions and migrate through the electrolyte toward the carbon anode where they combine with external electrons and are deposited between carbon layers as lithium atoms. This process is reversed during discharge.

The lithium polymer variation replaces the electrolyte with a plastic film that does not conduct electricity but allows ions to pass through it. The 60°C operating temperature requires a heater, reducing overall efficiency slightly.

Lithium ion batteries have a high energy density of about 0.72 MJ/L and have low internal resistance; so, they will achieve efficiencies in the 90% range and above. They have an energy density of around 0.72 MJ/kg. Their high energy efficiency and energy density make lithium-ion batteries excellent candidates for storage in all four applications we consider here: utility shaping, power quality, distributed generation, and automotive.

12.6.1.3 Nickel–Cadmium

Nickel–cadmium (NiCd) batteries operate according to the chemistry

$$2NiOOH + 2H_2O + Cd \leftrightarrow 2Ni(OH)_2 + Cd(OH)_2.$$

NiCd batteries are not common for large stationary applications. They have a specific energy of about 0.27 MJ/kg, an energy density of 0.41 MJ/L, and an efficiency of about 75%. Alaska's Golden Valley Electric Association commissioned a 40 MW/290 GJ nickel–cadmium battery in 2003 to improve reliability and to supply power for essentials during outages.[5] Resistance to cold and relatively low cost were among the deciding factors for choosing the NiCd chemistry.

Cadmium is a toxic heavy metal and there are concerns relating to the possible environmental hazards associated with the disposal of NiCd batteries. In November 2003, the European Commission adopted a proposal for a new battery directive that includes recycling targets of 75% for NiCd batteries. However, the possibility of a ban on rechargeable batteries made from nickel–cadmium still remains, and hence, the long-term viability and availability of NiCd batteries continues to be uncertain. NiCd batteries can also suffer from *memory effect*, where the batteries will only take full charge after a series of full discharges. Proper battery management procedures can help to mitigate this effect.

12.6.1.4 Nickel Metal Hydride

The nickel metal hydride (NiMH) battery operates according to the chemistry

$$MH + NiOOH \leftrightarrow M + Ni(OH)_2,$$

where M represents one of a large variety of metal alloys that serve to take up and release hydrogen. NiMH batteries were introduced as a higher energy density and more environmentally friendly version of the nickel–cadmium cell. Modern NiMH batteries offer up to 40% higher energy density than nickel–cadmium. There is potential for yet higher energy density, but other battery technologies (lithium ion in particular) may fill the same market sooner.

NiMH is less durable than nickel–cadmium. Cycling under heavy load and storage at high temperature reduces the service life. NiMH suffers from a higher self-discharge rate than the nickel–cadmium chemistry. NiMH batteries have a specific energy of 0.29 MJ/kg, an energy density of about 0.54 MJ/L, and an energy efficiency about 70%. These batteries have been an important bridging technology in the portable electronics and hybrid automobile markets. Their future is uncertain, because other battery chemistries promise higher energy storage potential and cycle life.

12.6.1.5 Sodium–Sulfur

A sodium–sulfur (NaS) battery consists of a liquid (molten) sulfur positive electrode and liquid (molten) sodium negative electrode, separated by a solid beta-alumina ceramic electrolyte (Figure 12.3). The chemistry is as follows:

$$2Na + xS \leftrightarrow Na_2S_x.$$

When discharging, positive sodium ions pass through the electrolyte and combine with the sulfur to form sodium polysulfides. x in the equation is 5 during early discharging, but once free sulfur has been exhausted, a more sodium-rich mixture of polysulfides with lower average values of x develops. This process is reversible as charging causes sodium polysulfides in the positive electrode to release sodium ions that migrate back through the electrolyte and recombine as elemental sodium. The battery operates at about 300°C. NaS batteries have a high energy density of around 0.65 MJ/L and a specific energy of up to 0.86 MJ/kg. These numbers would indicate an application in the automotive sector, but warm-up time and heat-related accident risk make its use there unlikely. The efficiency of this battery chemistry can be as high as 90% and would be suitable for bulk storage applications while simultaneously allowing effective power smoothing operations.[6]

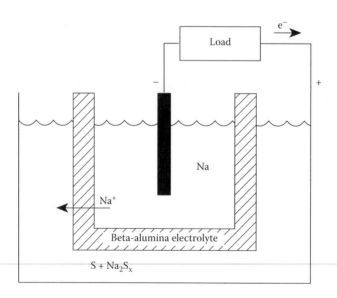

FIGURE 12.3
Sodium–sulfur battery showing discharge chemistry. The sodium (Na) and sulfur (S) electrodes are both in a liquid state and are separated by a solid, beta-alumina ceramic electrolyte that allows only sodium ions to pass. Charge is extracted from the electrolytes with metal contacts; the positive contact is the battery wall.

12.6.1.6 *Zebra*

Zebra is the popular name for the sodium nickel chloride battery chemistry:

$$NiCl_2 + 2Na \leftrightarrow Ni + 2NaCl.$$

Zebra batteries are configured similarly to sodium–sulfur batteries (see Figure 12.3), and also operate at about 300°C. Zebra batteries boast a greater than 90% energy efficiency, a specific energy of up to 0.32 MJ/kg, and an energy density of 0.49 MJ/L.[7] Its tolerance for a wide range of operating temperature and high efficiency, coupled with a good energy density and specific energy, make its most probable application the automobile sector, and as of 2003, Switzerland's MES-DEA is pursuing this application aggressively.[8] Its high energy efficiency also makes it a good candidate for the utility sector.

12.6.2 Flow Batteries

Most secondary batteries use electrodes both as an interface for gathering or depositing electrons, and as a storage site for the products or reactants associated with the battery's chemistry. Consequently, both energy and power density are tied to the size and shape of the electrodes. Flow batteries store and release electrical energy by means of reversible electrochemical reactions in two liquid electrolytes. An electrochemical cell has two compartments, one for each electrolyte, physically separated by an ion-exchange membrane. Electrolytes flow into and out of the cell through separate manifolds and undergo chemical reaction inside the cell, with ion or proton exchange through the membrane and electron exchange through the external electric circuit. The chemical energy in the electrolytes is turned into electrical energy and vice versa for charging. They all work in the same general way but vary in chemistry of electrolytes.[9]

There are some advantages to using the flow battery over a conventional secondary battery. The capacity of the system is scaleable by simply increasing the amount of solution. This leads to cheaper installation costs, as the systems get larger. The battery can be fully discharged with no ill effects and has little loss of electrolyte over time. Because the electrolytes are stored separately and in large containers (with a low surface area to volume ratio), flow batteries show promise to have some of the lowest self-discharge rates of any energy storage technology available.

Poor energy densities and specific energies remand these battery types to utility-scale power shaping and smoothing, though they might be adaptable for distributed generation use. There are three types of flow batteries that are closing in on commercialization: vanadium redox, polysulfide bromide, and zinc bromide. There is a fourth type of flow battery in early stages of R&D that may enable a significant increase in energy density over aqueous systems by enabling operation in a wider voltage window than allowed by water: nonaqueous flow batteries.

12.6.2.1 *Vanadium Redox*

The vanadium redox flow battery (VRB) was pioneered at the University of New South Wales, Australia, and has shown potentials for long cycle life and energy efficiencies of over 80% in large installations.[10] The VRB uses compounds of the element vanadium in both electrolyte tanks. The reaction chemistry at the positive electrode is

$$V^{5+} + e^- \leftrightarrow V^{4+}$$

and at the negative electrode is

$$V^{2+} \leftrightarrow V^{3+} + e^-.$$

Using vanadium compounds on both sides of the ion-exchange membrane eliminates the possible problem of cross-contamination of the electrolytes and makes recycling easier.[11] As of 2005, two small, utility-scale VRB installations are operating: one 2.9 GJ unit on King Island, Australia, and one 7.2 GJ unit in Castle Valley, Utah.

12.6.2.2 Polysulfide Bromide

The polysulfide bromide battery (PSB) utilizes two salt solution electrolytes, sodium bromide (NaBr) and sodium polysulfide (Na_2S_x). PSB electrolytes are separated in the battery cell by a polymer membrane that only passes positive sodium ions. Chemistry at the positive electrode is

$$NaBr_3 + 2Na^+ + 2e \leftrightarrow 3NaBr$$

and at the negative electrode is

$$2Na_2S_2 \leftrightarrow Na_2S_4 + 2Na^+ + 2e.$$

The PSB battery is being developed by Canada's VRB Power Systems, Inc.[12] This technology is expected to attain energy efficiencies of approximately 75%.[13] Though the salt solutions themselves are only mildly toxic, a catastrophic failure by one of the tanks could release highly toxic bromine gas. Nevertheless, the Tennessee Valley Authority released a finding of no significant impact for a proposed, 430 GJ facility and deemed it safe.[14]

12.6.2.3 Zinc Bromide

In each cell of a zinc bromide (ZnBr) battery, two different electrolytes flow past carbon-plastic composite electrodes in two compartments separated by a microporous membrane. Chemistry at the positive electrode follows the equation:

$$Br_2(aq) + 2e^- \leftrightarrow 2Br$$

and at the negative electrode:

$$Zn \leftrightarrow Zn^{2+} + 2e^-.$$

During discharge, Zn and Br combine into zinc bromide. During charge, metallic zinc is deposited as a thin film on the negative electrode. Meanwhile, bromine evolves as a dilute solution on the other side of the membrane, reacting with other agents to make thick bromine oil that sinks down to the bottom of the electrolytic tank. During discharge, a pump mixes the bromine oil with the rest of the electrolyte. The zinc bromide battery has an energy efficiency of nearly 80%.[15]

   Exxon developed the ZnBr battery in the early 1970s. Over the years, many GJ-scale ZnBr batteries have been built and tested. Meidisha demonstrated a 1 MW/14 GJ ZnBr battery in 1991 at Kyushu Electric Power Company. Some GJ-scale units are now available preassembled, complete with plumbing and power electronics.

12.6.2.4 Nonaqueous Redox Flow

Current flow battery technologies depend on aqueous electrodes, a choice that limits energy densities to <40 Wh/L as a consequence of low solubilities of redox species and operating voltages that are bounded by water electrolysis.[16] Employing nonaqueous solvents offers a wider window of electrochemical stability that enables cell operation at dramatically higher potentials. Higher cell voltage (>3 V) leads to higher energy density, and typically to higher roundtrip efficiency as well, which together reduce the total cost of energy. Compared to aqueous flow systems, fewer stack layers, lower flow velocities, smaller tanks, and fewer ancillaries are required, which significantly reduces hardware costs and enhances system reliability. In order to capitalize on the inherent benefits of nonaqueous redox flow, several fundamental science challenges must be overcome.

Nonaqueous flow batteries fall into two broad classes of early investigation: solution phase systems, in which electroactive materials are dissolved in the electrolyte,[17,18] and suspensions of active intercalant hosts, which are essentially fluidized versions of standard solid-state electrodes found in current lithium ion batteries.[19,20]

12.6.3 Electrolytic Hydrogen

Diatomic, gaseous hydrogen (H_2) can be manufactured with the process of electrolysis; an electric current applied to water separates it into components O_2 and H_2. The oxygen has no inherent energy value, but the HHV of the resulting hydrogen can contain up to 90% of the applied electric energy, depending on the technology.[21] This hydrogen can then be stored, and later combusted to provide heat or work, or power a fuel cell (see Chapter 9).

The gaseous hydrogen is low density and must be compressed to provide useful storage. Compression to a storage pressure of 350 bar, the value usually assumed for automotive technologies, consumes up to 12% of the hydrogen's HHV if performed adiabatically, though the loss approaches a lower limit of 5% as the compression approaches an isothermal ideal.[22] Alternatively, the hydrogen can be stored in liquid form, a process that costs about 40% of HHV using current technology, and that at best would consume about 25%. Liquid storage is not possible for automotive applications, because mandatory boiloff from the storage container cannot be safely released in closed spaces (i.e., garages).

Hydrogen can also be bonded into metal hydrides using an absorption process. The energy penalty of storage may be lower for this process, which requires pressurization to only 30 bar. However, the density of the metal hydride can be between 20 and 100 times the density of the hydrogen stored. Carbon nanotubes have also received attention as a potential hydrogen storage medium.[23]

12.7 Direct Electric Storage

12.7.1 Ultracapacitors

A capacitor stores energy in the electric field between two oppositely charged conductors. Typically, thin conducting plates are rolled or stacked into a compact configuration with a dielectric between them. The dielectric prevents arcing between the plates and allows the plates to hold more charge, increasing the maximum energy storage.

The ultracapacitor—also known as supercapacitor, electrochemical capacitor, or electric double layer capacitor (EDLC)—differs from a traditional capacitor in that it employs a thin electrolyte, in the order of only a few angstroms, instead of a dielectric. This increases the energy density of the device. The electrolyte can be made of either an organic or an aqueous material. The aqueous design operates over a larger temperature range, but has a smaller energy density than the organic design. The electrodes are made of a porous carbon that increases the surface area of the electrodes and further increases energy density over a traditional capacitor.

Ultracapacitors' ability to effectively equalize voltage variations with quick discharges makes them useful for power quality management and for regulating voltage in automotive systems during regular driving conditions. Ultracapacitors can also work in tandem with batteries and fuel cells to relieve peak power needs (e.g., hard acceleration) for which batteries and fuel cells are not ideal. This could help extend the overall life and reduce lifetime cost of the batteries and fuel cells used in HEV and EV. This storage technology also has the advantage of very high cycle life of greater than 500,000 cycles and a 10–12 year lifespan.[24] The limitations lie in the inability of ultracapacitors to maintain charge voltage over any significant time, losing up to 10% of their charge per day.

12.7.2 SMES

A superconducting magnetic energy storage (SMES) system is well suited in storing and discharging energy at high rates (high power.) It stores energy in the magnetic field created by direct current in a coil of cryogenically cooled, superconducting material. If the coil were wound using a conventional wire such as copper, the magnetic energy would be dissipated as heat due to the wire's resistance to the flow of current. The advantage of a cryogenically cooled, superconducting material is that it reduces electrical resistance to almost zero. The SMES recharges quickly and can repeat the charge–discharge sequence thousands of times without any degradation of the magnet. A SMES system can achieve full power within 100 ms.[25] Theoretically, a coil of around 150–500 m radius would be able to support a load of 18,000 GJ at 1000 MW, depending on the peak field and ratio of the coil's height and diameter.[26] Recharge time can be accelerated to meet specific requirements, depending on system capacity.

Because no conversion of energy to other forms is involved (e.g., mechanical or chemical), the energy is stored directly and round-trip efficiency can be very high.[5] SMES systems can store energy with a loss of only 0.1%; this loss is due principally to energy required by the cooling system.[6] Mature, commercialized SMES is likely to operate at 97%–98% round-trip efficiency and is an excellent technology for providing reactive power on demand.

12.8 Mechanical Energy Storage

12.8.1 Pumped Hydro

Pumped hydro is the oldest and largest of all of the commercially available energy storage technologies, with existing facilities up to 1000 MW in size. Conventional pumped hydro uses two water reservoirs, separated vertically. Energy is stored by moving water from the lower to the higher reservoir, and extracted by allowing the water to flow back to the lower reservoir. Energy is stored according to the fundamental physical principle of potential energy. To calculate the stored energy in joules, use the formula:

$$E_s = Vdgh,$$

where

V is the volume of water raised (m^3)

d is the density of water (1000 kg/m^3)

g is the acceleration of gravity (9.8 m/s^2)

h is the elevation difference between the reservoirs (m) and is often referred to as the *head*

Though pumped hydro is by nature a mechanical energy storage technology, it is most commonly used for electric utility shaping. During off-peak hours, electric pumps move water from the lower reservoir to the upper reservoir. When required, the water flow is reversed to generate electricity. Some high dam hydro plants have a storage capability and can be dispatched as pumped hydro storage. Underground pumped storage, using flooded mine shafts or other cavities, is also technically possible but probably prohibitively expensive. The open sea can also be used as the lower reservoir if a suitable upper reservoir can be built at close proximity. A 30 MW seawater pumped hydro plant was first built in Yanbaru, Japan, in 1999.

Pumped hydro is most practical at a large scale with discharge times ranging from several hours to a few days. There is over 90 GW of pumped storage in operation worldwide, which is about 3% of global electric generation capacity.[27] Pumped storage plants are characterized by long construction times and high capital expenditure. Its main application is for utility shaping. Pumped hydro storage has the limitation of needing to be a very large capacity to be cost effective, but can also be used as storage for a number of different generation sites.

Efficiency of these plants has greatly increased in the last 40 years. Pumped storage in the 1960s had efficiencies of 60% compared with 80% for new facilities. Innovations in variable speed motors have helped these plants to operate at partial capacity, and greatly reduced equipment vibrations, increasing plant life.

12.8.2 Compressed Air

A relatively new energy storage concept that is implemented with otherwise mature technologies is compressed air energy storage (CAES). CAES facilities must be coupled with a combustion turbine, so they are actually a hybrid storage/generation technology.

A conventional gas turbine consists of three basic components: a compressor, combustion chamber, and an expander. Power is generated when compressed air and fuel burned in the combustion chamber drive turbine blades in the expander. Approximately 60% of the mechanical power generated by the expander is consumed by the compressor supplying air to the combustion chamber.

A CAES facility performs the work of the compressor separately, stores the compressed air, and, at a later time, injects it into a simplified combustion turbine. The simplified turbine includes only the combustion chamber and the expansion turbine. Such a simplified turbine produces far more energy than a conventional turbine from the same fuel, because there is potential energy stored in the compressed air. The fraction of output energy beyond what would have been produced in a conventional turbine is attributable to the energy stored in compression.

The net efficiency of storage for a CAES plant is limited by the heat energy loss occurring at compression. The overall efficiency of energy storage is about 75%.[28]

CAES compressors operate on grid electricity during off-peak times, and use the expansion turbine to supply peak electricity when needed. CAES facilities cannot operate without combustion, because the exhaust air would exit at extremely low temperatures, causing trouble with brittle materials and icing. If 100% renewable energy generation is sought, biofuel could be used to fuel the gas turbines. There might still be other emissions issues, but the system could be fully carbon neutral.

The compressed air is stored in appropriate underground mines, caverns created inside salt rocks or possibly in aquifers. The first commercial CAES facility was a 290 MW unit built in Hundorf, Germany, in 1978. The second commercial installation was a 110 MW unit built in McIntosh, Alabama, in 1991. The third commercial CAES is a 2700 MW plant under construction in Norton, Ohio. This nine-unit plant will compress air to about 100 bar in an existing limestone mine 2200 ft (766 m) underground.[29] The natural synergy with geological caverns and turbine prime movers dictates that these be on the utility scale.

12.8.3 Flywheels

Most modern flywheel energy storage systems consist of a massive rotating cylinder (comprised of a rim attached to a shaft) that is supported on a stator by magnetically levitated bearings that eliminate bearing wear and increase system life. To maintain efficiency, the flywheel system is operated in a low vacuum environment to reduce drag. The flywheel is connected to a motor/generator mounted onto the stator that, through some power electronics, interacts with the utility grid.

The energy stored in a rotating flywheel, in joules, is given by

$$E = \tfrac{1}{2}I\omega^2,$$

where
   I is the flywheel's moment of inertia (kg-m^2)
   ω is its angular velocity (1/s^2)

I is proportional to the flywheel's mass, so energy is proportional to mass and the square of speed. In order to maximize energy capacity, flywheel designers gravitate toward increasing the flywheel's maximum speed rather than increasing its moment of inertia. This approach also produces flywheels with the higher specific energy.

Some of the key features of flywheels are low maintenance, a cycle life of better than 10,000 cycles, a 20-year lifetime, and environmentally friendly materials. Low speed, high mass flywheels (relying on I for energy storage) are typically made from steel, aluminum, or titanium; high speed, low mass flywheels (relying on ω for energy storage) are constructed from composites such as carbon fiber.

Flywheels can serve as a short-term ride-through before long-term storage comes online. Their low energy density and specific energy limit them to voltage regulation and UPS capabilities. Flywheels can have energy efficiencies in the upper 90% range, depending on frictional losses.

12.9 Direct Thermal Storage

Direct thermal technologies, though they are storing a lower grade of energy (heat, rather than electrical or mechanical energy), can be useful for storing energy from systems that provide heat as a native output (e.g., solar thermal, geothermal), or for applications where the energy's commodity value is heat (e.g., space heating, drying).

While thermal storage technologies can be characterized by specific energy and energy density like any other storage technology, they can also be characterized by an important, additional parameter, the delivery temperature range. Different end uses have more or less allowance for wide swings of the delivery temperature. Also, some applications require a high operating temperature that only some thermal storage media are capable of storing.

Thermal storage can be classified into two fundamental categories: sensible heat storage and latent heat storage. Applications that have less tolerance for temperature swings should utilize a latent heat technology.

Input to and output from heat energy storage is accomplished with heat exchangers. The following discussion focuses on the choice of heat storage materials; the methods of heat exchange will vary widely depending on properties of the storage material, especially its thermal conductivity. Materials with higher thermal conductivity will require a smaller surface area for heat exchange. For liquids, convection or pumping can reduce the need for a large heat exchanger. In some applications, the heat exchanger is simply the physical interface of the storage material with the application space (e.g., phase change drywall, see the following).

12.9.1 Sensible Heat

Sensible heat is the heat that is customarily and intuitively associated with a change in temperature of a massive substance. The heat energy E_s stored in such a substance is given by

$$E_s = (T_2 - T_1)cM,$$

where
  c is the specific heat of the substance (J/kg-°C)
  M is the mass of the substance (kg)
  T_1 and T_2 are the initial and final temperatures, respectively (°C)

The specific heat c is a physical parameter measured in units of heat per temperature per mass: substances with the ability to absorb heat energy with a relatively small increase in temperature (e.g., water) have a high specific heat, while those that get hot with only a little heat input (e.g., lead) have a low specific heat. Sensible heat storage is best accomplished with materials having a high specific heat.

12.9.1.1 Liquids

Sensible heat storage in a liquid is with very few exceptions accomplished with water. Water is unique among chemicals in having an abnormally high specific heat of 4186 J/kg-K, and furthermore has a reasonably high density. Water is also cheap and safe. It is the preferred choice for most nonconcentrating solar thermal collectors.

Liquids other than water may need to be chosen if the delivery temperature must be higher than 100°C, or if the system temperature can fall below 0°C. Water can be raised to temperatures higher than 100°C, but the costs of storage systems capable of containing the associated high pressures are usually prohibitive. Water can be mixed with ethylene glycol or propylene glycol to increase the useful temperature range and prevent freezing.

When a larger temperature range than that afforded by water is required, mineral, synthetic, or silicone oils can be used instead. The trade-offs for the increased temperature range are higher cost, lower specific heat, higher viscosity (making pumping more difficult), flammability, and, in some cases, toxicity.

For very high temperature ranges, salts are usually preferred, which balance a low specific heat with a high density and relatively low cost. Sodium nitrate has received the most prominent testing for this purpose, in the U.S. Department of Energy's *Solar Two* project located in Barstow, California.

Liquid sensible heat storage systems are strongly characterized not just by the choice of heat transfer fluid, but also by the system architecture. Two-tank systems store the cold and hot liquids in separate tanks (Figure 12.4). Thermocline systems use a single tank with cold fluid entering or leaving the bottom of the tank and hot fluid entering or leaving the top (Figure 12.5). Thermocline systems can be particularly low cost, because they minimize the required tank volume, but require careful design to prevent mixing of the hot and cold fluid.

One particularly interesting application of the thermocline concept is nonconvecting, salinity-gradient solar ponds, which employ the concept in reverse. Solar ponds are both an energy collection and energy storage technology. Salts are dissolved in the water to introduce a density gradient, with the densest (saltiest) water on the bottom and lightest (freshest) on top. Solar radiation striking the dark bottom of the pond heats the densest water, but convection of the heated water to the top cannot occur, because the density gradient prevents it. Salinity-gradient ponds can generate and store hot water at temperatures approaching 95°C.[30]

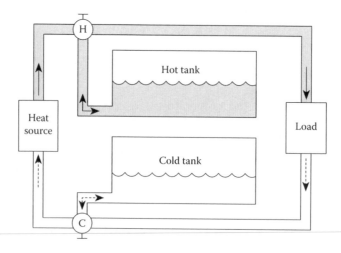

FIGURE 12.4

Two-tank thermal storage system; hot water is shown in gray and cold water is shown in white. When the heat source is producing more output than required for the load, valve H is turned to deposit hot liquid in the tank. When it is producing less than required for the load, the valve is turned to provide supplemental heat from the storage tank. Note that each tank must be large enough to hold the entire fluid capacity of the system.

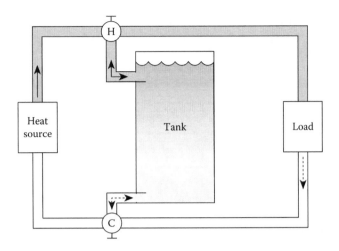

FIGURE 12.5
Thermocline storage tank. Thermocline storage tanks are tall and narrow to encourage the gravity-assisted separation of hot and cold fluid, and include design features (especially at the input/output connectors) to prevent mixing in the stored fluid.

12.9.1.2 Solids

Storage of sensible heat in solids is usually most effective when the solid is in the form of a bed of small units, rather than a single mass. The reason is that the surface-to-volume ratio increases with the number of units, so that heat transfer to and from the storage device is faster for a greater number of units. Energy can be stored or extracted from a thermal storage bed by passing a gas (such as air) through the bed. Thermal storage beds can be used to extract and store the latent heat of vaporization from water contained in flue gases.

Though less effective for heat transfer, monolithic solid storage has been successfully used in architectural applications and solar cookers.

12.9.2 Latent Heat

Latent heat is absorbed or liberated by a phase change or a chemical reaction, and occurs at a constant temperature. A phase change means the conversion of a homogenous substance among its various solid, liquid, or gaseous phases. One very common example is boiling water on the stovetop: though a substantial amount of heat is absorbed by the water in the pot, the boiling water maintains a constant temperature of 100°C. The latent heat E_s stored through a phase change is

$$E_s = lM,$$

where
  M is the mass of material undergoing a phase change (kg)
  l is the latent heat of vaporization (for liquid–gas phase changes) or the latent heat of fusion (for solid–liquid phase changes), in units of energy per mass (J/kg)

Conservation of energy dictates that the amount of heat absorbed in a given phase change is equal to the amount of heat liberated in the reverse phase change.

Though we use the term *phase change* to refer only to straightforward freezing and melting (Section 12.9.2.1), many sources use the term *phase change materials* (*PCMs*) to refer to any substance storing latent heat (including those described in Sections 12.9.2.2 and 12.9.2.3 as well).

12.9.2.1 Phase Change

Practical energy storage systems based on a material phase change are limited to solid–solid and solid–liquid phase changes. Changes involving gaseous phases are of little interest due to the expense associated with containing a pressurized gas, and difficulty of transferring heat to and from a gas.

Solid–solid phase changes occur when a solid material reorganizes into a different molecular structure in response to temperature. One particularly interesting example is lithium sulfate (Li_2SO_4) that undergoes a change from a monoclinic structure to a face-centered cubic structure at 578°C, absorbing 214 J/g in the process, more than most solid–liquid phase changes.[31]

Some common chemicals, their melting points, and heats of fusion are listed in Table 12.4. Fatty acids and paraffins received particular attention in the 1990s as candidate materials for the heat storage component of phase change drywall, a building material designed to absorb and release heat energy near room temperature for the purpose of indoor temperature stabilization.[32] In this application, solids in the drywall maintain the material's structural integrity even though the PCMs are transitioning between solid and liquid states.

TABLE 12.4

Melting Points and Heats of Fusion for Solid–Liquid Phase Changes

	Melting Point (°C)	Heat of Fusion (J/g)
Aluminum bromide	97	42
Aluminum iodide	191	81
Ammonium bisulfate	144	125
Ammonium nitrate	169	77
Ammonium thiocyanate	146	260
Anthracine	96	105
Arsenic tribromide	32	37
Beeswax	62	177
Boron hydride	99	267
Metaphosphoric acid	43	107
Naphthalene	80	149
Naphthol	95	163
Paraffin	74	230
Phosphoric acid	70	156
Potassium	63	63
Potassium thiocyanate	179	98
Sodium	98	114
Sodium hydroxide	318	167
Sulfur	110	56
Tallow	76	198
Water	0	335

Source: From Kreith, F. and Kreider J.F., *Principles of Solar Engineering*, Taylor & Francis, 1978. With permission.

12.9.2.2 Hydration–Dehydration Reactions

In this process, a salt or similar compound forms a crystalline lattice with water below a *melting point* temperature, and at the melting point, the crystal dissolves in its own water of hydration. Sodium sulfate (Na_2SO_4) is a good example, forming a lattice with ten molecules of water per molecule of sulfate ($Na_2SO_4 \cdot 10H_2O$) and absorbing 241 J/g at 32°C.[33]

Hydration–dehydration reactions have not found significant application in renewable energy systems, though they too have been a candidate for phase change drywall.

12.9.2.3 Chemical Reaction

A wide variety of reversible chemical reactions are available that release and absorb heat (e.g., Ref. [34]). The principal feature of this category of latent heat storage technologies is the ability to operate at extremely high temperatures, in some cases over 900°C. Extremely high temperature applications have focused primarily on fossil and advanced nuclear applications; to date, none of these chemical methods of heat storage have been deployed in commercial renewable energy applications.

12.10 Thermochemical Energy Storage

This section provides an overview of biomass storage technologies from an energetic perspective only.

12.10.1 Biomass Solids

Plant matter is a storage medium for solar energy. The input mechanism is photosynthesis conversion of solar radiation into biomass. The output mechanism is combustion of the biomass to generate heat energy.

Biologists measure the efficiency of photosynthetic energy capture with the metric *net primary productivity* (*NPP* , which is usually reported as a yield in units similar to dry Mg/ha-yr (dry metric tons per hectare per year.) However, to enable comparisons of biomass with other solar energy storage technologies, it is instructive to estimate a *solar efficiency* by multiplying the NPP by the biomass heating value (e.g., MJ/dry Mg) and then dividing the result by the average insolation at the crop's location (e.g., MJ/ha-yr). The solar efficiency is a unitless value describing the fraction of incident solar energy ultimately available as biomass heating value. Most energy crops capture between 0.2% and 2% of the incident solar energy in heating value of the biomass; Table 12.5 shows examples of solar efficiencies estimated for a number of test crops.

The principal method for extracting useful work or electricity from biomass solids is combustion. Hence, the solar efficiencies listed in Table 12.5 need to be multiplied by the efficiency of any associated combustion process to yield a net solar efficiency. For example, if a boiler-based electric generator extracts 35% of the feedstock energy as electricity, and the generator is sited at a switchgrass plantation achieving 0.30% solar capture efficiency on a mass basis, the electric plant has a net solar efficiency of 0.30% × 35% = 0.11%. Because biomass is a very low efficiency collector of solar energy, it is very land-intensive compared to photovoltaic or solar thermal collectors, which deliver energy at solar efficiencies over 20%. However, the capacity of land to store

TABLE 12.5

Primary Productivity and Solar Efficiency of Biomass Crops

Location	Crop	Yield (Dry Mg/ha-Year)	Average Insolation (W/m²)	Solar Efficiency (%)
Alabama	Johnsongrass	5.9	186	0.19
Alabama	Switchgrass	8.2	186	0.26
Minnesota	Willow and hybrid poplar	8–11	159	0.30–0.41
Denmark	Phytoplankton	8.6	133	0.36
Sweden	Enthropic lake angiosperm	7.2	106	0.38
Texas	Switchgrass	8–20	212	0.22–0.56
California	*Euphorbia lathyris*	16.3–19.3	212	0.45–0.54
Mississippi	Water hyacinth	11.0–33.0	194	0.31–0.94
Texas	Sweet sorghum	22.2–40.0	239	0.55–0.99
Minnesota	Maize	24.0	169	0.79
West Indies	Tropical marine angiosperm	30.3	212	0.79
Israel	Maize	34.1	239	0.79
Georgia	Subtropical saltmarsh	32.1	194	0.92
Congo	Tree plantation	36.1	212	0.95
New Zealand	Temperate grassland	29.1	159	1.02
Marshall Islands	Green algae	39.0	212	1.02
New South Wales	Rice	35.0	186	1.04
Puerto Rico	*Panicum maximum*	48.9	212	1.28
Nova Scotia	Sublittoral seaweed	32.1	133	1.34
Colombia	Pangola grass	50.2	186	1.50
West Indies	Tropical forest, mixed ages	59.0	212	1.55
California	Algae, sewage pond	49.3–74.2	218	1.26–1.89
England	Coniferous forest, 0–21 years	34.1	106	1.79
Germany	Temperate reedswamp	46.0	133	1.92
Holland	Maize, rye, two harvests	37.0	106	1.94
Puerto Rico	*Pennisetum purpurcum*	84.5	212	2.21
Hawaii	Sugarcane	74.9	186	2.24
Java	Sugarcane	86.8	186	2.59
Puerto Rico	Napier grass	106	212	2.78
Thailand	Green algae	164	186	4.90

Source: From Klass, D.L., *Biomass for Renewable Energy, Fuels, and Chemicals*, Academic Press, San Diego, CA, 1998. With permission.

standing biomass over time is extremely high, with densities up to several hundred Mg/ha (and therefore several thousand GJ/ha), depending on the forest type. Standing biomass can serve as very long-term storage, though multiple stores need to be used in order to accommodate fire risk. For short-term storage, woody biomass may be dried, and is frequently chipped or otherwise mechanically treated to create a fine and homogenous fuel suitable for burning in a wider variety of combustors.

12.10.2 Ethanol

Biomass is a more practical solar energy storage medium if it can be converted to liquid form. Liquids allow for more convenient transportation and combustion, and enable extraction on demand (through reciprocating engines) rather than through a less dispatchable, boiler- or turbine-based process. This latter property also enables its use in automobiles.

Biomass grown in crops or collected as residue from agricultural processes consists principally of cellulose, hemicellulose, and lignin. The sugary or starchy by-products of some crops such as sugarcane, sugar beet, sorghum, molasses, corn, and potatoes can be converted to ethanol through fermentation processes, and these processes are the principal source of ethanol today. Starch-based ethanol production is low efficiency, but does succeed in transferring about 16% of the biomass heating value to the ethanol fuel.[35]

When viewed as a developing energy storage technology, ethanol derived from cellulose shows much more promise than the currently prevalent starch-based ethanol.[36] Cellulosic ethanol can be manufactured with two fundamentally different methods: either the biomass is broken down to sugars using a hydrolysis process, and then the sugars are subjected to fermentation, or the biomass is gasified (see the following), and the ethanol is subsequently synthesized from this gas with a thermochemical process. Both processes show promise to be far cheaper than traditional ethanol manufacture via fermentation of starch crops, and will also improve energy balances. For example, it is estimated that dry sawdust can yield up to 224 L/Mg of ethanol, thus recovering about 26% of the HHV of the sawdust.[37] Since the ethanol will still need to be combusted in a heat engine, the gross, biomass-to-useful-work efficiency will be well below this. In comparison, direct combustion of the biomass to generate electricity (per the discussion in Section 12.10.1) makes much more effective use of the biomass as an energy storage medium. Hence, the value of ethanol as an energy storage medium lies mostly in the convenience of its liquid (rather than solid) state.

12.10.3 Biodiesel

As starch-based ethanol is made from starchy by-products, most biodiesel is generated from oily by-products. Some of the most common sources are rapeseed oil, sunflower oil, and soybean oil. Biodiesel yields from crops like these range from about 300 to 1000 kg/ha-year, but the crop as a whole produces about 20 Mg/ha-year, meaning that the gross solar capture efficiency for biodiesel from crops ranges between 1/20 and 1/60 the solar capture efficiency of the crop itself. Because of this low solar capture efficiency, biomass cannot be the principal energy storage medium for transportation needs.[38]

Biodiesel can also be manufactured from waste vegetable or animal oils; however, in this case, the biodiesel is not functioning *per se* as a solar energy storage medium, so it is not further treated in this work.

12.10.4 Syngas

Biomass can be converted to a gaseous state for storage, transportation, and combustion (or other chemical conversion).[39] Gasification processes are grouped into three different classes: *pyrolysis* is the application of heat in anoxic conditions; *partial oxidation* is combustion occurring in an oxygen-starved environment; *reforming* is the application of heat in the presence of a catalyst. All three processes form *syngas*, a combination of methane, carbon monoxide, carbon dioxide, and hydrogen. The relative abundances of the gaseous products can be controlled by adjusting heat, pressure, and feed rates. The HHV of the resulting gas can contain up to 78% of the original HHV of the feedstock, if the feedstock is dry.[40] Compositions and heating values of two example syngases are listed in Table 12.3.

The equivalent of up to 10% of the gas HHV will be lost when the gas is pressurized for transportation and storage. Even with this loss, gasification is a considerably more efficient method than ethanol manufacture for transferring stored solar energy to a nonsolid medium.

References

1. Srinivasan, V., Batteries for vehicular applications, *Physics of Sustainable Energy, AIP Conference Proceedings*, 1044, 283–296, 2008.
2. Goodenough, J.B., Abruna, H.D., Buchanan, M.V., *Basic Research Needs in Electrical Energy Storage*, Bethesda, MD, 2007. Basic energy sciences workshop report. web.anl.gov/energy-storage-science/publications/ees.rpt.pdf.
3. Grid-scale rampable intermittent dispatchable storage, U.S. Department of Energy—Advanced Research Projects Agency—Energy (ARPA-E), Funding Opportunity Number: DE-FOA-0000290, 2010.
4. EA Technology, Review of electrical energy storage technologies and systems and of their potential for the UK, United Kingdom Department of Trade and Industry, London, U.K., 2004 (URN 04/1876).
5. DeVries, T., World's biggest battery helps stabilize Alaska, *Modern Power Systems*, 22, 40, 2002.
6. Nourai, A., NaS battery demonstration in the USA, *Electricity Storage Association Spring Meeting*, Electricity Storage Association, Morgan Hill, CA, 2003.
7. Sudworth, J.L., Sodium/nickel chloride (Zebra) battery, *Journal of Power Sources*, 100, 149, 2001.
8. The Zebra, *Fleets & Fuels*, February 17, 2003, p. 8.
9. Price, A., Technologies for energy storage—Present and future: Flow batteries, *2000 Power Engineering Society Meeting*, Seattle, WA, July 16–20, 2000.
10. Skyllas-Kazacos, M., *Recent Progress with the Vanadium Redox Battery*, University of New South Wales, Sydney, Australia, 2000.
11. Menictas, C. et al., Status of the vanadium battery development program, *Proceedings of the Electrical Engineering Congress*, Sydney, Australia, 1994.
12. VRB Power Systems, Inc., http://www.vrbpower.com.
13. Wilks, N., Solving current problems, *Professional Engineering*, 13, 27, 2000.
14. Scheffler, P., *Environmental Assessment for the Regenesys Energy Storage System*, Tennessee Valley Authority, 2001.
15. Lex, P. and Jonshagen, B., The zinc/bromide battery system for utility and remote applications, *Power Engineering Journal*, 13, 142, 1999.
16. Weber, A.Z., Mench, M.M., Meyers, J.P., Ross, P.N., Gostick, J.T., and Liu, Q., Redox flow batteries: A review, *Journal of Applied Electrochemistry*, 41(10), 1137–1164, 2011.
17. Brushett, F.R., Jansen, A.N., Vaughey, J.T., and Zhang, Z., Exploratory research of non-aqueous flow batteries for renewable energy storage, *220th Electrochemical Society Meeting*, Boston, MA, October 9–14, 2011.
18. Huskinson, B., Marshak, M.P., Suh, C., Er, S., Gerhardt, M.R., Galvin, C.J., Chen, X., Aspuru-Guzik, A., Gordon, R.G., and Aziz, M.J., A metal-free organic–inorganic aqueous flow battery, *Nature*, 505, 195–198, 2014.
19. Duduta, M., Ho, B.Y., Wood, V.C., Limthongkul, P., Brunini, V.E., Carter, W.C., and Chiang, Y.M., Semi-solid lithium rechargeable flow battery, *Advanced Energy Materials*, 1, 511–516, 2011.
20. Ho, B.Y., An experimental study on the structure-property relationship of composite fluid electrodes for use in high energy density semi-solid flow cells, PhD thesis, MIT, Cambridge, MA, February 2012.

21. Kruger, P., Electric power requirement for large-scale production of hydrogen fuel for the world vehicle fleet, *International Journal of Hydrogen Energy*, 26, 1137, 2001.
22. Bossel, U., Eliasson, B., and Taylor, G., The future of the hydrogen economy: Bright or bleak?, *European Fuel Cell Forum*, Oberrohrdorf, Switzerland, 2003.
23. Dillon, A. et al., Storage of hydrogen in single-walled carbon nanotubes, *Nature*, 386, 377, 1997.
24. Linden, D. and Reddy, T.B., *Handbook of Batteries*, 3rd edn., McGraw-Hill, New York, 2002.
25. Luango, C.A., Superconducting storage systems: An overview, *IEEE Transactions on Magnetics*, 32, 1996.
26. Cheung, K.Y.C. et al., *Large Scale Energy Storage Systems*, Imperial College London, London, U.K., 2003 (ISE2 2002/2003).
27. Donalek, P., Advances in pumped storage, *Electricity Association Spring Meeting*, Chicago, IL, 2003.
28. Kondoh, J. et al., Electrical energy storage systems for energy networks, *Energy Conservation and Management*, 41, 1863, 2000.
29. van der Linden, S., The case for compressed air energy system, *Modern Power Systems*, 22, 19, 2002.
30. Hull, J. et al., *Salinity Gradient Solar Ponds*, CRC Press, 1988.
31. Sørensen, B., *Renewable Energy: Its Physics, Engineering, Environmental Impacts, Economics & Planning*, 3rd edn., Elsevier Academic Press, Burlington, MA, 2004.
32. Neeper, D.A., Thermal dynamics of wallboard with latent heat storage, *Solar Energy*, 68, 393, 2000.
33. Goswami, D.Y., Kreith, F., and Kreider, J.F., *Principles of Solar Engineering*, 2nd edn., Taylor & Francis, Philadelphia, PA, 2000.
34. Hanneman, R., Vakil, H., and Wentorf Jr., R., Closed loop chemical systems for energy transmission, conversion and storage, *Proceedings of the Ninth Intersociety Energy Conversion Engineering Conference*, American Society of Mechanical Engineers, New York, 1974.
35. Shapouri, H., Duffield, J.A., and Wang, M., The energy balance of corn ethanol: An update, USDA, Office of Energy Policy and New Uses, Agricultural Economics, 2002 (Rept. No. 813).
36. Hammerschlag, R., Ethanol's energy return on investment: A survey of the literature 1990–present, *Environmental Science & Technology* (submitted).
37. Klass, D.L., *Biomass for Renewable Energy, Fuels, and Chemicals*, Academic Press, San Diego, CA, 1998.
38. Bockey, D. and Körbitz, W., *Situation and Development Potential for the Production of Biodiesel—An International Study*, Union zur Forderung von Oel- und Proteinpflanzen, e.V., Berlin, Germany, 2003.
39. Bridgwater, A.V., The technical and economic feasibility of biomass gasification for power generation, *Fuel*, 74, 631, 1995.
40. Klass, *Biomass for Renewable Energy, Fuels, and Chemicals*, 302.
41. U.S. Department of Energy, Properties of fuels, Alternative Fuels Data Center, 2004.
42. Kreith, F. and Kreider J.F., *Principles of Solar Engineering*, Taylor & Francis, 1978.

13

Demand-Side Management

Clark W. Gellings and Kelly E. Parmenter

CONTENTS

13.1 Introduction

Since the mid-1980s, demand-side management has been an important element of the electric utility planning approach referred to as *integrated resource planning*. At that time, annual demand-side management expenditures in the United States were measured in billions of dollars, energy savings were measured in billions of kWh, and peak load reductions were stated in thousands of MW. While activities nationally slowed during the couple of decades that followed the 1980s, activity has again been on the rise since the turn of the millennium. Expenditures for utility and third-party administered electric efficiency programs were nearly $6 billion in 2012, up from a low of $0.9 billion in 1998; in addition, estimated electric savings nationwide exceeded 22 billion kWh in 2011 (Hayes et al. 2013). Therefore, demand-side management practices have continued to persevere and have recently regained a considerable and growing influence on the demand for energy resources. This chapter defines demand-side management, describes the role demand-side management plays in integrated resource planning, discusses the main elements of demand-side management programs, and summarizes the key best practices for program design and delivery. It then presents case studies of four successful demand-side management programs.

13.2 What Is Demand-Side Management?

The term *demand-side management* is the result of a logical evolution of planning processes used by utilities in the late 1980s. One of the first terms, *demand-side load management* was introduced by the author, Clark W. Gellings, in an article for *IEEE Spectrum* in 1981. Shortly after the publication of this article, at a meeting of The Edison Electric Institute (EEI) Customer Service and Marketing Executives in 1982, Mr. Gellings altered the term to *demand-side planning*. This change was made to reflect the broader objectives of the planning process. Mr. Gellings coined the term *demand-side management* and continued to popularize the term throughout a series of more than 100 articles since that time, including the five volume set *Demand-Side Management* which is widely recognized as a definitive and practical source of information on the demand-side management process.

Perhaps the most widely accepted definition of demand-side management is the following:

> Demand-side management is the planning, implementation, and monitoring of those utility activities designed to influence customer use of electricity in ways that will produce desired changes in the utility's load shape, i.e., changes in the time pattern and magnitude of a utility's load. Utility programs falling under the umbrella of demand-side management include: load management, new uses, strategic conservation, electrification, customer generation, and adjustments in market share (Gellings 1984–1988).

However, demand-side management is even more encompassing than this definition implies because it includes the management of all forms of energy at the demand-side, not just electricity. In addition, groups other than just electric utilities (including natural gas suppliers, government organizations, nonprofit groups, and private parties) implement demand-side management programs.

In general, demand-side management embraces the following critical components of energy planning:

1. Demand-side management *will influence customer use.* Any program intended to influence the customer's use of energy is considered demand-side management.

2. Demand-side management *must achieve selected objectives.* To constitute a desired load-shape change, the program must further the achievement of selected objectives; that is, it must result in reductions in average rates, improvements in customer satisfaction, achievement of reliability targets, etc.

3. Demand-side management *will be evaluated against non-demand-side management alternatives.* The concept also requires that selected demand-side management programs further these objectives to at least as great an extent as non-demand-side management alternatives, such as generating units, purchased power, or supply-side storage devices. In other words, it requires that demand-side management alternatives be compared to supply-side alternatives. It is at this stage of evaluation that demand-side management becomes part of the integrated resource planning process.

4. Demand-side management *identifies how customers will respond.* Demand-side management is pragmatically oriented. Normative programs (we ought to do this) do not bring about the desired; positive efforts (if we do this, that will happen) are required. Thus, demand-side management encompasses a process that identifies how customers will respond not how they should respond.

5. Demand-side management *value is influenced by load shape*. Finally, this definition of demand-side management focuses upon the load shape. This implies an evaluation process that examines the value of programs according to how they influence costs and benefits throughout the day, week, month, and year.

Subsets of these activities have been referred to in the past as *load management, strategic conservation*, and *marketing*.

13.3 Demand-Side Management and Integrated Resource Planning

A very important part of the demand-side management process involves the consistent evaluation of demand-side to supply-side alternatives and vice versa. This approach is referred to as *integrated resource planning*. Figure 13.1 illustrates how demand-side management fits into the integrated resource planning process. For demand-side management to be a viable resource option, it has to compete with traditional supply-side options.

13.4 Demand-Side Management Programs

A variety of programs have been implemented since the introduction of demand-side management in the early 1980s. Mr. Gellings and the EPRI (Electric Power Research Institute) have been instrumental in defining a framework for utilities and other implementers to follow when planning demand-side management programs (Gellings 1984–1988; Evans et al. 1993; Gellings and Chamberlin 1993; Gellings 2002; Siddiqui et al. 2008). This section describes the main elements of the demand-side management planning framework. It then discusses the types of end-use sectors, buildings, and end-use technologies targeted during program development. It also lists the various entities typically responsible for implementing programs, along with several program implementation methods. Lastly, this section summarizes specific characteristics of successful program design and delivery.

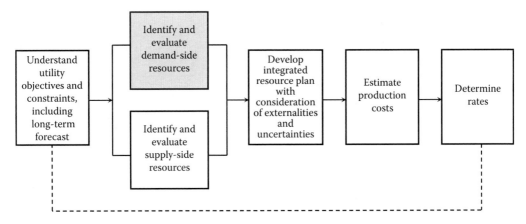

FIGURE 13.1
How demand-side management fits into integrated resource planning.

13.4.1 Elements of the Demand-Side Management Planning Framework

Figure 13.2 illustrates the five main elements of the demand-side management planning framework. These five elements are summarized as follows:

1. *Set objectives*: The first step in demand-side management planning is to establish overall organizational objectives. These strategic objectives are quite broad and generally include examples such as conserving energy resources, reducing peak demand (thereby deferring need to build new power plants), decreasing

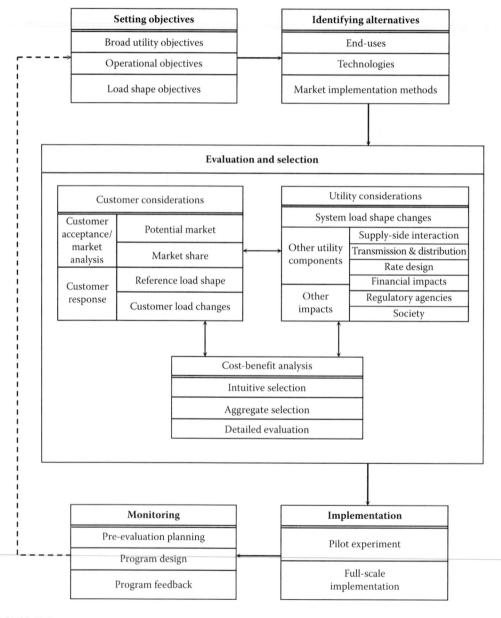

FIGURE 13.2
Elements of the demand-side management planning framework.

greenhouse gas emissions, reducing dependence on foreign imports, improving cash flow, increasing earnings, and improving customer and employee relations. In this level of the formal planning process, the planner needs to operationalize broad objectives to guide policymakers to specific actions. It is at this operational level or tactical level that demand-side management alternatives should be examined and evaluated. For example, an examination of capital investment requirements may show periods of high investment needs. Postponing the need for new construction through a demand-side management program may reduce investment needs and stabilize the financial future of an energy company, or a utility and its state or country. Specific operational objectives are established on the basis of the conditions of the existing energy system—its system configuration, cash reserves, operating environment, and competition. Once designated, operational objectives are translated into desired demand-pattern changes or load-shape changes that can be used to characterize the potential impact of alternative demand-side management programs. Although there is an infinite combination of load-shape-changing possibilities, six have been illustrated in Figure 13.3 to show the range of possibilities, namely peak clipping, valley filling, load shifting, strategic conservation, strategic load growth, and flexible load shape. These six are not mutually exclusive, and may frequently be employed in combinations.

2. *Identify alternatives*: The second step is to identify alternatives. The first dimension of this step involves identifying the appropriate end-uses whose peak load and energy consumption characteristics generally match the requirements of the load-shape objectives established in the previous step. In general, each

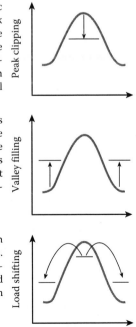

Peak clipping—or the reduction of the system peak loads, embodies one of the classic forms of load management and is now commonly referred to as *demand response*. Peak clipping is generally considered as the reduction of peak load by using time-based rate options or incentive-based strategies, with or without enabling technologies. While many utilities consider this as a means to reduce peaking capacity or capacity purchases and consider strategies only during the most probable days and times of system peak, these strategies can be used to reduce operating cost and dependence on critical fuels by economic dispatch.

Valley filling—is the second classic form of load management and applies to both gas and electric systems. Valley filling encompasses building off-peak loads. This may be particularly desirable where the long-run incremental cost is less than the average price of energy. Adding properly priced off-peak load under those circumstances decreases the average price. Valley filling can be accomplished in several ways, one of the most popular of which displaces loads served by fossil fuels with electric loads that are operated during off-peak periods (e.g., water heating and/or space heating).

Load shifting—is the last classic form of load management and also applies to both gas and electric systems. This involves shifting load from on-peak to off-peak periods. Popular applications include use of storage water heating, storage space heating, coolness storage (the most common type of thermal energy storage), and customer load shifts. The load shift from storage devices involves displacing what would have been conventional appliances.

FIGURE 13.3

Six generic load-shape objectives that can be considered during demand-side management planning. (*Continued*)

Strategic conservation—is the load-shape change that results from programs directed at end-use consumption. Not normally considered load management, the change reflects a modification of the load shape involving a reduction in consumption as well as a change in the pattern of use. In employing energy conservation, the planner must consider what conservation actions would occur naturally and then evaluate the cost-effectiveness of possible intended programs to accelerate or stimulate those actions. Examples include weatherization and appliance efficiency improvement.

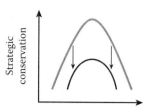

Strategic load growth—is the load-shape change that refers to a general increase in sales beyond the valley filling described previously. Load growth may involve increased market share of loads that are or can be, served by competing fuels, as well as economic development. Load growth may include electrification. Electrification is the term being employed to describe the new emerging electric technologies surrounding electric vehicles, industrial process heating, and automation. These have a potential for increasing the electric energy intensity of the industrial sector. This rise in intensity may be motivated by reduction in the use of fossil fuels and raw materials resulting in improved overall productivity.

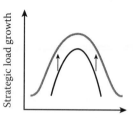

Flexible load shape—is a concept related to electric system reliability, a planning constraint. Once the anticipated load shape, including demand-side activities, is forecast over some horizon, the power supply planner studies the final optimum supply-side options. Among the many criteria he or she uses is reliability. Load shape can be flexible—if customers are presented with options as to the variations in quality of service that they are willing to allow in exchange for various incentives. The program involved can be variations of interruptible or curtailable load; concepts of pooled, integrated energy management systems; or individual customer load control devices offering service constraints.

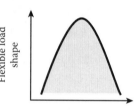

FIGURE 13.3 (*Continued*)
Six generic load-shape objectives that can be considered during demand-side management planning.

end-use (e.g., residential space heating, commercial lighting) exhibits typical and predictable demand or load patterns. The extent to which load pattern modification can be accommodated by a given end-use is one factor used to select an end-use for demand-side management. The second dimension of demand-side management alternatives involves choosing appropriate technology alternatives for each target end-use. This process should consider the suitability of the technology for satisfying the load-shape objective. Even though a technology is suitable for a given end-use, it may not produce the desired results. For example, although water-heater wraps are appropriate for reducing domestic water-heating energy consumption, they are not appropriate for load shifting. In this case, an option such as electric water-heating direct load control would be a better choice. The third dimension involves investigating market implementation methods (see Section 13.4.5 for a description of potential implementation methods).

3. *Evaluate and select program(s)*: The third step balances customer considerations, supplier considerations, and cost/benefit analyses to identify the most viable demand-side management alternative(s) to pursue. Although customers and suppliers act independently to alter the pattern of demand, the concept of demand-side management implies a supplier/customer relationship that produces mutually beneficial results. To achieve that mutual benefit, suppliers must carefully consider such

factors as the manner in which the activity will affect the patterns and amount of demand (load shape), the methods available for obtaining customer participation, and the likely magnitudes of costs and benefits to both supplier and customer prior to attempting implementation.

4. *Implement program(s)*: The fourth step is to implement the program(s), which takes place in several stages. As a first step, a high level, demand-side management project team should be created with representation from the various departments and organizations, and with the overall control and responsibility for the implementation process. It is important for implementers to establish clear directives for the project team, including a written scope of responsibility, project team goals and time frame. When limited information is available on prior demand-side management program experiences, a pilot experiment may precede the program. Pilot experiments can be a useful interim step toward making a decision to undertake a major program. Pilot experiments may be limited either to a subregion or to a sample of consumers throughout an area. If the pilot experiment proves cost-effective, then the implementers may consider initiating the full-scale program.

5. *Monitor program(s)*: The fifth step is to monitor the program(s). The ultimate goal of the monitoring process is to identify deviations from expected performance and to improve both existing and planned demand-side management programs. Monitoring and evaluation processes can also serve as a primary source of information on customer behavior and system impacts, foster advanced planning and organization within a demand-side management program, and provide management with the means of examining demand-side management programs as they develop.

13.4.2 Targeted End-Use Sectors/Building Types

The three broad categories of end-use sectors targeted for demand-side management programs are residential, commercial, and industrial. Each of these broad categories includes several subsectors. In some cases, the program will be designed for one or more broad sectors; in other cases, it may be designed for a specific subsector. For example, the residential sector can be divided into several subsectors including single family homes, multifamily homes, mobile homes, low-income homes, etc. In addition, the commercial sector can be split into subsets, such as offices, restaurants, health care facilities, educational facilities, retail stores, grocery stores, hotels/motels, etc. There are also numerous specific industrial end-users that may be potentially targeted for a demand-side management program. Moreover, the program designer may want to target a specific type or size of building within the chosen sector. The program could focus on new construction, old construction, renovations and retrofits, large customers, small customers, or a combination. Crosscutting programs target multiple end-use sectors and/or multiple building types. Figure 13.4 illustrates the broad types of end-use sectors and building types and how they relate to other aspects of demand-side management program planning.

13.4.3 Targeted End-Use Technologies/Program Types

There are several end-use technologies or program types targeted in demand-side management programs (see Figure 13.4). Some programs are comprehensive, and crossover between end-use technologies. Other programs target specific end-use equipment, such as lighting,

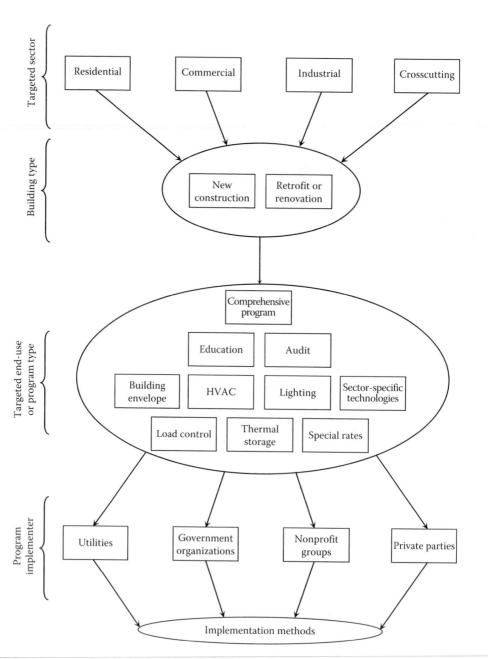

FIGURE 13.4
Relationship between end-use sectors, building types, end-use programs, and program implementers.

air conditioners, dishwashers, etc. Still others target load control measures, such as demand response programs whereby customers temporarily curtail loads in response to peak demand events or programs that permanently shift loads to off-peak hours (e.g., via thermal energy storage). Figure 13.4 shows representative end-use technologies or program types and how they relate to other aspects of demand-side management program planning.

13.4.4 Program Implementers

Implementers of demand-side management programs are often utilities. However, other possible implementers include government organizations, nonprofit groups, private parties, or a collaboration of several entities (see Figure 13.4). Utilities and governments, in particular, have a special interest in influencing customers' demand—treating it not as fate but as choice—in order to provide better service at lower cost while increasing their own profits and reducing their business risks. Energy planners can choose from a wide range of market push and pull methods designed to influence consumer adoption and reduce barriers, as discussed in the next section.

13.4.5 Implementation Methods

Among the most important dimension in the characterization of demand-side alternatives is the selection of the appropriate market implementation methods. Planners and policy makers can select from a variety of methods for influencing customer adoption and acceptance of demand-side management programs. The methods can be broadly classified into six categories. Table 13.1 lists examples for each category of market implementation method. The categories include the following:

1. *Customer education*: Many energy suppliers and governments have relied on some form of customer education to promote general customer awareness of programs. Websites, brochures, bill inserts, information packets, clearinghouses, educational curricula, and direct mailings are widely used. Customer education is the most basic of the market implementation methods available and should be used in conjunction with one or more other market implementation method for maximum effectiveness.

2. *Direct customer contact*: Direct customer contact techniques refer to face-to-face communication between the customer and an energy supplier or government representative to encourage greater customer acceptance of programs. Energy suppliers have for some time employed marketing and customer service representatives to provide advice on appliance choice and operation, sizing of heating/cooling systems, lighting design, and even home economics. Direct customer contact can be accomplished through energy audits, specific program services (e.g., equipment servicing), store fronts where information and devices are displayed, workshops, exhibits, on-site inspection, etc. A major advantage of these methods is that they allow the implementer to obtain feedback from the consumer, thus providing an opportunity to identify and respond to major customer concerns. They also enable more personalized marketing, and can be useful in communicating interest in and concern for controlling energy costs.

TABLE 13.1

Examples of Market Implementation Methods

Market Implementation Method	Illustrative Objective	Examples
Customer education	• Increase customer awareness of programs • Increase perceived value of energy services	• Websites • Bill inserts • Brochures • Information packets • Displays • Clearinghouses • Direct mailings
Direct customer contact	• Through face-to-face communication, encourage greater customer acceptance and response to programs	• Energy audits • Direct installation • Store fronts • Workshops/energy clinics • Exhibits/displays • Inspection services
Trade ally cooperation (i.e., architects, engineers, appliance dealers, heating/cooling contractors)	• Increase capability in marketing and implementing programs • Obtain support and technical advice on customer adoption of demand-side technologies	• Cooperative advertising and marketing • Training • Certification • Selected product sales/service
Advertising and promotion	• Increase public awareness of new programs • Influence customer response	• Mass media (Internet, radio, TV, and newspaper) • Point-of-purchase advertising
Alternative pricing	• Provide customers with pricing signals that reflect real economic costs and encourage the desired market response	• Demand rates • Time-of-use rates • Real-time pricing • Critical peak pricing • Off-peak rates • Seasonal rates • Inverted rates • Variable levels of service • Promotional rates • Conservation rates
Direct incentives	• Reduce up-front purchase price and risk of demand-side technologies to the customer • Increase short-term market penetration • Provide incentives to employees to promote demand-side management programs	• Low- or no-interest loan • Cash grants • Subsidized installation/modification • Rebates • Buyback programs • Rewards to employees for successful marketing of demand-side management programs

3. *Trade ally cooperation*: Trade ally cooperation and support can contribute significantly to the success of many demand-side management programs. A trade ally is defined as any organization that can influence the transactions between the supplier and its customers or between implementers and consumers. Key trade ally groups include home builders and contractors, local chapters of professional societies, technology/product trade groups, trade associations, and associations representing wholesalers and retailers of appliances and energy consuming devices. Depending on the type of trade ally organization, a wide range of services are performed, including

development of standards and procedures, technology transfer, training, certification, marketing/sales, installation, maintenance, and repair. Generally, if trade ally groups believe that demand-side management programs will help them (or at least not hinder their business), they will likely support the program.

4. *Advertising and promotion*: Energy suppliers and government energy entities have used a variety of advertising and promotional techniques. Advertising uses various media to communicate a message to customers in order to inform or persuade them. Advertising media applicable to demand-side management programs include the Internet, radio, television, magazines, newspapers, outdoor advertising and point-of-purchase advertising. Promotion usually includes activities to support advertising, such as press releases, personal selling, displays, demonstrations, coupons, and contest/awards.

5. *Alternative pricing*: Pricing as a market-influencing factor generally performs three functions: (a) transfers to producers and consumers information regarding the cost or value of products and services being provided, (b) provides incentives to use the most efficient production and consumption methods, and (c) determines who can afford how much of a product. These three functions are closely interrelated. Alternative pricing, through innovative schemes can be an important implementation technique for utilities promoting demand-side options. For example, rate incentives for encouraging specific patterns of utilization of electricity can often be combined with other strategies (e.g., direct incentives) to achieve electric utility demand-side management goals. Pricing structures include time-of-use rates, real-time pricing, critical peak pricing, inverted rates, seasonal rates, variable service levels, promotional rates, off-peak rates, etc. A major advantage of alternative pricing programs over some other types of implementation techniques is that the supplier has little or no cash outlay. The customer receives a financial incentive, but over a period of years, so that the implementer can provide the incentives as it receives the benefits.

6. *Direct incentives*: Direct incentives are used to increase short-term market penetration of a cost control/customer option by reducing the net cash outlay required for equipment purchase or by reducing the payback period (i.e., increasing the rate of return) to make the investment more attractive. Incentives also reduce customer resistance to options without proven performance histories or options that involve extensive modifications to the building or the customer's lifestyle. Direct incentives include cash grants, rebates, buyback programs, billing credits, and low-interest or no-interest loans. One additional type of direct incentive is the offer of free, or very heavily, subsidized, equipment installation or maintenance in exchange for participation. Such arrangements may cost the supplier more than the direct benefits from the energy or demand impact, but can expedite customer recruitment, and allow the collection of valuable empirical performance data.

Energy suppliers, utilities, and government entities have successfully used many of these marketing strategies. Typically, multiple marketing methods are used to promote demand-side management programs. The selection of the individual market implementation method or mix of methods depends on a number of factors, including the following:

- Prior experience with similar programs
- Existing market penetration
- The receptivity of policy makers and regulatory authorities

- The estimated program benefits and costs to suppliers and customers
- Stage of buyer readiness
- Barriers to implementation

Some of the most innovative demand-side marketing programs started as pilot programs to gauge consumer acceptance and evaluate program design before large-scale implementation.

The objective of the market implementation methods is to influence the marketplace and to change the customer behavior. The key question for planners and policy makers is the selection of the market implementation method(s) to obtain the desired customer acceptance and response. Customer acceptance refers to customer willingness to participate in a market implementation program, customer decisions to adopt the desired fuel/appliance choice and efficiency, and behavior change as encouraged by the supplier, or state. Customer response is the actual load-shape change that results from customer action, combined with the characteristics of the devices and systems being used.

Customer acceptance and responses are influenced by the demographic characteristics of the customer, income, knowledge, and awareness of the technologies and programs available, and decision criteria, such as cash flow and perceived benefits and costs, as well as attitudes and motivations. Customer acceptance and response are also influenced by other external factors, such as economic conditions, energy prices, technology characteristics, regulation, and tax credits.

13.4.6 Characteristics of Successful Programs

Numerous demand-side management programs are implemented in the United States yearly by various organizations including utilities, state agencies, and third-party implementers. In addition, some businesses implement their own internal or corporate-wide energy efficiency programs. Over the years, groups such as the EPRI, American Council for an Energy-Efficient Economy (ACEEE), Energy Trust of Oregon, and California Best Practices Project Advisory Committee with contractor, Quantum Consulting Inc., have reviewed and compared demand-side management programs to identify exemplary programs and best practices. The results of those studies are included in an assortment of reports (Peters 2002; Quantum Consulting 2004; Wikler et al. 2008; Nowak et al. 2013; Young and Mackres 2013).

EPRI conducted an assessment of best practices in demand-side management programs (Wikler et al. 2008). The study included interviews with utility program staff, an online survey of utility representatives, and a comprehensive review of recent best practices studies in the literature. The study revealed a number of elements associated with successful program design and delivery. Some of the elements vary by customer segment and program type, while others are common across multiple program types. Table 13.2 lists the challenges faced by selected types of Residential and Commercial & Industrial (C&I) programs and identifies the key factors that have contributed to programs' success. The table also includes examples of programs that have applied these best practices. The following subsections distill some of the more generally applicable best practices for program design and delivery.

TABLE 13.2

Synthesis of Key Challenges and Success Factors for Selected Energy Efficiency and Load Management Programs

Program Type	Key Challenges	Key Success Factors	Program Example
Residential conservation	• Technology developments • Easy access to pay stations	• Enhancing customer knowledge and control over their consumption	Salt River Project—M Power
Residential lighting	• Low customer motivation, especially if electricity prices are low • Utility resource allocation toward marketing and outreach efforts • Overcome customer experience or perception of poor quality products • CFL disposal issues	• Aggressive customer awareness and outreach efforts • Strong publicity campaigns • Leverage ENERGY STAR® brand • Build and maintain relationships with manufacturers and retailers	Georgia Power— ENERGY STAR Change a Light (CAL) Campaign Snohomish County PUD— Residential CFL Program
Residential load management	• Obtain customer trust to install equipment • Address customer complaints unrelated to the program • Low level of customer motivation if electricity prices are low	• Simple program design • Widespread customer awareness and outreach efforts • Establish system reliability • Maintain continuity in customer participation • Build strong customer relationships • Provide customer choice and control on electricity usage	Florida Power and Light— Residential On Call Gulf Power— Good Cents Select
Residential new construction	• Generate interest in the program by all parties • Maintain consistency with builders and energy raters	• ENERGY STAR name recognition • Build and maintain relationships with builders and raters	Center Point Energy— ENERGY STAR New Home Oncor Electric Delivery— ENERGY STAR Homes Program
Low income	• Build customer trust and confidence • Rapid turnaround in contractor workforce responsible for delivery	• Deliver nonenergy benefits, such as comfort and safety • Engender participant trust through program stability and continuity • Successful education and outreach efforts • Partner with community-based organizations for effective delivery	Pacific Gas and Electric—Energy Partners
C&I energy efficiency	• Incorporate flexible incentives to fit different customer requirements, while keeping program design simple	• Simplicity in program design • Strong financial incentives • Keep customer's financial bottom-line in mind • Strong customer relationships/maintaining close contact with customers directly and through contractors	Alliant Energy— Shared Savings Program

(Continued)

TABLE 13.2 (*Continued*)

Synthesis of Key Challenges and Success Factors for Selected Energy Efficiency and Load
Management Programs

Program Type	Key Challenges	Key Success Factors	Program Example
C&I new construction	• Generate interest from designers, builders, and owners • Shortage of qualified staff	• Build broad awareness • Start with a pilot program and expand • Staff development and training • Flexible approach • Full energy simulation capability	We Energies—C/I New Construction Program
C&I retrofit	• Need to conform to new/ increased codes and standards • Interactions with regulators	• Flexibility in accommodating different measures • Stakeholder collaborative process in designing programs • Highly skilled utility staff • Contractor capacity building • Contractor–customer relationships	Connecticut Light and Power - Energy Opportunities
C&I niche	• Accommodate industry requirements • Obtain customer interest for participation • Differences in using the new technology	• Target niche, high-growth industries • Dynamic program design to fit market requirements • Form partnerships and collaborations with related groups • Integrate energy efficiency into customers' business strategies • Achieve more than just the energy benefits	Pacific Gas and Electric—High Tech Energy Efficiency Program Salt River Project—Pre Rinse Spray Valves
Small business	• Develop web-based delivery infrastructure • Bring vendors up to speed	• Make program participation as simple as possible • Build a strong network of vendors	Southern California Edison—Express Efficiency Rebate Program

Source: Wikler, G. et al., Best practices in energy efficiency and load management programs, 1016383, EPRI, Palo Alto, CA, 2008.

13.4.6.1 Key Elements of Program Design

Successful program design effectively translates the program design theories into a practical program structure with actionable policies and procedures aimed at meeting program goals and objectives. Key elements of success include the following:

- Maintain simplicity in program design
- Design incentive structure to fit customer requirements
- Maintain flexibility in program design to accommodate measures
- Maintain dynamism in program design to fit market requirements in specific industries

- Develop sound performance tracking mechanisms
- Incorporate customer choice and control features in program design
- Ensure resource allocation is commensurate with program tasks
- Obtain stakeholder support right at the design stage
- Maintain high quality of products
- Establish program branding
- Undertake program improvements over time

13.4.6.2 Key Elements of Program Delivery

Program delivery consists of marketplace actions to promote demand-side management practices and to increase adoption of technologies and measures. Specific delivery activities encompass the implementation methods discussed in Section 13.5. There are several key elements of successful program delivery:

- Foster trade ally relationships and partnerships
- Undertake contractor capacity building efforts
- Build networks and alliances with other relevant parties and groups
- Undertake program publicity campaigns
- Establish strong customer education and outreach efforts
- Foster utility–customer relationships
- Coordinate with other utilities and program administrators
- Build customer–contractor relationships
- Maintain strong in-house capabilities
- Integrate energy efficiency into customer's business strategy
- Deliver nonenergy benefits
- Maintain consistency over time
- Enhance program delivery through collaborative efforts

13.5 Case Studies

This section examines case studies of four successful demand-side management programs and a smart grid demonstration project. Each of the four programs targets a different sector and set of end-uses. In addition, each program represents a different U.S. geographical region and has a different implementation structure. The smart grid demonstration project is a widespread effort involving a collaboration of over 20 utilities in the United States and abroad with the common goal of advancing the integration of distributed energy resources into the electric power grid.

Table 13.3 presents the first case study, which is a comprehensive industrial energy efficiency program administered by Bonneville Power Administration (BPA). The second

TABLE 13.3

Case Study #1—2012 Bonneville Power Administration (BPA) Energy Smart Industrial (ESI) Program

---

Description:

BPA's ESI program was designed and rolled out in 2009 with partner Cascade Energy to help BPA meet aggressive industrial energy savings goals. The program targets the industrial sector in BPA's service territory in the Pacific Northwest and incorporates a wide range of retrofit and operation and maintenance measures as well as continuous improvement options aimed at behavioral changes. BPA's participating utility customers can choose from this assortment of measures and several different market delivery options to serve their industrial customers. Customers receive incentives for projects and many are eligible for subsidized technical consulting services. A unique program aspect is the use of a qualified ESI partner (ESIP) assigned by the ESI program to act as a single point of contact for the program to the utilities and industrial customers. In addition, a pilot feature of the program incorporates three innovative Energy Management options to improve operation and maintenance and management aspects of energy use so that savings persist: (1) co-funding for an Energy Project Manager to serve as a staff resource to promote energy efficiency projects at qualifying facilities, (2) support for no-cost/low-cost operation and maintenance improvements including tools for data collection and tracking and incentive funding (referred to as Track and Tune), and (3) use of behavior-based and continuous improvement approaches to train industrial facilities to incorporate energy management into core business practices (referred to as High Performance Energy Management). The program's success is reflected in its high performance over a short period of time, cost-effectiveness, utility and industrial customer satisfaction with offerings, and simplified communications approach.

Targeted sector/building type:

Industrial customers of BPA's multiple utility customers in the Pacific Northwest.

Any industrial customer of a participating utility is eligible.

Targeted industries include pulp and paper, wood products, food processing, and water and wastewater.

Targeted end-use technology/program type:

Targets multiple end-users through various measure types: custom and prescriptive retrofit projects; no-cost/low-cost operation and maintenance improvements; and behavior-based measures

Program administrator/implementer:

Administered by Bonneville Power Administration

Implemented by Cascade Energy

Program expenditures for 2012 (estimated):

$15.2 million

Program results for 2012 (estimated):

Net energy savings: 91,980 MWh

Net demand savings: 10.50 aMW

Participation: 105 enrolled utilities; 86 engaged utilities; 478 end-users

---

Source: Nowak, S. et al., Leaders of the pack: ACEEE's third national review of exemplary energy efficiency programs, Report No. U132, ACEEE, Washington, DC, 2013.

Personal communication with Jennifer Eskil, Agriculture/Industrial Sector Lead, Bonneville Power Administration, August 20, 2013.

case study (Table 13.4) is a retrocommissioning (RCx) program offered to the commercial sector by Commonwealth Edison (ComEd) with a focus on controls optimization and other operational energy efficiency improvements. The third case study (Table 13.5) is a Southern California Edison (SCE) pilot program that was offered to water utilities and targeted both water and energy savings. These first three programs received the distinction of *exemplary* by ACEEE in 2013 for their excellence in program design and delivery. Table 13.6 presents the fourth case study, a smart grid-enabled demand response program offered to residential and small commercial customers by Oklahoma Gas and

TABLE 13.4

Case Study #2—2011/2012 Commonwealth Edison (ComEd) Smart Ideas for Your Business Retrocommissioning (RCx) and Monitoring-Based Commissioning (MBCx) Program

---

Description:

In ComEd's Smart Ideas for Your Business RCx and MBCx programs, approved engineering firms (Service Providers) provide onsite assessments and analysis of building energy systems that focus on controls optimization and other operational improvements. ComEd pays the Service Providers to do the study, at no cost to the customer. The customer in exchange agrees to invest at least a minimum amount on operational improvements with paybacks of 18 months or less. A newer feature of the program is the inclusion of natural gas measures in the analysis, which is enabled by partnerships with local gas companies. Investigating electric and gas measures simultaneously optimizes the Service Providers' time at the site and provides additional value to the customers and utilities. Another enhancement to the program is the addition of an MBCx option in which the customer receives an incentive to help offset the cost of installing advanced building automation software to monitor energy use and identify opportunities for operational improvements over a period of 18 months or more. Then, the customer is paid an additional incentive for verified energy savings resulting from these improvements. Some of the reasons for the program's success include the use of preapproved Service Providers as a sales channel, the partnerships with natural gas utilities and ability to identify electric and gas measures, and the program's continued efforts to enhance processes and adapt to reach more customers and savings opportunities.

Targeted sector/building type:

Commercial sector in Northern Illinois: retail and office buildings; commercial real estate; hospitals; education; hospitality; and other buildings with > 150,000 sq. ft. of air-conditioned floor space

Targeted end-use technology/program type:

RCx and MBCx programs

Targets operation and maintenance improvements such as economizer and ventilation control, equipment scheduling, and fan optimization and air distribution modifications

Include electric and gas saving measures

Program administrator/implementer:

Administrated by ComEd in partnership with Nicor Gas, North Shore Gas, and Peoples Gas

Implemented by Nexant

Program expenditures for program year 4 (2011/2012):

$4.84 million

Program results for program year 4 (2011/2012):

Net electricity savings: 25,021 MWh

Participation: 50 customers

Source: Nowak, S. et al., Leaders of the pack: ACEEE's third national review of exemplary energy efficiency programs, Report No. U132, ACEEE, Washington, DC, 2013.

Personal communication with Rick Tonielli, Sr. Energy Efficiency Program Manager, ComEd, August 14, 2013.

Electric (OG&E). OG&E was recognized as Utility of the Year in 2011 by *Electric Light & Power*, in part due to OG&E's successful demand management programs and their high customer satisfaction rating during a period of very high Smart Meter deployment. The smart grid demonstration project (Table 13.7) is a large-scale EPRI initiative focused on demonstrating grid integration of distributed energy resources, such as demand response technologies, electric vehicles, thermal energy storage, electric storage, solar photovoltaics, wind generation, conservation voltage reduction, and distributed generation. The project just completed its fifth year of a 7-year effort. As Tables 13.3 through 13.7 show, all of the programs and initiatives presented in these case studies have yielded impressive results. Even more importantly, their innovative approaches have contributed to the advancement of demand-side management practices.

TABLE 13.5

Case Study #3—Southern California Edison (SCE) Leak Detection Pilot Program

Description:

The SCE Leak Detection Pilot Program was one of nine *Embedded Energy in Water Pilots* established by the California Public Utility Commission (CPUC) to study alternatives for reducing water-related energy use. The pilot program involved conducting audits and repairing leaks in water distribution systems. Saving water through leak repair also reduces the embedded energy requirements associated with the supply, conveyance, treatment, and distribution of water to end-users. The pilot consisted of three partners: SCE, participating water agencies, and the audit implementation contractor, Water Systems Optimization Inc. The program demonstrated the largest energy savings potential for relatively low cost compared to the CPUC pilots and was successful as measured by the satisfaction of program partners. Success was due to a combination good planning, dedicated involvement by the SCE program manager, good communication between partners, and professional and high quality work. The potential for larger scale deployment of the program hinges on the CPUC and California Energy Commission. Of particular relevance is whether or not the CPUC will publish a rule allowing electric utilities to claim embedded energy savings arising from water saving measures.

Targeted sector/building type:

Municipal water utilities in Southern California

Targeted end-use technology/program type:

Targets leaks in water distribution systems

Pilot program that includes audits of distribution systems and repairs of leaks

Program administrator/implementer:

SCE

Program expenditures for an 18-month period (July 08–December 09):

$300,000

Program results:

Annual energy savings: 498 MWh

Annual water savings: 83 million gallons

Participation: three water agencies (Las Virgenes Municipal Water District, Apple Valley Ranchos Water Company, and Lake Arrowhead Community Services District)

Source: Young, R. and Mackres, E., Tackling the nexus: Exemplary programs that save both energy and water, Report No. E131, ACEEE, Washington, DC, 2013.

Personal communication with Gene Rodrigues, Director of DSM Strategy, Portfolio Oversight and Technical Support, Southern California Edison Company, August 13, 2013.

TABLE 13.6

Case Study #4—2012 OG&E SmartHours Program

Description:

OG&E's SmartHours Program is a multilevel voluntary dynamic pricing program designed to encourage participants to reduce on-peak energy use by offering them lower rates during off-peak periods. The on-peak hours are defined as 2:00–7:00 p.m. on weekdays during the summer from June 1 through September 30, excluding holidays. The rest of the time is considered off-peak. During the on-peak, there are several pricing levels based on system demand and weather conditions: low, standard, high, and critical. Pricing events are called via e-mail, text message, and/or voice mail the day ahead for the first three levels. For the critical level, pricing events can be called a day ahead or there is also a provision for a critical price overcall event (CPE) in which participants receive a minimum of 2-h notice and then the price rises to the critical level. A CPE lasts 2–8 h and the maximum number of CPE hours per year is 80.

OG&E has deployed Smart Meters throughout its service territory. All SmartHours participants have Smart Meters and access to their energy information through a website. Program participants are also offered a free programmable communicating thermostat (PCT) to help them automate their response to pricing events (62% of 2012 participants accepted the PCT). OG&E does not directly control the thermostat; instead, customers set thresholds based on pricing events and their personal preferences.

Aspects of the program that have contributed to its success include free enabling technology and services, customer empowerment of thermostat control and overall energy use, pricing that reflects the true cost of electricity, and the no-lose proposition provided to customers for the first year to guarantee they will not pay more on the SmartHours rate than they would have on the standard flat rate.

Targeted sector/building type:

Residential and small commercial customers

Targeted end-use technology/program type:

Program targets reduction in on-peak loads through time-based pricing and smart grid products and services

Program implementer:

OG&E

Program expenditures for 2012 (approximate):

$20 million

Program results for 2012:

Overall maximum demand reduction: 51.4 MW on day of the system peak at hour-ending 4:00 p.m.

System peak demand reduction: 44.1 MW on day of system peak at system peak hour (hour-ending 5:00 p.m.)

Participation: 35,144 participants as of September 2012

Participants without PCTs responded significantly less than those with PCTs

Source: Marrin, K. and Williamson, C., *2012 Evaluation of OG&E SmartHours Program*, Prepared for Oklahoma Gas & Electric, EnerNOC, Walnut Creek, CA, 2012.

Personal communication with Mike Farrell, Director of Customer Programs, Oklahoma Gas and Electric, October 1, 2013.

TABLE 13.7

Case Study #5—EPRI Smart Grid Demonstration Initiative

Overview of initiative:

EPRI's Smart Grid Demonstration Initiative is a 7-year collaborative research effort involving two dozen utilities from Australia, Canada, France, Ireland, Japan, and the United States. The focus of the initiative is to design, deploy, and evaluate ways to integrate distributed energy resources into the electric grid and into market operations. The goal of the collaborative effort is to leverage electric utility Smart Grid investments and share research and demonstration results with smart grid technologies and applications to create a *smarter* grid. Each project that is undertaken by a member of the initiative contributes to the collective knowledge. With the initiative recently completing its fifth year, many projects addressing a wide range of research questions have been completed or are underway. EPRI and utility members of the initiative have created hundreds of reports and case studies describing the results (see www.smartgrid.epri.com).

Collaborators:

The collaborators consist of host sites and non-host sites. Host site collaborators are conducting the majority of the field projects.

Host sites: American Electric Power, Con Edison, Duke Energy, Electricité de France, Ergon, ESB Networks, Exelon (ComEd/PECO), First Energy, Hawaiian Electric Company, Hydro Québec, Kansas City Light and Power, PNM Resources, Sacramento Municipal Utility District, Southern California Edison, Southern Company

Non-host sites: Ameren, Central Hudson Gas and Electric, CenterPoint Energy, Entergy, Salt River Project, Tennessee Valley Authority, Tokyo Electric Power Company, Wisconsin Public Service

Distributed energy resources under investigation:

The initiative is studying grid integration of eight different types of distributed energy resource technologies: demand response technologies, electric vehicles, thermal energy storage, electric storage, solar photovoltaics, wind generation, conservation voltage reduction, and distributed generation

Communications and standards under investigation:

Customer Domain (SEP, WiFi, etc.), Distribution (DNP3, IEC 61850, etc.), Cyber Security, Advanced Metering Infrastructure or Automated Meter Reading (AMI or AMR), Radio Frequency (RF) Mesh or Tower, Public or Private Internet, Cellular 3G (GPRS, CDMA, etc.), and Cellular 4G (WiMAX, LTE, etc.)

Grid management under investigation:

Volt/VAR optimization (VVO) and conservation voltage reduction (CVR), Distribution Automation, and Grid Management System (DMS, DERMS, DRMS)

Demand-side management programs under investigation:

Price-based (time of use, critical peak pricing, real-time pricing, etc.)
Incentive-based (direct load control, interruptible, etc.)

Operations and planning under investigation:

System Operations Integration, System Planning Integration, and Modeling and/or Simulation Tools

Source: Wakefield, M. and Horst, G., EPRI smart grid demonstration initiative: 5 year update, 3002000778, EPRI, Palo Alto, CA, 2013.

13.6 Conclusions

Since the early 1970s, economic, political, social, technological, and resource supply factors have combined to change the energy industry's operating environment and its outlook for the future. Many utilities are faced with staggering capital requirements for new plants, significant fluctuations in demand and energy growth rates, declining financial performance, and political or regulatory and consumer concern about rising prices and the environment. While demand-side management is not a cure-all for these difficulties, it does provide for a great many additional alternatives that have myriad nonenergy benefits as well as the more obvious energy-related benefits. These demand-side alternatives are equally appropriate for consideration by utilities, energy suppliers, energy-service suppliers, and government entities. Implementation of demand-side measures not only benefits the implementing organization by influencing load characteristics, delaying the need for new energy resources, and in general improving resource value, but also provides benefits to customers such as reduced energy bills and/or improved performance from new technological options. In addition, society as a whole receives economic, environmental, and national security benefits. For example, since demand-side management programs can postpone the need for new power plants, the costs and emissions associated with fossil-fueled electricity generation are avoided. Demand-side management programs also tend to generate more jobs and expenditures within the regions where the programs are implemented, boosting local economies. Moreover, demand-side management programs can help reduce a country's dependence on foreign oil imports, improving national security. Demand-side management alternatives will continue to hold an important role in resource planning in the United States and abroad, and will be a critical element in the pursuit of a sustainable energy future.

References

Evans, M., P. Meagher, A. Faruqui, and J. Chamberlin. 1993. Principles and practice of demand-side management. TR-102556. Palo Alto, CA: EPRI.

Gellings, C. W. 1984–1988. *Demand-Side Management: Volumes 1–5*. Palo Alto, CA: EPRI.

Gellings, C. W. 2002. Using demand-side management to select energy efficient technologies and programs. In *Efficient Use and Conservation of Energy*, C. W. Gellings (ed.). Encyclopedia of Life Support Systems (EOLSS). Oxford, U.K.: EOLSS Publishers. http://www.eolss.net (accessed March 3, 2015).

Gellings, C. W. and J. H. Chamberlin. 1993. *Demand-Side Management: Concepts and Methods*, 2nd edn. Lilburn, GA: The Fairmont Press, Inc.

Hayes, S., N. Baum, and G. Herndon. 2013. Energy efficiency: Is the United States improving? White Paper. Washington, DC: ACEEE.

Marrin, K. and C. Williamson. 2012. *2012 Evaluation of OG&E SmartHours Program*. Prepared for Oklahoma Gas & Electric. Walnut Creek, CA: EnerNOC.

Nowak, S., M. Kushler, P. Witte, and D. York. 2013. Leaders of the pack: ACEEE's third national review of exemplary energy efficiency programs. Report No. U132. Washington, DC: ACEEE.

Peters, J. 2002. *Best Practices from Energy Efficiency Organizations and Programs*. Portland, OR: Energy Trust of Oregon.

Quantum Consulting. 2004. *National Energy Efficiency Best Practices Study.* Multiple Volumes. Submitted to California Best Practices Project Advisory Committee. Berkeley, CA: Quantum Consulting Inc. http://www.eebestpractices.com (accessed March 3, 2015).

Siddiqui, O., P. Hurtado, K. Parmenter et al. 2008. Energy efficiency planning guidebook: Energy efficiency initiative. 1016273. Palo Alto, CA: EPRI.

Wakefield, M. and G. Horst. 2013. EPRI smart grid demonstration initiative: 5 year update. 3002000778. Palo Alto, CA: EPRI.

Wikler, G., D. Ghosh, K. Smith, and O. Siddiqui. 2008. Best practices in energy efficiency and load management programs. 1016383. Palo Alto, CA: EPRI.

Young, R. and E. Mackres. 2013. Tackling the nexus: Exemplary programs that save both energy and water. Report No. E131. Washington, DC: ACEEE.

Index

3M energy management program, 254

A

A-lamps, 188
Absorption chillers, 106
Absorption cycle heat pumps, 243–244
AC motors
    electronic adjustable speed drives in,
        342–344
    high-performance speed control of, 351–352
ACEEE
    energy efficiency report, 250–251
    energy-efficient industrial technologies
        study, 326–327
    Fan and Pump study, 323
Actuators, 139
    linear, 134
    rotary, 134
Adaptive control, 126, 170
Adjustable speed drives. *See* ASDs
Advanced control systems, design topics,
        168–175
Advanced ground transportation, use of
        biomass fuel for, 17–18
Advanced window technologies, 186
Advertising, implementation of demand-side
        management, 397
AHUs, optimization of operation, 166–167
Air compressors
    energy and demand balances, 262–263
    energy management strategies for, 292–293
Air conditioner efficiency, 105
Air conditioning
    energy and demand balances, 262
    energy efficient design for, 223–224
    residential, 216–217
Air conditioning expansion valves
    model predictive control for, 174–175
    nonlinear compensation for, 170–171
Air handling units. *See* AHUs
Air preheaters, 317–318
Air-cooled chillers, 106
Air-source heat pumps, 232
    cold climate, 234
    dual fuel, 234
    premium efficiency, 232–233

Air-to-air heat pumps, 101
AIRMaster+, 270
Algae, derivation of biofuels from, 9
All-electric vehicles, 22
Alternative fuels, 8–10
    feedstocks for, 9
Alternative pricing, implementation of
        demand-side management, 397
American Council for an Energy-Efficient
        Economy. *See* ACEEE
American Society of Heating, Refrigerating,
        and Air-Conditioning Engineers.
        See ASHRAE
AMOPs, derivation of biofuels from, 9
Analog control, 125–126
Anodizing, 275
Appliances
    clothes dryers, 218
    clothes washers, 218
    cooktops and ovens, 219
    cost effectiveness of efficient
        designs for, 226
    dishwashers, 219
    efficient designs of, 219
    electricity use by in residential sector, 87
    energy use efficiency, 117–118
    furnaces and boilers, 216
    heat pumps, 217–218
    refrigerator-freezers and freezers, 215
    residential energy use, 97–98, 118–120
    U.S. production of, 220
Aquatic microbial oxygenic photoautotrophs.
        See AMOPs
Armstrong Ceilings DC Flexzone system, 186
Artificial light, 180
    architectural design and, 185–186
Artificial neural networks, use of to verify
        energy savings, 80
ASDs, 345
    cost-effectiveness of, 355
    electronic, 342–344
    energy analysis of, 268–269
    energy-saving applications of, 349–352
    high-performance applications, 351–352
    potential savings by sector and end use,
        353–355
    unit and installation costs, 355